Mechanisms of Secondary Brain Damage

Current State

Edited by

A. Baethmann, O. Kempski, L. Schürer

Acta Neurochirurgica
Supplementum 57

Springer-Verlag Wien New York

Professor Dr. Alexander Baethmann
Institut für Chirurgische Forschung, Ludwig-Maximilians-Universität,
Klinikum Grosshadern, München, Federal Republic of Germany

Professor Dr. Oliver Kempski
Institut für Neurochirurgische Pathophysiologie,
Johannes-Gutenberg-Universität, Mainz, Federal Republic of Germany

Dr. Ludwig Schürer
Neurochirurgische Klinik, Ludwig-Maximilians-Universität,
Klinikum Grosshadern, München, Federal Republic of Germany

With 76 Figures

Softcover reprint of the hardcover 1st edition 1993

Printed on acid-free paper

ISSN 0065-1419
ISBN-13: 978-3-7091-9268-9 e-ISBN-13: 978-3-7091-9266-5
DOI: 10.1007/ 978-3-7091-9266-5

Preface

Great progress has been made in the understanding and prevention of secondary brain damage from acute cerebral disorders, such as trauma and ischemia. Advances may be concerned in particular with better organization and logistics of preclinical emergency care, including rapid arrival of well-trained medical staff on the scene of an accident and of transportation to a competent hospital. Nevertheless, it is a safe assumption that development of secondary brain damage from both intra- and extracranial causes still represents a major factor for the final outcome in severe head injury. Thus, exchanges of experiences and information between various disciplines involved with this important clinical problem – trauma still assumes the number one position as a cause of morbidity and mortality up to an age of 45 years – may provide a basis for in-depth analysis of remaining problems as well as of methods of their solution. This exactly is the purpose of the present publication on concepts and findings pertinent for the general subject of secondary brain damage from various experimental as well as clinical viewpoints.

An internationally high-ranking group of experts has been contributing to this collection of reviews on cerebral trauma and ischemia and its adverse sequelae, including cerebral exploration by most modern technologies, such as NMR spectroscopy or PET scanning, among others. Of course, the pathophysiological basis of secondary brain damage continues to remain a focus of such an endeavor, as demonstrated by up-to-date analyses on tissue acidosis, the present understanding of the neuroexcitotoxicity of glutamate or, conversely, the significance of endogenous defense mechanisms afforded by adenosine. Highlights of clinical analyses are studies on the impressive potential of brain tissue to recover from ischemia, or the advanced level of multifactorial treatment modalities developed for cerebral resuscitation after cardiac arrest, employing among others cardiopulmonary bypass circulation, hypertensive hemodilution, and hypothermia. Further, the volume provides an update on the management of patients with severe head injury after arrival in a trauma center, focusing in particular on intracranial hypertension, or brain tissue acidosis. Due to the continuous improvement of cerebral monitoring we may witness in the near future application of something like tailormade treatment protocols, which are matching the specific pathophysiological course of a patient.

Important information is included on the current level of organization and practice of emergency care in trauma patients on a regional as well as supra-regional basis. Thereby, one arrives at the unavoidable conclusion that a major potential of prevention of secondary brain damage in patients with head injury is not yet fully exploited. A variety of factors might be involved, such as the level of training of the lay public and emergency personnel, or the limited availability of specialized neurotrauma centers, often making necessary administration of primary care by a general hospital, which is not qualified enough to diagnose and treat cerebral injuries, particularly if neurosurgery is required. An interdisciplinary discussion as provided herewith may be helpful to find solutions for this important problem.

It is our understanding that currently available methods of specific inhibition of pathomechanisms causing secondary brain damage might have little chances to salvage a patient in jeopardy from trauma or ischemia, if avoidable factors, such as deficits in emergency or intensive care leading to secondary hypoxia or circulatory failure, are not more efficiently prevented. Specific treatment modalities, such as glutamate receptor blockers or adenosine agonists, among others, may protect the cerebrum far better, if the brain primarily injured by trauma or ischemia is not suffering in addition from secondary insults. The present publication discussing at a highly competent level mechanisms of secondary brain damage from an experimental as well as clinical point of view may make obvious that progress in its prevention might be enhanced, if the various disciplines involved are interacting more effectively than before. Examples are given where collaboratory efforts appear particularly promising.

We would like to take the opportunity to express our gratitude to FIDIA, Munich for their generous support, to Springer, Vienna, for publishing this synopsis, and for the competent secretarial assistance provided by Ulrike Goerke, Isolde Juna, Helga Kleylein, and Monika Stucky in completing this volume.

A. Baethmann
O. Kempski
L. Schürer

Munich and Mainz, 1992

Contents

Brain Damage Studied by NMR and Other Methods

Gadian, D. G., Williams, S. R., Bates, T. E., Kauppinen, R. A.: NMR Spectroscopy: Current Status and Future Possibilities ... 1

Behar, K. L.: Cerebral Metabolic Studies *in vivo* by Combined $^{1}H/^{31}P$ and $^{1}H/^{13}C$ NMR Spectroscopic Methods ... 9

Hossmann, K.-A., Behar, K. L., Rothman, D. L.: NMR-Spectroscopic Investigation of Cerebral Reanimation After Prolonged Ischemia ... 21

Bulte, J. W. M., de Jonge, M. W. A., Kamman, R. L., Zuiderveen, F., The, T. H., de Leij, L., Go, K. G.: Magnetite as a Potent Contrast-Enhancing Agent in Magnetic Resonance Imaging to Visualize Blood-Brain Barrier Disruption ... 30

Gjedde, A., Kuwabara, H.: Absent Recruitment of Capillaries in Brain Tissue Recovering from Stroke ... 35

Graham, D. I., Adams, J. H., Doyle, D., Ford, I., Gennarelli, T. A., Lawrence, A. E., Maxwell, W. L., McLellan, D. R.: Quantification of Primary and Secondary Lesions in Severe Head Injury ... 41

Gennarelli, T. A., Tipperman, R., Maxwell, W. L., Graham, D. I., Adams, J. H., Irvine, A.: Traumatic Damage to the Nodal Axolemma: An Early, Secondary Injury ... 49

Heye, N., Campos, A., Sampaolo, S., Cervos-Navarro, J.: Morphometrical Evaluation of Triflusal in Brain Infarction ... 53

Mediators and Antagonism in Secondary Brain Damage

Plum, F.: *In vivo* and *in vitro* Regulation of Acid-Base Control of Brain Cells During Ischemic and Selective Acidic Exposure ... 57

Wahl, M., Schilling, L., Unterberg, A., Baethmann, A.: Mediators of Vascular and Parenchymal Mechanisms in Secondary Brain Damage ... 64

McCulloch, J., Ozyurt, E., Kun Park, C., Nehls, D. G., Teasdale, G. M., Graham, D. I.: Glutamate Receptor Antagonists in Experimental Focal Cerebral Ischaemia ... 73

Schubert, P., Kreutzberg, G. W.: Cerebral Protection by Adenosine ... 80

James, H. E., Schneider, S.: Effects of Acute Isotonic Saline Administration on Serum Osmolality, Serum Electrolytes, Brain Water Content and Intracranial Pressure ... 89

Diemer, N. H., Johansen, F. F., Benveniste, H., Bruhn, T., Berg, M., Valente, E., Jørgensen, M. B.: Ischemia as an Excitotoxic Lesion: Protection Against Hippocampal Nerve Cell Loss by Denervation ... 94

Symon, L.: Recovery of Brain Function Following Ischemia ... 102

Safar, P., Sterz, F., Leonov, Y., Radovsky, A., Tisherman, S., Oku, K.: Systematic Development of Cerebral Resuscitation After Cardiac Arrest. Three Promising Treatments: Cardiopulmonary Bypass, Hypertensive Hemodilution, and Mild Hypothermia 110

Functional Consequences of Cerebral Lesions

Pöppel, E.: Taxonomy of Subjective Phenomena: a Neuropsychological Basis of Functional Assessment of Ischemic or Traumatic Brain Lesions 123

Hartmann, A., Dettmers, C., Lagreze, H., Tsuda, Y.: Blood Flow and Clinical Course in Patients with Ischemic Stroke without Cerebrospecific Therapy 130

Emergency Care and Treatment in Acute Cerebral Insults

Bouillon, B., Schweins, M., Lechleuthner, A., Vorweg, M., Troidl, H.: Assessment of Emergency Care in Trauma Patients 137

Sefrin, P.: Current Level of Prehospital Care in Severe Head Injury – Potential for Improvement 141

Garcia, J. H.: Prehospital Management of Head Injuries: International Perspectives 145

Miller, J. D., Piper, I. R., Dearden, N. M.: Management of Intracranial Hypertension in Head Injury: Matching Treatment with Cause 152

Marmarou, A., Holdaway, R., Ward, J. D., Yoshida, K., Choi, S. C., Muizelaar, J. P., Young, H. F.: Traumatic Brain Tissue Acidosis: Experimental and Clinical Studies 160

Subject Index 165

Acta Neurochir (1993) [Suppl] 57: 1–8

Brain Damage Studied by NMR and Other Methods

NMR Spectroscopy: Current Status and Future Possibilities

D. G. Gadian, S. R. Williams, T. E. Bates, and **R. A. Kauppinen**

Hunterian Institute, Royal College of Surgeons of England, London, U.K..

Abstract

Nuclear magnetic resonance (NMR) spectroscopy is now established as a non-invasive method of studying metabolism in living systems, ranging from cellular suspensions to man. With respect to clinical applications, recent developments include the successful implementation of new techniques for spatial localisation, and in particular the acquisition of excellent ^{1}H spectra from selected regions of the human brain. Localised ^{1}H spectroscopy opens the way to monitoring a wide range of compounds that are inaccessible to ^{31}P NMR, and should add considerably to the information that is available from ^{31}P studies. NMR spectroscopy does, however, have its limitations, which arise primarily from the fact that it is an insensitive technique. This lack of sensitivity limits the spatial resolution for metabolic studies, and means that metabolites must be present at fairly high concentrations in order to produce detectable signals. In this article, we illustrate the scope and limitations of NMR spectroscopy by describing a few examples of studies undertaken on animals and humans.

Keywords: ^{1}H-spectroscopy; spectral editing; *in vivo* metabolism; ischaemia

Abbreviations

NMR: nuclear magnetic resonance
PCr: phosphocreatine
PME: phosphomonoester
PDE: phosphodiester

Introduction

Following the initial ^{31}P NMR studies of intact tissue that were carried out 15 years ago (Hoult *et al.*, 1974), NMR is now established as a method of studying metabolism in living systems, ranging from cellular suspensions to man. The non-invasive study of cerebral metabolism in small animals became feasible with the development of the surface coil (Ackerman *et al.*, 1980), which is a type of detecting coil that can be placed adjacent to a superficial region of interest. Extension to human studies (firstly of limbs, then of the neonatal brain) followed shortly afterwards with the construction of larger magnets (Ross *et al.*, 1981; Cady *et al.*, 1983). With the development of whole body spectroscopy systems and additional or alternative methods of spatial localisation, studies of the adult brain soon followed. The development of magnetic resonance spectroscopy for metabolic studies has proceeded in parallel with analogous developments in magnetic resonance imaging, which now has widespread use as a clinical imaging technique. The emphasis in this article is on metabolism and hence on spectroscopy rather than imaging, but it should be stressed that there is increasing interaction between these two aspects of NMR. There are many texts covering the theory and applications of NMR spectroscopy and imaging (Gadian, 1982; Morris, 1986; Foster and Hutchison, 1987; Stark and Bradley, 1988), and this article assumes a basic knowledge of the NMR technique.

The Nuclei that are Used for Metabolic Studies

^{31}P NMR

The nucleus that has been used most widely for metabolic studies is ^{31}P, which is the naturally occurring phosphorus nucleus. ^{31}P spectra of the brain include characteristic signals from ATP, phosphocreatine (PCr), phosphodiesters (PDE), inorganic phosphate (P_i) and phosphomonesters (PME). The relative areas of the signals provide information about the relative concentrations of the metabolites, while the frequency (chemical shift) of the P_i signal can be used for the measurement of intracellular pH.

Several interesting points emerged from the first surface coil studies of the rat brain (Ackerman *et al.*, 1980). (i) From the chemical shifts of the three ATP signals, it could be concluded that the ATP was predominantly complexed to Mg^{2+} ions. (ii) The PCr/ATP ratio was on the high side of values commonly obtained by rapid freezing techniques. (iii) The concentrations of P_i and of free cytoplasmic ADP are low, and this has several important implications in relation to the control and energetics of cellular metabolism. The latter two points illustrate one of the major advantages of NMR; namely that being non-invasive it gives a direct monitor of tissue energy status *in vivo*. It should be stressed that narrow signals are only observed from relatively mobile compounds; highly immobilised species such as membrane phospholipids will normally give very broad signals that show up as a baseline "hump". Thus it is commonly assumed that NMR spectroscopy monitors cytoplasmic metabolites, although there has been one report in which a second P_i contribution was attributed to mitochondrial P_i (Garlick *et al.*, 1983).

The major contribution to the PME signal in brain spectra is believed to be from phosphorylethanolamine (Gyulai *et al.*, 1984; Prenton *et al.*, 1985; Pettegrew *et al.*, 1986). This signal is particularly large in the neonatal brain, both in humans (Cady *et al.*, 1983) and in rats (Tofts and Wray, 1985). However, many other metabolites (including sugar phosphates) can contribute to this region of the spectrum, and in particular could contribute to any changes that occur in disease. Studies on the characterisation of the PDE signal in the dog brain have shown that there are contributions from glycerophosphorylcholine and glycerophosphorylethanolamine, but that the major component of the signal may be from a relatively mobile fraction of membrane phospholipids (Cerdan *et al.*, 1986). The contribution of the membrane phospholipids may be field-dependent, being greater at the lower field because of a field-dependent relaxation mechanism (see Bates *et al.*, 1989b). This would be consistent with the common observation of a large PDE signal in clinical spectroscopy, which is typically performed at field strengths of 1.5–2 Tesla.

^{31}P NMR has been widely used to investigate the

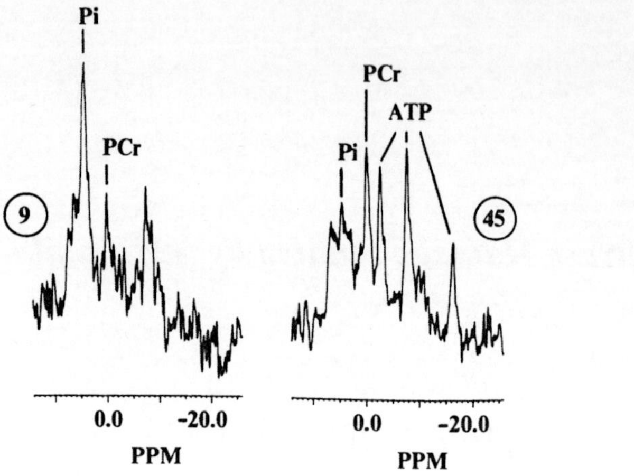

Fig. 1. ^{31}P spectra and flow measurements recorded from the left and right hemispheres of the gerbil brain during unilateral occlusion. The flow measurements (in ml/100 g/min) are circled adjacent to the corresponding spectra. (Adapted from Crockard *et al.*, 1987)

metabolic changes associated with ischaemia; provided that the regional cerebral blood flow falls sufficiently, there is a decline in PCr and ATP, an increase in P_i and a fall in intracellular pH, and these changes have been monitored in several animal models, as discussed below.

^{1}H NMR

The development of ^{1}H NMR for metabolic studies *in vivo* has been rather slower than ^{31}P NMR. This is partly for technical reasons – although ^{1}H NMR is about 15 times more sensitive than ^{31}P NMR, it is technically more difficult because of the need to suppress the large signals from water and, in some cases, from fats, and because of the large number of metabolites that produce signals in a relatively narrow chemical shift range. However, techniques for solvent suppression and spectral "editing" are now sufficiently well developed to permit the non-invasive monitoring of several metabolites of interest, including lactate, alanine, N-acetylaspartate, glutamine, glutamate, and total creatine (i.e. creatine + PCr). Following the first report by Behar *et al.* (1983), the use of ^{1}H NMR for studies of brain metabolism is becoming increasingly widespread, and on the basis of recent results described by several groups, applications to human brain metabolism appear very promising.

Other Nuclei

^{19}F is an excellent NMR nucleus, offering 85% of the sensitivity of ^{1}H NMR. There are no naturally occurring fluorine-containing compounds, and so the the role of ^{19}F NMR is in the monitoring of exogenous ^{19}F containing molecules.

Studies of fluorocarbon blood substitutes provide an opportunity for studying tissue perfusion and oxygenation, as the relaxation time T_1 of some fluorocarbons depends on O_2 tension (Clark *et al.*, 1984; Eidelberg *et al.*, 1988). A related application of ^{19}F NMR involves the use of fluorinated molecules such as Freon-22 and Freon-23 as blood flow tracers (Eleff *et al.*, 1988; Ewing *et al.*, 1989). The physiological effects of Freon-22 may limit its use as a CBF indicator, but these animal studies demonstrate the feasibility of using ^{19}F NMR for blood flow measurements, and show promise for future use in humans.

Metabolism of 2-fluoro-2-deoxy-D-glucose (2FDG) can be followed directly by ^{19}F NMR (Nakada *et al.*, 1986), and it has been shown that at relatively low doses (by NMR standards) of 20 mg/kg 2FDG can undergo extensive metabolism to compounds other than 2FDG-6-phosphate in the rat brain (Berkowitz *et al.*, 1987). In fact, it appears from these studies and from ^{31}P NMR observations of 2DG metabolism (Deuel *et al.*, 1985) that the metabolism of labelled 2DG is more complex than has often been assumed, which would have some influence on quantitative analysis of 2FDG studies using positron emission tomography. Another use of ^{19}F NMR is in the measurement of intracellular cation concentrations. Fluorinated NMR indicators have been synthesised as intracellular cation probes for H^+ (Deutsch *et al.*, 1982) and Ca^{2+} (Smith *et al.*, 1983), and a recent report has described intracellular Ca^{2+} measurements by ^{19}F NMR in superfused brain slices using fluoro-BAPTA (Bachelard *et al.*, 1988).

^{23}Na NMR spectra *in vivo* show just a single peak representing Na^+ ions in both the intracellular and extracellular spaces; thus the different environments of the Na^+ ions cannot be distinguished on the basis of chemical shift effects (and it seems unlikely that paramagnetic "shift reagents" will prove effective for studies of the brain *in vivo*). However, the relaxation properties of the ^{23}Na nucleus may differ in the two environments, permitting information to be obtained about oedema and the distribution of Na^+ ions between the intracellular and extracellular spaces.

^{13}C NMR provides a method of studying metabolism that has analogies with the use of ^{14}C for radioactive tracer studies, with the advantages that well-resolved signals can be attributed to different carbons within individual molecules (Alger and Shulman, 1984). In view of the poor sensitivity of ^{13}C relative to ^1H NMR, there are advantages in following the ^{13}C nuclei indirectly, by observing protons that are coupled to the ^{13}C nuclei. This approach is discussed in detail in the accompanying article by Behar *et al.* (this volume)

Experimental Animal Models

In this section, we shall discuss briefly some animal studies of ischaemia and of acute liver failure, as illustrative examples of the type of information NMR can provide.

Ischaemia

NMR spectroscopy *in vivo* has been used to follow changes in high energy phosphates, P_i, intracellular pH, lactate and changes in Na^+ distribution in various animal models of ischaemia, anoxia and hypoxia. From the earliest experiments it became apparent that NMR is best used in conjunction with other techniques for the investigation and characterisation of ischaemia. For example, in 1982 Thulborn and co-workers (Thulborn *et al.*, 1982) observed changes using ^{31}P NMR in one ischaemic hemisphere of the gerbil and correlated these with subsequent histology and specific gravity measurements to determine oedema and the extent of ischaemic damage. More recently, correlations of changes in ^{31}P and ^1H spectra during ischaemia have been made with changes in cerebral blood flow determined either simultaneously with the NMR measurements using H_2 flow (Gadian *et al.*, 1987; Crockard *et al.*, 1987; Allen *et al.*, 1988), or immediately after the NMR experiment using ^{14}C-iodoantipyrine (Naritomi *et al.*, 1988).

Our combined flow and metabolism studies have been carried out on the anaesthetised gerbil. The NMR measurements were made with two surface coils, one on each hemisphere, each coil being doubly tuned to the ^1H and ^{31}P frequencies. Changes in metabolic state, together with H_2 clearance measurements of regional flow, were monitored during and following 60 min of unilateral carotid occlusion or 30 min of bilateral carotid occlusion. During ischaemia, no changes were detected in the ^1H and ^{31}P spectra provided that the regional flow remained above 25 ml/100 g/min, whereas if the flow fell to about 20 ml/100 g/min or below, large metabolic changes characteristic of energy failure were observed; i.e. there was a decrease in high energy phosphates and intracellular pH, and an increase in P_i and lactate (Crockard *et al.*, 1987; see Fig. 1). The flow threshold value is similar to that at which, in a variety of species, electrical activity ceases, and is also similar to the flow level at which water accumulates in the gerbil brain. These results therefore provide evidence suggesting that the thresholds for electrical function and oedema are a direct consequence of energy failure. The model could prove to be of value in studying modifications to flow and metabolism produced by pharmaceuticals or other forms of therapy.

We have also studied the recovery processes that take place immediately following 15 or 30 min periods of ischaemia (Allen *et al.*, 1988). Among our results, we find that there is a period during recirculation when the lactate is still elevated, while the ^{31}P spectrum and intracellular pH are close to normal. This observation provides additional evidence for a mechanism of H^+ extrusion from the intracellular region, possibly via Na^+/H^+ exchange, as has been discussed by Siesjö (1985). In addition, these results indicate that ^{1}H NMR spectroscopy of lactate may be a more sensitive monitor than ^{31}P NMR of a previous ischaemic episode, or of repeated transient ischaemia. This is especially so in view of the high concentration to which lactate can accumulate, coupled with the higher sensitivity of ^{1}H NMR (see below).

It should be pointed out that in many NMR studies, including our studies of cerebral ischaemia, caution is necessary in interpreting spectra from regions where the metabolic state may be heterogeneous. For example, at a flow value close to the threshold of 20 ml/100 g/min the ^{31}P spectrum may contain similar amounts of PCr, ATP and P_i, the P_i indicating a pH of 6.7. A simple interpretation of such a spectrum is that all the cells contributing signal have a similar metabolic state. An alternative explanation is that there are two populations of cells, one of which has high P_i and low PCr and ATP with a pH of 6.7, while the other cells are healthy and have a normal pH but insufficient P_i to contribute significantly to the observed spectrum. This illustrates the point that, as the pH is measured from the P_i resonance, those regions with high levels of P_i will tend to dominate the measurement. These will, of course, be the ischaemic, acidotic regions. In general, therefore, NMR will not measure the volume-averaged pH, but will provide a pH value weighted towards the ischaemic regions. This is one reason why it is difficult to detect a "split P_i" signal during ischaemia, in which two distinguishable P_i resonances corresponding to normoxic and ischaemic tissue pH compartments are seen. Unless the proportion of normoxic tissue is considerably larger than the ischaemic proportion, the small "normoxic" P_i will be swamped by the ischaemic signal.

It is also important to point out that the acute recovery of ATP does not necessarily imply recovery of function (Naruse *et al.*, 1984). It is possible that a better indication of damage will be provided by longer term measurements of metabolism using combined ^{31}P and ^{1}H NMR.

Acute Liver Failure

We have used NMR spectroscopy to investigate metabolism in the liver and brain of rats with acute liver failure. Although this article is concerned with brain metabolism, one aspect of the liver data should be mentioned as it is of general relevance to *in vivo* NMR. This relates to the measurement of absolute concentrations. In rats with acute liver failure, the ratio of hepatic ATP to P_i determined *in vivo* by NMR was similar to that in controls, whereas studies of tissue extracts showed that the ATP level was reduced by a factor of 2–3 in the animals with acute liver failure (Bates *et al.*, 1988). It was concluded that the ^{31}P signals were observed only from the relatively healthy cells (which gave a normal ATP/P_i ratio), and not from severely damaged cells. It was suggested that in the damaged cells, any P_i that was generated subsequently leaked out and was washed away, so that by the time the spectra were accumulated (48 h after administration of the hepatotoxin), these cells did not contribute any signal. If this is the case, then the possibility arises that in many disease states (in particular in chronic disease where there is ample time for P_i to be washed out), the ratios of phosphorus metabolites may appear normal but the absolute concentrations averaged over the tissue volume will be reduced. This is consistent with the observations made by Bottomley *et al.* (1986) in chronic adult cerebral infarctions, as discussed below.

For the brain studies (Bates *et al.*, 1989), acute liver failure was induced by a single intragastric dose of carbon tetrachloride, resulting in a plasma ammonia level of over 500 μM. We showed that the forebrain is capable of maintaining normal phosphorus energy metabolite ratios and intracellular pH despite the metabolic challenge associated with the elevated ammonia. This is consistent with the ^{31}P NMR observations of Deutz *et al.* (1988a) who concluded, with a different rat model, that a change in high energy phosphates is not an important pathophysiological mechanism during the development of acute hepatic encephalopathy. Our

^{1}H NMR studies showed a significant increase in the brain glutamine level and a concomitant decrease in glutamate during hyperammonaemia, and the brain lactate concentration increased twofold. These results are also consistent with data obtained by other groups (Deutz *et al.*, 1988b; Behar *et al.*, this volume), and add weight to the view that ^{1}H NMR may prove to be a more sensitive indicator of disease than ^{31}P NMR. Further information about the effects of ammonia can be obtained using brain slice preparations (Brooks *et al.*, 1989).

Studies of the Human Brain

Neonates

Birth asphyxia is the commonest cause of impaired neurological function in full-term infants (Hope and Reynolds, 1985), and hypoxic-ischaemic injury is likely to be the main factor involved. Because of the importance of understanding the mechanisms of hypoxic-ischaemic injury in newborns, ^{31}P NMR spectroscopy of neonates was started almost as soon as magnets of large enough dimensions became available (Cady *et al.*, 1983; Hope *et al.*, 1984; Younkin *et al.*, 1984). Reynolds and colleagues have demonstrated that the PCr/P_i ratio is a good prognostic indicator of the likely clinical outcome following birth asphyxia; below a certain value for the ratio infants either die, or survive with impaired neurological function (Hope and Reynolds, 1985). A feature of the timecourse of the ^{31}P spectra is that an apparently normal spectrum (in terms of metabolite ratios) is recorded soon after the birth asphyxia episode (8–17 hours), but over the next few days the energy status declines (Hope and Reynolds, 1985). As Hope and Reynolds state, this raises the exciting possibility that intervention during the intermediate "normal" stage may be able to ameliorate the clinical outcome.

The Adult Brain

Studies of the adult brain require improved methods of spatial localisation. The development of improved techniques is an area of active research, and several methods have now been implemented for ^{31}P studies (Bottomley *et al.*, 1984; Oberhaensli *et al.*, 1986; Segebarth *et al.*, 1987; Bottomley *et al.*, 1988; Coutts *et al.*, 1989). However, regardless of the method that is employed, the spatial resolution is limited by the intrinsically poor sensitivity of the NMR technique, so that typically the resolution for ^{31}P NMR of the human brain will be about 3–6 cm. One advantage of ^{1}H spectroscopy is that ^{1}H NMR is intrinsically more sensitive than ^{31}P NMR, and this, together with the high concentration to which lactate can accumulate in disease (note that a concentration of 10 mmol/kg wet wt is equivalent to 30 mmol/kg wet wt in methyl protons) means that better spatial resolution should be available for ^{1}H spectroscopy in many clinical situations.

Relatively few ^{31}P studies of ischaemic disease have so far been reported (see Bottomley, 1989 for review). In chronic adult infarctions, Bottomley *et al.* (1986) have seen decreases of up to 40% in the total ^{31}P NMR metabolite signals when compared with normal contralateral regions of the patients' brains. However, there were no accompanying changes in metabolite ratios or intracellular pH. One has to be cautious in interpreting absolute signal intensities, particularly when using surface coils. Nevertheless, it was possible to attribute these observations to a reduction in the total number of metabolically active brain cells.

Welch and his colleagues have carried out ^{31}P NMR studies on a series of patients with acute ischaemic stroke due to major cerebral vessel occlusion (Welch *et al.*, 1988). All patients had focal neurological deficits, and they were serially studied at 18 hours (acute), between 32 and 72 hours after the onset of the clinical deficit (subacute), 7–9 days after stroke (intermediate) and 10–40 days after stroke (late). Overall, Welch and his colleagues observed distinct metabolic changes in the stroke patients that were greatest during the acute or subacute stages. No significant abnormalities in high-energy phosphates were measured in the intermediate or later stages despite persistent neurological deficit and evidence for infarction from X-ray CT or NMR images. It was suggested that, apart from partial recovery of some neurons, the return of high energy phosphates may also originate from glial cells that are relatively more resistant to ischaemia or macrophages that infiltrate the infarct. As in the studies of Bottomley *et al.* (1986) discussed above, this serves to emphasise the importance of measuring absolute signal intensities in addition to signal ratios.

So far, the major focus for clinical spectroscopy has been in the investigation of tumour metabolism, with particular interest in whether NMR can provide a useful indication of early response to therapy. Much of the emphasis in the ^{31}P spectroscopy is on the phosphomonester and phosphodiester signals, as these are the signals that tend to display the greatest abnormalities (Oberhaensli *et al.*, 1986; Segebarth *et al.*, 1987). The compounds that make the greatest contributions to these signals (see above) appear to be more closely related to membrane metabolism than to energy metabolism. One further point of interest is that the pH that is measured by ^{31}P NMR in human tumours is not acid; in general it is normal or somewhat alkaline, and it has been suggested (Oberhaensli *et al.*, 1986) that this may reflect a particularly effective Na^+/H^+ exchange mechanism.

The study of human brain metabolism by ^1H NMR has lagged behind ^{31}P studies, primarily for technical reasons, as discussed above. However, excellent ^1H spectra of the human brain have now been reported (Hanstock *et al.*, 1988; Bomsdorf *et al.*, 1988; Luyten *et al.*, 1988a,b; Bruhn *et al.*, 1989), showing striking abnormalities in some patients with intracranial tumours (Bruhn *et al.*, 1988; Luyten *et al.*, 1988a) and cerebrovascular disease (Luyten *et al.*, 1988b; Bruhn *et al.*, 1989). For example, in the spectra obtained from a patient with a 4-day old infarct, there was no detectable N-acetylaspartate signal in the damaged region, and the lactate signal was highly elevated (Bruhn *et al.*, 1989). On the basis of these results, together with the animal studies referred to above, it seems likely that ^1H NMR spectroscopy will prove to be at least as useful as ^{31}P NMR, and probably more so, in many studies of brain disease. It is interesting to note that N-acetylaspartate is believed to be located primarily in neurons (Nadler and Cooper, 1972; Koller *et al.*, 1984), and recent studies of biopsy samples from astrocytomas (Gill *et al.*, submitted) add weight to the possibility that the ^1H NMR signal from this compound could provide an endogenous neuronal marker.

Acknowledgements

We thank the Rank Foundation and the Wolfson Foundation for their support. Dr Kauppinen was on leave of absence from the Department of Clinical Neurophysiology, University Central Hospital of Kuopio, Finland, and was supported by a joint grant from the Sigrid Juselius Foundation and the Wellcome Trust.

References

1. Ackerman JJH, Grove TH, Wong GG, Gadian DG, Radda GK (1980) Mapping of metabolites in whole animals by ^{31}P NMR using surface coils. Nature 283: 167–170
2. Alger JR, Shulman RG (1984) Metabolic applications of ^{13}C nuclear magnetic resonance spectroscopy. Br Med Bull 40: 160–164
3. Allen K, Busza AL, Crockard HA, Frackowiak RSJ, Gadian DG, Proctor E, Ross Russell RW, Williams SR (1988) Acute cerebral ischaemia: concurrent changes in cerebral blood flow, energy metabolites, pH and lactate measured with hydrogen clearance and ^{31}P and ^1H nuclear magnetic resonance spectroscopy: III Changes following ischaemia. J Cereb Blood Flow Metabol 8: 816–821
4. Bachelard HS, Badar-Goffer RS, Brooks KJ, Dolin SJ, Morris PG (1988) Measurement of free intracellular calcium in the brain by ^{19}F-nuclear magnetic resonance spectroscopy. J Neurochem 51: 1311–1313
5. Bates TE, Williams SR, Busza AL, Gadian DG, Proctor E (1988) A ^{31}P nuclear magnetic resonance study *in vivo* of metabolic abnormalities in rats with acute liver failure. NMR Biomed 1: 67–73
6. Bates TE, Williams SR, Kauppinen RA, Gadian DG (1989) Observation of cerebral metabolites in an animal model of acute liver failure *in vivo*. A ^1H and ^{31}P nuclear magnetic resonance study. J Neurochem 53: 102–110
7. Bates TE, Williams SR, Gadian DG (1989) Phosphodiesters in the liver: the effect of field strength on the ^{31}P signal. Magn Reson Med 12: 145–150
8. Behar KL, den Hollander JA, Stromski ME, Ogino T, Shulman RG, Petroff OAC, Prichard JW (1983) High resolution ^1H NMR study of cerebral hypoxia *in vivo*. Proc Natl Acad Sci USA 80: 4945–4948
9. Berkowitz BA, Song S-K, Deuel RK, Ackerman JJH (1987) 2-Fluoro-2-deoxy-D-glucose cerebral metabolism at low NMR dose (20 mg/kg) in the conscious rat *in situ*: ^{19}F-(^1H) NMR investigation. Proc Soc Magn Reson Med 6th annual meeting New York, p 109
10. Bomsdorf H, Hetzel T, Kunz D, Roeschmann P, Tschendel O, Wieland J (1988) Spectroscopy and imaging on a 4T whole-body magnetic resonance system. NMR Biomed 1: 151–158
11. Bottomley PA (1989) Human *in vivo* NMR spectroscopy in diagnostic medicine: clinical tool or research probe? Radiology 170: 1–15
12. Bottomley PA, Foster TB, Darrow RD (1984) Depth-resolved surface-coil spectroscopy (DRESS) for *in vivo* ^1H, ^{31}P, and ^{13}C NMR. J Magn Reson 59: 338–342
13. Bottomley PA, Drayer BP, Smith LS (1986) Chronic adult cerebral infarction studied by phosphorus NMR spectroscopy. Radiology 160: 763–766
14. Bottomley PA, Charles HC, Roemer PB, Flamig D, Engeseth H, Edelstein WA, Mueller OM (1988) Human *in vivo* phosphate metabolite imaging. Magn Reson Med 7: 319–336
15. Brenton DP, Garrod PJ, Krywawych S, Reynolds EOR, Bachelard HS, Cox DW, Morris PG (1985) Phosphoethanolamine is major constituent of phosphomonoester peak detected by ^{31}P NMR in newborn brain. Lancet i: 115
16. Brooks KJ, Kauppinen RA, Williams SR, Bachelard HS, Bates TE, Gadian DG (1989) Ammonia causes a drop in intracellular pH in metabolising cortical brain slices. A ^{31}P and ^1H nuclear magnetic resonance study. Neuroscience 33: 185–192
17. Bruhn H, Frahm J, Gyngell ML, Merboldt KD, Hanicke W, Sauter R (1988) Localized proton spectroscopy of tumours *in vivo*: patients with primary and secondary tumours. Proc Soc Magn Reson Med, 7th annual meeting San Francisco, p 253
18. Bruhn H, Frahm J, Gyngell ML, Merboldt KD, Hanicke W, Sauter R (1989) Cerebral metabolism in man after acute stroke: new observations using localized proton NMR spectroscopy. Magn Reson Med 9: 126–131
19. Cady EB, Costello AMdeL, Dawson MJ, Delpy DT, Hope PL, Reynolds EOR, Tofts PS, Wilkie DR (1983) Non-invasive investigation of cerebral metabolism in newborn infants by phosphorus nuclear magnetic resonance spectroscopy. Lancet i: 1059–1062
20. Cerdan S, Harihara Subramanian V, Hilberman M, Cone J, Egan J, Chance B, Williamson JR (1986) ^{31}P NMR detection of mobile dog brain phospholipids. Magn Reson Med 3: 432–439

21. Clark LC, Ackerman JL, Thomas SR, Millard RW (1984) High-contrast tissue and blood oxygen imaging based on fluorocarbon ^{19}F NMR relaxation times. Magn Reson Med 1: 135–136
22. Coutts GA, Cox IJ, Gadian DG, Sargentoni J, Bryant DJ, Collins AG (1989) Phosphorus-31 magnetic resonance spectroscopy of the normal human brain: approaches using four dimensional chemical shift imaging and phase mapping techniques. NMR Biomed 1: 190–197
23. Crockard HA, Gadian DG, Frackowiak RSJ, Proctor E, Allen K, Williams SR, Ross Russell RW (1987) Acute cerebral ischaemia: concurrent changes in cerebral blood flow, energy metabolites, pH and lactate measured with hydrogen clearance and ^{31}P and ^{1}H nuclear magnetic resonance spectroscopy. II. Changes during ischaemia. J Cereb Blood Flow Metabol 7: 394–402
24. Deuel RK, Yue GM, Sherman WR, Schnicker DJ, Ackerman JJH (1985) Monitoring the time course of cerebral deoxyglucose metabolism by ^{31}P nuclear magnetic resonance spectroscopy. Science 228: 1329–1331
25. Deutsch C, Taylor JS, Wilson DF (1982) Regulation of intracellular pH by human peripheral blood lymphocytes as measured by ^{19}F NMR. Proc Natl Acad Sci USA 79: 7944–7948
26. Deutz NEP, Chamuleau RAFM, de Graaf AA, Bovee WMMJ, de Beer R (1988a) *In vivo* ^{31}P NMR spectroscopy of the rat cerebral cortex during acute hepatic encephalopathy. NMR Biomed 1: 101–106
27. Deutz NEP, de Graaf AA, de Haan JG, Bovee WMMJ, de Beer R (1988b) *In vivo* brain ^{1}H NMR spectroscopy during acute hepatic encephalopathy. In: Soeters PB (ed) Advances in ammonia metabolism and acute hepatic encephalopathy. Elsevier, Amsterdam, pp 439–446
28. Eidelberg D, Johnson G, Barnes D, Tofts PS, Delpy D, Plummer D, McDonald WI (1988) ^{19}F NMR imaging of blood oxygenation in the brain. Magn Reson Med 6: 344–352
29. Eleff SM, Schnall MD, Ligetti L, Osbakken M, Subramanian VH, Chance B, Leigh JS (1988) Concurrent measurements of cerebral blood flow, sodium, lactate and high-energy phosphate metabolism using ^{19}F, ^{23}Na, ^{1}H and ^{31}P NMR spectroscopy. Magn Reson Med 7: 412–424
30. Ewing JR, Branch CA, Helpern JA, Smith MB, Butt SM, Welch KMA (1989) Cerebral blood flow measured by NMR indicator dilution technique in cats. Stroke 20: 259–267
31. Foster MA, Hutchison JMS (eds) (1987) Practical NMR Imaging. IRL Press, Oxford Washington
32. Gadian DG (1982) Nuclear magnetic resonance and its applications to living systems. Oxford University Press, Oxford
33. Gadian DG, Frackowiak RSJ, Crockard HA, Proctor E, Allen K, Williams SR, Ross Russell RW (1987) Acute cerebral ischaemia: concurrent changes in cerebral blood flow, energy metabolites, pH and lactate measured with hydrogen clearance and ^{31}P and ^{1}H nuclear magnetic resonance spectroscopy. I. Methodology. J Cereb Blood Flow Metabol 7: 199–206
34. Garlick PB, Brown TR, Sullivan RH, Ugurbil K (1983) Observation of a second phosphate pool in the perfused heart by ^{31}P NMR; is this the mitochondrial phosphate? J Mol Cell Cardiol 15: 855–858
35. Gill SS, Small RK, Thomas DGT, Patel P, Porteous R, Van Bruggen N, Gadian DG, Kauppinen RA, Williams SR (1989) Brain metabolites as ^{1}H NMR markers of neuronal and glial disorders. NMR Biomech 2: 196–200
36. Gyulai L, Bolinger L, Leigh JS, Barlow C, Chance B (1984) Phosphorylethanolamine: the major constituent of the phosphomonoester peak observed by ^{31}P NMR on developing dog brain. FEBS Lett 178: 137–142
37. Hanstock CC, Rothman DL, Prichard JW, Jue T, Shulman RG (1988) Spatially localized ^{1}H NMR spectra of metabolites in the human brain. Proc Natl Acad Sci USA 85: 1821–1825
38. Hope PL, Costello AMdeL, Cady EB, Delpy DT, Tofts PS, Chu A, Hamilton PA, Reynolds EOR, Wilkie DR (1984) Cerebral energy metabolism studied with phosphorus NMR spectroscopy in normal and birth-asphyxiated infants. Lancet ii: 366–370
39. Hope PL, Reynolds EOR (1985) Investigation of cerebral energy metabolism in newborn infants by ^{31}P NMR spectroscopy. Clin Perinatology 12: 261–275
40. Hoult DI, Busby SJW, Gadian DG, Radda GK, Richards RE, Seeley PJ (1974) Observations of tissue metabolites using ^{31}P nuclear magnetic resonance. Nature 252: 285–287
41. Koller KJ, Zaczek R, Coyle JT (1984) N-Acetyl-aspartyl-glutamate: regional levels in rat brain and the effects of brain lesions as determined by a new HPLC method. J Neurochem 43: 1136–1142
42. Luyten PR, den Hollander JA, Segebarth C, Baleriaux D (1988a) Localized ^{1}H NMR spectroscopy and spectroscopic imaging of human brain tumours *in situ*. Proc Soc Magn Reson Med, 7th annual meeting San Francsisco, p 252
43. Luyten PR, van Rijen PC, Berkelbach vd Sprenkel JW, Tulleken CAF, den Hollander JA (1988b) ^{1}H NMR spectroscopic detection of cerebral metabolic alterations in patients with hemodynamically significant cerebrovascular disease. Proc Soc Magn Reson Med, 7th annual meeting, San Francisco, p 619
44. Morris PG (1986) Nuclear magnetic resonance imaging in medicine and biology. Clarendon Press, Oxford
45. Nadler JV, Cooper JR (1972) N-acetyl-L-aspartic acid content of human neural tumours and bovine peripheral tissues. J Neurochem 19: 313–319
46. Nakada T, Kwee IL, Conboy CB (1986) Noninvasive *in vivo* demonstration of 2-fluoro-2-deoxy-D-glucose metabolism beyond the hexokinase reaction in rat brain by ^{19}F nuclear magnetic resonance spectroscopy. J Neurochem 46: 198–201
47. Naritomi H, Sasaki M, Kanashiro M, Kitani M, Sawada T (1988) Flow thresholds for cerebral energy disturbance and Na^+ pump failure as studied by ^{31}P and ^{23}Na NMR spectroscopy. J Cereb Blood Flow Metabol 8: 16–23
48. Naruse S, Horikawa Y, Tanaka C, Hirakawa K, Nishikawa H, Watari H (1984) *In vivo* measurement of energy metabolism and the concomitant monitoring of electroencephalogram in experimental cerebral ischemia. Brain Res 296: 370–372
49. Oberhaensli RD, Bore PJ, Rampling RP, Hilton-Jones D, Hands J, Radda GK (1986) Biochemical investigation of human tumours *in vivo* with phosphorus-31 magnetic resonance spectroscopy. Lancet ii: 8–11
50. Pettegrew JW, Kopp SJ, Dadok J, Minshew NJ, Feliksik JM, Glonek T, Cohen MM (1986) Chemical characterization of a prominent phosphomonoester resonance from mammalian brain. ^{31}P and ^{1}H NMR analysis at 4.7 and 14.1 T. J Magn Reson 67: 443–450
51. Ross BD, Radda GK, Gadian DG, Rocker G, Esiri M, Falconer-Smith J(1981) Examination of a case of suspected McArdle's syndrome by ^{31}P nuclear magnetic resonance. N Engl J Med 304: 1338–1342
52. Segebarth CM, Baleriaux DF, Arnold DL, Luyten PR, den Hollander JA (1987) Image-guided localised ^{31}P MRS spectroscopy of human brain tumours *in situ*: effect of treatment. Radiology 165: 215–219
53. Siesjö BK (1985) Acid-base homeostasis in the brain: physiology, chemistry, and neurochemical pathology. Prog Brain Res 63: 121–154
54. Smith GA, Hesketh RT, Metcalfe JC, Feeney J, Morris PG (1983) Intracellular calcium measurements by ^{19}F NMR of fluorine-labelled chelators. Proc Natl Acad Sci USA 80: 7178–7182
55. Stark DD, Bradley WG (eds) (1988) Magnetic resonance imaging. CV Mosby, St Louis

56. Thulborn KR, du Boulay GH, Duchen LW, Radda GK (1982) A ^{31}P nuclear magnetic resonance *in vivo* study of cerebral ischaemia in the gerbil. J Cereb Blood Flow Metabol 2: 299–306
57. Tofts P, Wray S (1985) Changes in brain phosphorus metabolites during the post-natal development of the rat. J Physiol 359: 417–429
58. Welch KMA, Gross B, Licht J, Levine SR, Glasberg M, Smith MB, Helpern JA, Bueri J, Gorell JM (1988) Magnetic resonance spectroscopy of neurologic diseases. In: SH Appel (ed) Current neurology, Vol 8, pp 295–331
59. Younkin DP, Delivora-Papadopoulos M, Leonard JC, Subramanian VH, Eleff S, Leigh JS, Chance B (1984) Unique aspects of human newborn cerebral metabolism evaluated with phosphorus nuclear magnetic resonance spectroscopy. Ann Neurol 6: 581–586

Correspondence and Reprints: David G. Gadian, Department of Physics in Relation to Surgery, Hunterian Institute, Royal College of Surgeons of England, 35–43 Lincoln's Inn Fields, London WC2A 3PN, U.K.

Acta Neurochir (1993) [Suppl] 57: 9–20

Cerebral Metabolic Studies *in vivo* by Combined $^{1}H/^{31}P$ and $^{1}H/^{13}C$ NMR Spectroscopic Methods

K. L. Behar

Department of Molecular Biophysics and Biochemistry, Yale University, New Haven, U.S.A.

Abstract

Intracellular pH and ammonium ion concentration are potent modulators of cerebral amino acid metabolism. Furthermore, intracellular acidosis and hyperammonemia accompany conditions such as ischemic encephalopathy and seizures and may contribute to the pathological sequelae observed. *In vivo* NMR spectroscopy permits multiple, non-destructive measurements of important cerebral metabolic intermediates in the same animal. We describe here the use of ^{1}H, and ^{31}P NMR spectroscopy to investigate the effects of acute changes in intracellular pH and ammonium ions on cerebral glutamate, glutamine, and lactate levels *in vivo*. We then show how ^{1}H NMR can be used to indirectly follow the flow of ^{13}C label from [1–^{13}C] glucose into the cerebral glutamate pool, allowing us to measure cerebral TCA activity in normal and chronically hyperammonemic rats.

Male Sprague-Dawley rats (160–210 gm), fasted 24-hours, were tracheotomized, paralyzed and ventilated on 30% O_2 / 70% N_2O. NMR spectroscopy was performed at a field strength of 8.4 Tesla using a Bruker AM-360 wide bore spectrometer. An elliptical surface-coil (8 × 12 mm) was double-tuned to either the ^{1}H and ^{31}P or ^{1}H and ^{13}C frequencies. After retraction of extracranial tissues, the coil was positioned over the skull 2 mm posterior to the bregma. Tail arteries and veins were cannulated allowing periodic measurements of PO_2, pCO_2, pH and glucose in arterial blood and intravenous infusions.

Respiratory acidosis was induced in rats by the addition of CO_2 to the ventilation gas mixture. Arterial pCO_2 increased within 5 min from a pre-hypercarbic value of 36.4 ± 6.1 mm Hg to 200–220 mm Hg and was maintained at this level for over 1 hour. Hypercarbia led to rapid cerebral acidification. Intracellular pH decreased from 7.18 ± 0.08 (pre-hypercarbic period) to 6.68 ± 0.06 (n = 4) at 10 min and remained stable throughout the NMR observation period. Glutamate decreased to 53 ± 4% of control after 60 min of hypercarbia, while glutamine increased to 126 ± 7% of control.

Acute hyperammonemia was produced by a programmed intravenous infusion of 250 mM ammonium acetate, which rapidly raised and maintained the concentration of ammonium ions in the blood at approximately 500 μM. Shortly after the start of the infusion (10–20 min), the levels of glutamine and lactate rose continuously throughout the experiment, reaching levels of 170 ± 25% and 260 ± 60% of control, respectively (n = 12) after 50 min. Glutamate decreased during the same time interval to 80 ± 4% of control (n = 12). These changes were not observed in animals infused with sodium acetate (control group, n = 6). No changes were observed in intracellular pH or high energy phosphates in the ^{31}P spectrum as a result of either the ammonium acetate or sodium acetate infusions.

In the final study we used ^{1}H-observed, ^{13}C-decoupled spectroscopy to measure the flow of ^{13}C label from [1–^{13}C] glucose into cerebral glutamate (and glutamine) *in vivo* in normal and chronically hyperammonemic (4-week portacaval shunted) rats. The flow of label into glutamate-C4 allowed us to measure the steady-state cerebral TCA cycle activity *in vivo*. An infusion protocol was developed to rapidly increase and maintain the enrichment of [1–^{13}C] glucose in arterial blood. The metabolism of [1–^{13}C] glucose through glycolysis and the TCA cycle results initially in the incorporation of the label in the C4 of α-ketoglutarate (α-KG) and glutamate, with which α-KG is in rapid equilibrium. Subsequent turns of the TCA cycle will lead to ^{13}C label incorporation in C2, C3 and later in C1 of glutamate. In control animals the rate of incorporation of ^{13}C into glutamate-C4 followed first order kinetics with a rate constant of 0.130 ± 0.012 min^{-1}; glutamate-C3 was much slower (0.026 ± 0.004 min^{-1}). In contrast, the rate constant for glutamate-C4 labeling in 4-week portacaval shunted rats was only 0.038 ± 0.008 min^{-1}, suggesting a significant reduction in TCA cycle activity in chronic hyperammonemia.

These studies indicate

1) acute cerebral acidosis leads to a large reduction in cerebral glutamate and a smaller increase in glutamine;

2) acute hyperammonemia does not alter cerebral pH, but leads to large increases in glutamine and lactate, and a smaller decrease in glutamate;

3) chronic hyperammonemia causes a significant reduction in the rate of label incorporation from [1–^{13}C] glucose into glutamate, probably due to a fall in cerebral TCA cycle activity.

Keywords: Intracellular pH; hyperammonemia; ^{1}H-observed ^{13}C-decoupled spectroscopy; label flow.

Introduction

Developments in *in vivo* applications of ^{1}H NMR spectroscopy have progressed at a rapid pace in the few short years since its introduction. An extremely powerful method to study cerebral metabolism, ^{1}H NMR is intrinsically more sensitive than ^{31}P or ^{13}C

NMR. However, ^{1}H spectroscopy is also more difficult due to the need to suppress intense signals arrising from water and fats and attain good peak resolution in a spectrum where metabolite resonances are crowded into a spectral bandwidth of $\leq$7 ppm. These requirements, although demanding in terms of spectroscopy hardware and software capabilities, have been met in recent years leading to numerous experimental and clinical applications in the study of cerebral metabolism.

The assignment of resonances in the ^{1}H spectrum of brain are based on analysis of brain tissue and its acid extract (Behar *et al.*, 1983; 1986; Middlehurst *et al*, 1984a, b; Arus *et al*, 1985; Cerdan *et al.*, 1985; Fan *et al.*, 1986; Petroff *et al.*, 1988a, b; Richards and Budinger, 1988). The majority of low molecular weight organic acids present in the ^{1}H spectrum of mammalian brain have been identified (see Table 1 and Fig. 1). Studies of rat brain at high magnetic field strength with homonuclear spectral editing techniques, indicate that lactate, alanine, glutamate/glutamine, GABA, and taurine resonances can be selected and quantitated (Rothman *et al*, 1984; Hetherington *et al.*, 1985). The resonances of N-acetylaspartate, creatine/phosphocreatine, trimethylamines, and lactate are amenable to analysis at low field strengths. Amino acid resonances have also been identified in ^{1}H spectra of human brain at the lower magnetic field strengths of currently available large bore diameter magnets (1.5 to 2.1 Tesla) (Hanstock *et al.*, 1989; Frahm *et al.*, 1989; Luyten *et al.*, 1989). Lactic acid metabolism has been studied in the basal state and under pathophysiologic conditions in animal and human brain (Fitzpatrick *et al.*, 1988; Hanstock *et al.*, 1988a, d; Petroff *et al.*, 1986).

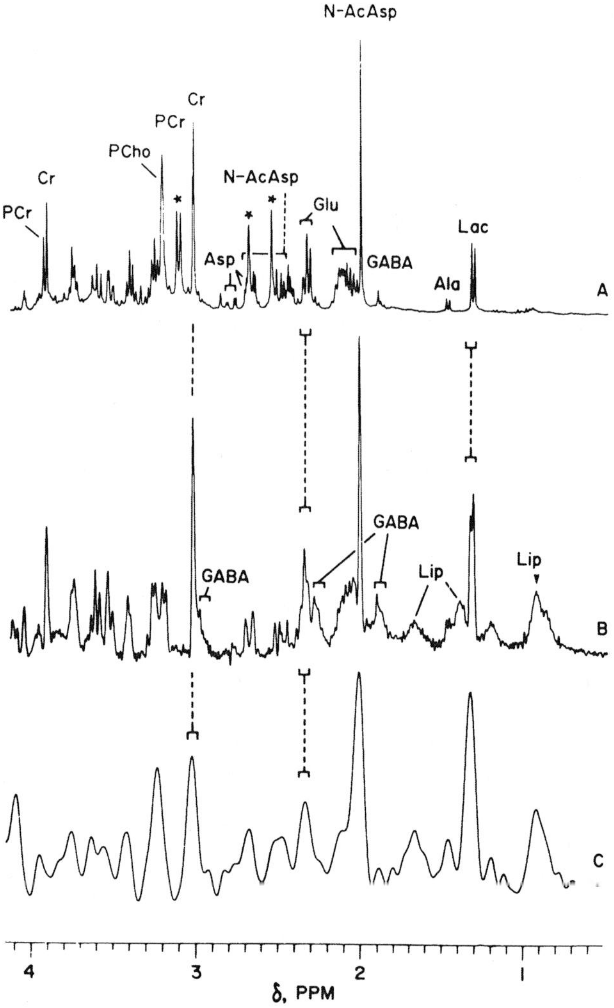

Fig. 1. ^{1}H NMR spectra of rat brain obtained at 8.4 Tesla. (A) Acid extract obtained from rat brain frozen *in situ* with liquid N_2. (B) Excised brain tissue. (C) Spectrum obtained *in vivo* with surface coil following hypoxia. *PCr* phosphocreatine; *Cr* creatine; *PCho* phosphatidyl choline and other trimethyl amine groups; *Asp* aspartate; *N-AcAsp* N-acetylaspartate; *Glu* glutamate; *GABA* γ-aminobutyric acid; *Ala* alanine; *Lac* lactate; *Lip* lipid (and proteins); * contaminants. Reproduced from Behar *et al.* (1983)

This review is not comprehensive and is meant to serve only as an overview of a rapidly expanding field. Impressive advances have been made, recently, in the attainment of well-resolved ^{1}H NMR spectra of localized volumes from human brain. For this reason I have included several studies presented as abstracts at the 8th annual meeting of the Society of Magnetic Resonance in Medicine which was held in Amsterdam, The Netherlands on August 12–18, 1989.

Studies of Lactic Acid Metabolism

Hypoxia

Lactate was the first cerebral metabolite to be studied by ^{1}H NMR in the brain *in vivo* under conditions of hypoxia (Behar *et al*, 1983; Behar *et al*, 1984). The rise in cerebral lactate, as represented in the ^{1}H spectrum, was shown to depend critically on the arterial PO_2. By combining both ^{1}H and ^{31}P NMR techniques it has been possible to correlate changes in brain lactate, pH_i, and energy metabolism simultaneously within the same animal. Figures 2–4 show the effects of hypoxia on rat brain (Behar, 1985). Intracellular pH was not affected during the first 5 min of hypoxia; a time when rapid changes were observed in lactate and phosphocreatine. Some variation was observed in the amount of lactate that accumulated before a significant decrease in pH_i was observed; increases in lactate concentration to ~ 4–8 μmol/g wet wt were observed before the decrease

Table 1. *The Major Compounds Observed in ^{1}H NMR Spectra of the Brain*

	Chemical shift (ppm)	
Lac	[3] CH_3	1.33
	[2] CH	4.12
Ala	[3] CH_3	1.48
	[2] CH	3.78
GABA	[2] CH_2	2.31
	[3] CH_2	1.91
	[4] CH_2	3.02
N-AcAsp	[2] CH	4.40
	[3] CH_2	2.52
	[3′] CH_2	2.70
	– CH_3	2.02*
Glutamate	[2] CH	3.76
	[3] CH_2	2.05–2.10
	[4] CH_2	2.35
Glutamine	[2] CH	3.76
	[3] CH_2	2.14
	[4] CH_2	2.46
Taurine	SCH_2	3.43
	NCH_2	3.27
Inositol	[1,3] CH	3.56
	[2] CH	4.06
	[4,6] CH	3.62
	[5] CH	3.28
Phosphocreatine/ creatine	– CH_2 –	3.93
	– CH_3	3.04
Phosphatidyl choline	– N $(CH_3)_3$	3.21

* All chemicals shifts *in vivo* are referenced to the methyl group of N-AcAsp at 2.02 ppm.

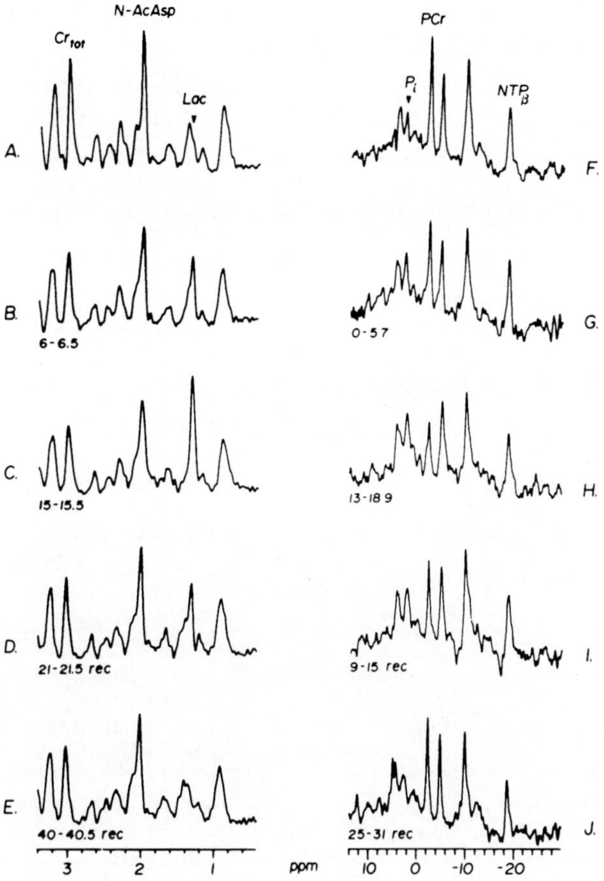

Fig. 2. Effects of hypoxia and recovery upon lactate and high energy phosphates in the rat brain as revealed by sequential ^{1}H and ^{31}P NMR spectra. Rats were tracheotomized and ventilated on 25–33% O_2 in nitrous oxide and the scalp was retracted as described in Behar *et al.* (1983). ^{1}H NMR spectra are presented along the left column and ^{31}P NMR spectra along the right column. The time intervals, in min, is indicated beneath each spectrum; zero time represents the induction of hypoxia (5–6%O_2). Spectra A and F were obtained prior to hypoxia. Recovery was obtained by administration of 25% O_2 at time zero; this is indicated by *rec* and is followed by the elapsed recovery time (in min). For labels in the ^{1}H spectrum, see Fig. 1; Cr_{tot}, total creatine (creatine + phosphocreatine) – CH_3. Labels in the ^{31}P spectrum: P_i, inorganic orthophosphate; PCr, phosphocreatine; NTP_β, nucleoside triphosphate β. ^{1}H spectra: interpulse interval, 2.2 sec (total time per spectrum of 70.4 sec; ^{1}H spectrum A is the sum of 32 scans while spectra B–E are the sum of 2–16 scan spectra of 35.2 sec each. ^{31}P spectra: interpulse interval, 0.6 sec (total time of 4.0 min; ^{31}P spectra G–I are the sum of 2–200 scan spectra of 2.0 min each; spectra F and J are the sum of 600 scans each). Reproduced from Behar (1985)

in pH_i was > 0.05 pH units. The magnitude of the changes in lactate and phosphocreatine generally reflected the intensity of the hypoxic stress. During the first 10 min of hypoxia (10% O_2), lactate increased at an average rate of ~ 0.5 μmol/min/g wet wt (n = 3). No significant changes in nucleoside triphosphate (NTP_β) were detected in ^{31}P-NMR spectra acquired throughout the hypoxic periods, an observation that is in accord with other NMR studies of hypoxia (Prichard *et al.*, 1983; Gyulai *et al.*, 1987) and results obtained from *in situ* freeze-extraction studies (Gurdjian *et al.*, 1944; Duffy *et al.*, 1972).

During hypoxia the electroencephalogram (EEG) revealed a wave pattern that progressed from high to low frequencies and increased in amplitude with time. This response of the EEG to hypoxia is well-known (Gurdjian *et al.*, 1949). Progressive EEG slowing during hypoxia correlated with increasing brain lactate (Fig. 5), an observation first made by Gurdjian *et al.* (1949) in their studies of hypoxia in dog brain. Isoelectricity was associated with the highest levels of lactate during hypoxia, ≈ 18–23 μmol/g wet wt. However, repeated descents into hypoxia in the same animal often led to an isoelectric EEG before lactate had accumulated beyond 4–5 μmol/g.

The complete recovery of lactate to pre-hypoxic levels took 20–30 min and was longer than that of PCr. The concentration of lactate (Δ[Lac], converted to μmol/g i.c. water) was plotted against pH_i (Fig. 6). If the major changes in [H^+] during the post-hypoxic recovery period are assumed to be due to changes in [lac] and [P_i] alone, and $\Delta[P_i] = \Delta[PCr]$, then an estimate of the change in buffer base concentration (Δ[BB]), which is the sum of the conjugate base buffer species, can be plotted against pH_i according to the

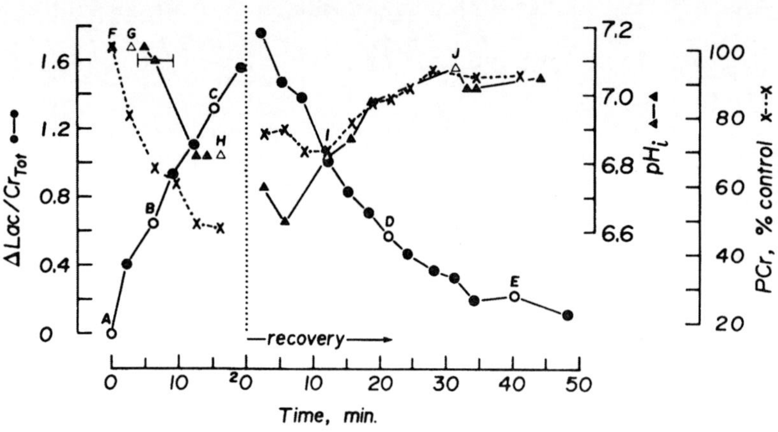

Fig. 3. Time-dependent changes in lactate, phosphocreatine, and pH_i during hypoxia and recovery in the rat brain. Time-course data were taken from the ^1H and ^{31}P spectra presented in Fig. 2. The change in lactate (Δlac) is presented as the ratio to total creatine (Cr_{tot}). Phosphocreatine (PCr) is plotted as a percentage of the control value. The hypoxic interval (FiO_2 = 4.8–6.1%; 0–20 min) is followed by recovery (FiO_2 = 25%); this is delineated by the dotted vertical line. Intracellular pH, pH_i. Reproduced from Behar (1985)

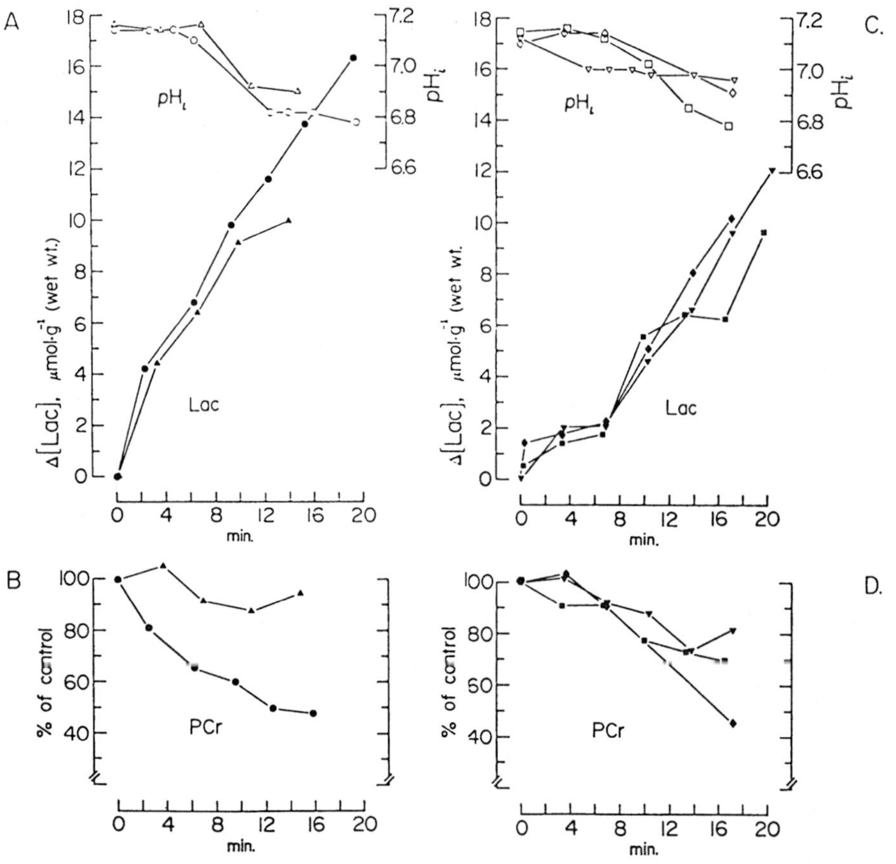

Fig. 4. Changes in cerebral pH_i, lactate, and phosphocreatine during hypoxia in the rat. (A, B) FiO_2 = 4.8–6.2% (O, O) and 7% (Δ, Δ), respectively. (C, D) FiO_2 = 10% (n = 3). The time coordinate for each datum point represents the mid-point time of spectrum accumulation. Different symbols represent the response of individual animals; filled symbols refer to axes on the left side of each figure, while unfilled symbols in (A) and (C) refer to the pH_i on the right. Mean rectal temperatures measured during hypoxia were between 35.5–36.6°C for all animals except that represented by (•, ∘) in (A) and (B), where it fell to 34.6°C. Reproduced from Behar (1985)

following relation (modified from Nilsson and Siesjö, 1971):

$$-[BB] = \Delta[Lac] - (K_2'/(K_2' + [H^+])) * \Delta[PCr] \quad (1)$$

where K_2' is the second apparent ionization constant for phosphoric acid. The relationship between $\Delta[BB]$ and pH_i was used to obtain an apparent buffer value (β') for rat brain *in vivo*. Although blood gases were not measured in these animals, the arterial pCO_2 would be expected to have a major affect on intracellular buffering and the value of β'. Arterial pCO_2 decreases during hypoxia in unanesthetized, spontaneously breathing rats (Lewis *et al.*, 1973) and this (including a decrease in cerebral tissue CO_2) occurs even when ventilation is constant (Norberg and Siesjö, 1975), as in the present experiments. The effect that changes in the pCO_2 would have on the relationship between pH_i and $\Delta[BB]$ and β (where $\beta = [BB]/\Delta pH_i$) was calculated from an equation that mimics the intracellular physiochemical buffering of rat brain (Siesjö, 1973). The NMR-derived *in vivo* data were observed to parallel roughly the 40 mmHg pCO_2 isobar (Fig. 6). The NMR-derived data for pH_i and $\Delta[BB]$ were fitted by the least squares method to a linear function of $\Delta[BB]$ giving a slope and intercept (at $\Delta[BB] = 0$) of 0.0248 and 7.13, respectively ($r^2 = 0.90$). The reciprocal of the slope ($\Delta[BB]/\Delta pH_i$), which is a measure of the apparent buffer value (β'), was 40.3 μmol/g i.c. water/ pH unit or 32 μmol/g wet wt/pH unit assuming an intracellular water content of 0.8 g i.c. water/g wet tissue wt. This value is not significantly different from buffer values measured during ischemia (see below). The intercept value of pH_i of 7.13 at $\Delta[BB] = 0$ did not differ significantly from the normal average pH_i of 7.14 (n = 4 rats) obtained before hypoxia; therefore, no changes occurred in the steady state pH_i that were not also reversible within this period.

The rapid clearance of lactate may be due to metabolic removal through the oxidation of pyruvate in the TCA cycle (Drews and Gilboe, 1973). The initial decay rates of lactate following reoxygenation ranged between 0.5–1.4 μmol/g/min. These values are far in excess of the transport V_{max} estimated for lactate between brain and blood of 0.08–0.12 μmol/g/min (Pardridge *et al.*, 1975) indicating that this transport carrier must be saturated with lactate. The exponential decay rate of the clearance curve suggests that transport is not responsible and that a pathway exists with a higher capacity for lactate removal than transport. If pyruvate is being oxidized during recovery at a rate

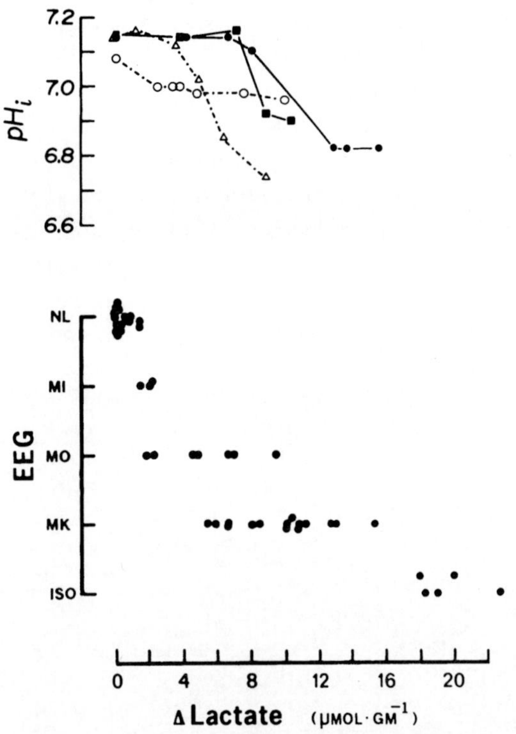

Fig. 5. Association between changes in cerebral lactate and the electroencephalogram during hypoxia in the rat. The patterns of electrical activity displayed in the EEG at a given lactate concentration are given at the bottom of the figure and the symbols are defined as following: *NL* normal with high frequencies; *MI* mild slowing; *MO* moderate slowing with increased amplitude; *MK* marked slowing with increased amplitude; *ISO* isoelectric. Reproduced from Behar (1985)

that is commensurate with its pre-hypoxic value, then the net disappearance of lactate must involve an increase in O_2 consumption.

Ischemia

Incomplete cerebral ischemia has been investigated by ^1H and ^{31}P NMR in newborn piglets (Corbett *et al.*, 1988), young lambs (Hope *et al.*, 1987; 1988), gerbils (Gadian *et al.*, 1987; Crockard *et al.*, 1987; Allen *et al.*, 1988), rats (Richards *et al.*, 1987) and cats (Gyulai *et al.*, 1987; Ligeti *et al.*, 1987). Most studies to date have used a combination of bilateral carotid artery ligation and hemorraghic hypotension to induce partial but reversible ischemia. During ischemia lactate and pH_i change reciprocally: lactate may reach concentrations of 14–40 μmol/g while pH_i decreases to 5.9–6.4. Although previous studies have shown that brain lactate levels during cardiovascular arrest or profound hypoxia depend on the preischemic concentration of blood glucose (Myers and Yamaguchi, 1976), studies of incomplete ischemia by ^1H NMR are less clear. Hope *et al.* (1988) found no effect of glucose administration (arterial blood glucose concentrations up to 18 mM) on lactate levels or pH_i during partial ischemia or recirculation in lamb brain. Poor correlation was also found between arterial glucose concentrations and brain lactate or ΔpH_i *during* partial ischemia in neonatal piglet brain (Corbett *et al.*, 1988); however, a stronger correlation was found when comparing the preischemic values with lactate or ΔpH_i measured at the end of hypotension or death (complete ischemia).

The change in pH_i in response to a given change in the concentration of lactate has been measured during ischemia; in the adult brain, changes in [lactate] and pH_i are linear between 0 and 20 μmol/g, yielding a slope of 29 ± 4 μmol lactate/pH unit (Gyulai *et al.*, 1987). Corbett *et al.* (1988) has shown that a linear relation of 30 ± 2 μmol lactate/pH unit exists between [buffer base] and pH_i to beyond 40 μmol/g in the neonatal piglet (Corbett *et al.*, 1988). A somewhat lower value of ≈10 μmol lactate/pH unit can be calculated from the data of Allen *et al.* (1988) for the period between 10 and 20 min of recirculation following ischemia in the gerbil brain; although significantly

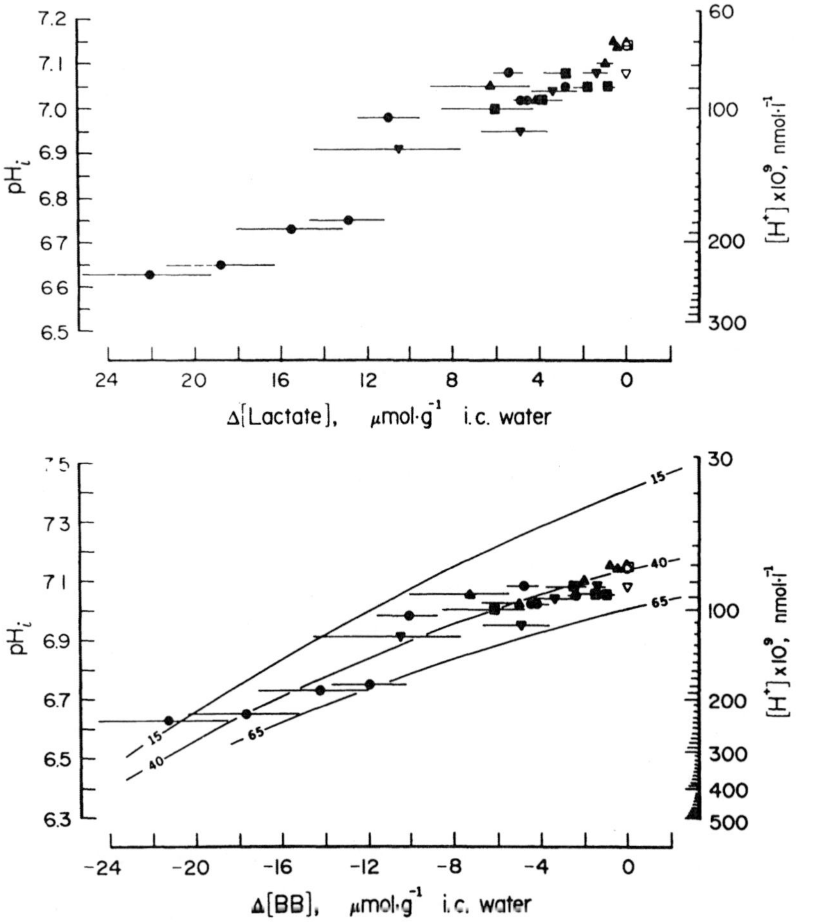

Fig. 6. The relationship between cerebral pH_i, lactate, and buffer base concentration during recovery from hypoxia. (Upper) Intracellular pH was determined from the ^{31}P spectrum. Lactate and total creatine amplitudes were measured in the ^1H spectrum collected alternately between ^{31}P spectra, and expressed as the ratio ($\Delta lac/Cr_{tot}$). The values were fit to a first order exponential function of time and the slope and intercept calculated. Lactate values were calculated for the midpoint and endpoint times that correspond to the time interval over which the pH measurement was made; horizontal error bars correspond to these endpoints and are larger for high values of *Δ[lactate]* where lactate is changing most rapidly. Lactate concentrations are reported as μmol/g intracellular (i.c.) water. Unfilled symbols give the average cerebral pH_i over the pre-hypoxic control period for the respective animal represented by that symbol. (Lower) The relationship between the cerebral pH_i and the change in buffer base concentration during recovery from hypoxia. Buffer base changes were calculated from Δ[lac] and Δ[PCr] according to Eq. [1] in the text and were expressed as μmol/g i.c. water to allow comparison with an ideal buffer model equation. The PCO_2 isobars (curving solid lines) were constructed from an equation that relates *Δ[BB]* to the ionization constants of CO_2 and 9 other arbitrary non-carbonic acid buffers of 6 mM each (total buffer concentration 54 mM). The pK_a's of the arbitrary buffers were chosen to be equally spaced between 5.3 and 7.7 (Siesjö, 1973). Reproduced from Behar (1985).

lower than other reported *in vivo* values, the lactate clearance rate of 0.35 mM/min was nearly the same as the value reported for the brains of larger animal species. Gyulai *et al.* (1987) measured a clearance rate of 0.36 ± 0.02 μmol/min/g in cat brain for lactate concentrations above 4 μmol/g. The kinetics was zero order and independent of the severity of the insult. As discussed above, the high rate of lactate clearance is probably due to metabolic removal; Chang *et al.* (1989) have demonstrated marked increases in the rate of lactate recovery following complete global ischemia following activation of pyruvate dehydrogenase complex by dichloroacetate (100 mg/kg i.v.).

Recovery of lactate and pH_i during recirculation after prolonged complete cerebral ischemia differs from that of hypoxia. Following recirculation after 1 hr complete cerebral ischemia in the cat, pH_i decreases transiently but does not renormalize for an additional 20 min, although there is a substantial recovery of NTP and PCr (Behar *et al.*, 1989b). Lactate levels recover more slowly than pH_i or high energy metabolites.

Rapid progress has been made over the last two years in ^{1}H spectroscopy of the human brain. Volume localized techniques have yielded ^{1}H spectra of high resolution and quality from brain volumes between 14 to 27 cm³ and, as low as, 1 cm³ (Hanstock *et al.*, 1988b,d; Frahm *et al.*, 1989). Lactate concentrations between 5 and 16 mM were measured in patients after cerebral infarction (Berkelbach van der Sprenkel *et al.*, 1988; Bruhn *et al.*, 1989). Bruhn *et al.* (1989) reported a complete loss of N-acetylaspartate and a significant loss of creatine and choline-containing molecules within the ischemic region 4 days after an acute infarction; these losses may reflect both edema and leakage of the metabolites from necrotic tissue.

Seizures and Hypocarbia

^{1}H NMR studies of bicuculline-induced seizures in rabbits (Petroff *et al.*, 1986), cats (Schnall *et al.*, 1988), fluorothyl and bicuculline-induced seizures in neonatal dogs (Young *et al.*, 1987; 1989) and electroshock-

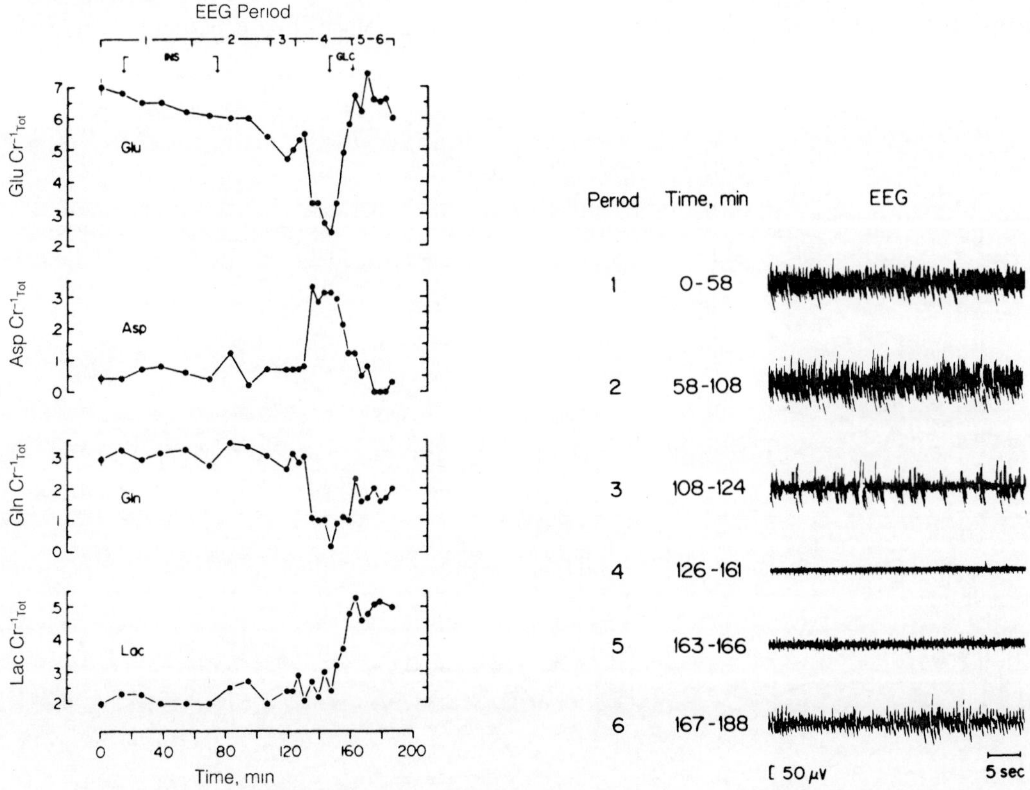

Fig. 7. Changes in amino acids, lactate, and the EEG in the rat brain during insulin-induced hypoglycemia and recovery following glucose administration. Resonance amplitudes were measured directly from *in vivo* ^{1}H spectra (32 scans; 1.18 min). Changes in metabolites are expressed as a ratio to total creatine (Cr_{tot}). Insulin and glucose administration are indicated by downward pointing arrows marked by *"Ins"* and *"Glc"*, respectively. Each EEG trace represents a given interval of time along the time axis and are described as follows: *1* 0–58 min, normal and fast; *2* 58–108 min, mild to moderate slow; *3* 108–124 min, marked slow; *4* 126–161 min, isoelectric (the periodic spikes are respirator artifacts); *5* 163–166 min, low amplitude; *6* 167–188 min, moderate to marked slow. Reproduced from Behar *et al.* (1985)

induced seizures (Prichard *et al.*, 1987) in rabbits have all reported increased levels of cerebral lactate. Following bicuculline administration, brain pH_i decreases to ≈ 6.8 after several minutes (Petroff *et al.*, 1984, 1986; Schnall *et al.*, 1988). Petroff *et al.* (1986) reported a rapid increase in lactate which stabilized after 30 min (pseudo-first order time constant of 12 min) and remained nearly constant for the next 1–2 hours in seizures that lasted more than 1 hour. For seizure discharges lasting only 10–20 min, the elevation of lactate persisted even after pH_i and the EEG had recovered to normal. Lactate levels decrease slowly in rabbit brain after electroshock-induced seizures to the cortex or optic nerve (Prichard *et al.*, 1987) and this pool is metabolically active, as ^{13}C is incorporated rapidly from [1–^{13}C]glucose into lactate (Petroff *et al.*, 1989).

Intracerebral alkalosis of 0.1–0.2 pH units and increased lactate have been observed together during hyperventilation-induced hypocapnia in rabbit (Petroff *et al.*, 1985) and human brain by combined ^{1}H and ^{31}P NMR techniques. In both cases, high energy phosphates (PCr and NTP) were unaffected indicating the probable lack of vasoconstriction-induced tissue hypoxia as an explanation for the increase in lactate. Cady *et al.* (1987) found no significant change in cerebral pH_i during increased volume ventilation (pCO_2 of 16 mmHg) in the new-born lamb; this may relate to specific differences in buffering (Corbett *et al.,* 1988) and/or regulation of pH_i in hypocapnia between mature and immature brain.

Amino Acid Metabolism

Metabolic encephalopathies of the brain often lead to disturbances in TCA cycle and amino acid metabolism. Important amino acids, such as glutamate, aspartate, glutamine, and GABA are detected in the ^{1}H spectrum of the brains of experimental animals (Behar *et al.*, 1983; Rothman *et al.*, 1984) and human brain by volume selection techniques for sufficiently short spin echo delay times (Hanstock *et al.*, 1988b,c, 1989; Frahm *et al.*, 1989). Because many ^{1}H resonances are crowded into a narrow spectral band, both homo- and heteronuclear resonance editing techniques have been applied to detect and quantitate brain amino acid resonances (Rothman *et al.*, 1984, 1985; Hetherington *et al.*, 1985).

Substrate oxidation in the TCA cycle is tightly coupled to energy metabolism, so it is particularly advan-

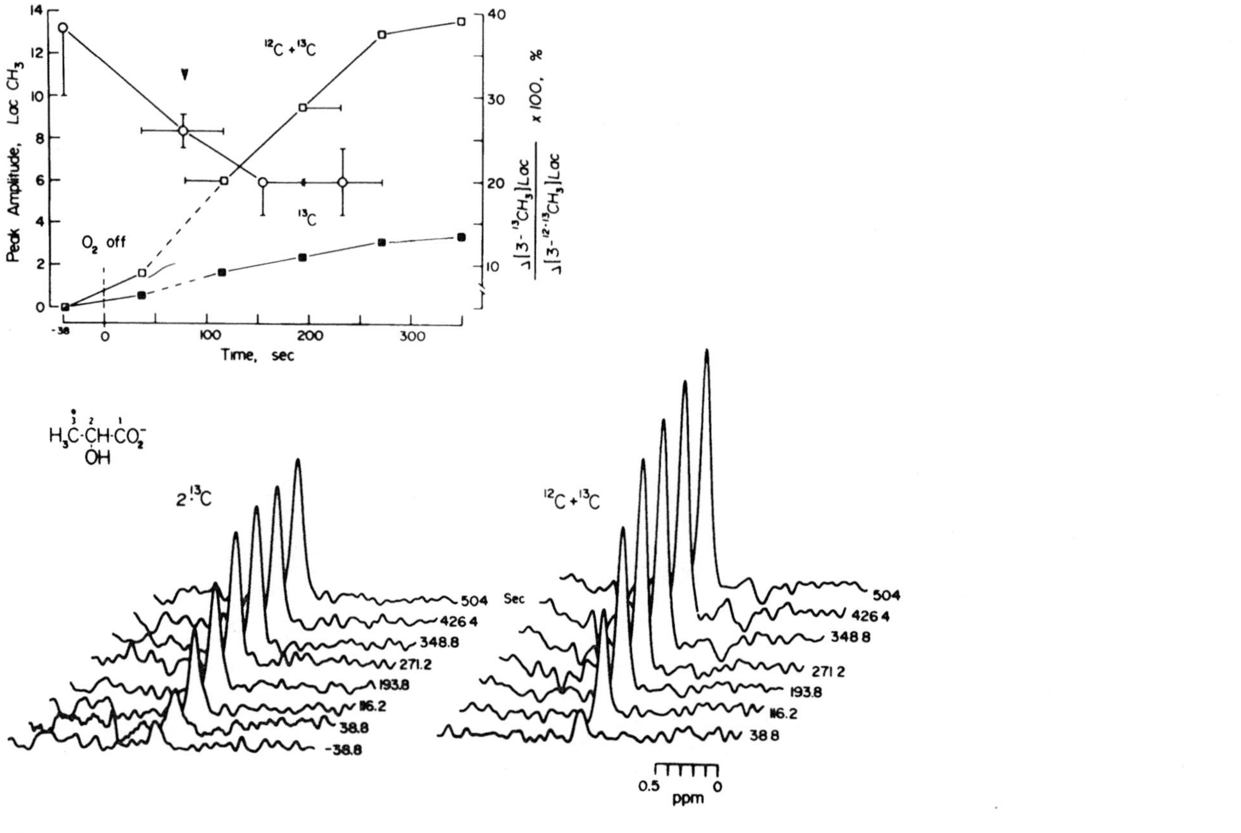

Fig. 8. Proton-observed, carbon-decoupled difference spectrum of [3–$^{13}CH_2$] lactate during ischemia following an infusion of [1–^{13}C] glucose. Reproduced from Rothman *et al.* (1985)

tageous to combine spectroscopic observations by ^1H and ^{31}P NMR. These techniques were applied to study the effects of insulin-induced hypoglycemia (Behar *et al.*, 1985), hypercapnia (Behar *et al.*, 1987, 1989), hepatic encephalopathy and hyperammonemia (Deutz, 1988; Fitzpatrick *et al.*, 1989a,b) on amino acids, pH_i, and high energy phosphates of rat brain.

Hypoglycemia

The cerebral metabolic effects of insulin-induced hypoglycemia in the rat was studied by ^1H and ^{31}P NMR (Behar *et al.*, 1985). Although no changes were observed in high energy phosphates (PCr and NTP) or pH_i during the period of EEG slowing, energy failure occurred rapidly when the EEG became isoelectric. These changes were followed by an equally rapid decrease in glutamate and glutamine and an increase in aspartate (Fig. 7). These effects were reversed after glucose administration and metabolite levels recovered over a time period of ≈ 20 min; however, cerebral electrical activity did not reappear until glutamate and aspartate had recovered to 90% of their prehypoglycemic levels. Following the administration of glucose lactate levels increased indicating that the rate of glycolysis and pyruvate oxidation in the TCA cycle were not balanced.

Hypercapnia

Hypercapnia produces rapid intracellular acidification which has a marked effect on the concentrations of glycolytic and TCA cycle intermediates. Amino acids linked to TCA cycle intermediates by transaminases and dehydrogenases are also affected by changes in pH and net changes in their steady state levels have been proposed as a means of pH regulation (Siesjö and Messeter, 1971). Decreases in glutamate have been observed in ^1H spectra of rat brain during CO_2 inhalation (Behar *et al.*, 1987), and the timecourse of changes in both glutamate and glutamine has been defined in normal and portacaval shunted rats (Behar and Fitzpatrick, 1989). Following an increase in the pCO_2 to 200 mmHg, pH_i decreased to ≈ 6.72 and remained stable over a period of 60 min. The rate of glutamate disappearance appeared biphasic – a rapid decrease for the first 20 min was followed by a more gradual decline between 20–60 min. The slower rate of decrease of glutamate between 20–60 min matched the slow rate of increase of glutamine over the entire 60 min of observation. The reason why glutamate levels decrease during hypercarbia is not well understood but the increase in glutamine may follow an increase in the ammonium ion concentration.

Hyperammonemia and Hepatic Encephalopathy

Increased concentrations of blood and brain ammonium ions occur during hypoxia/ischemia, seizures, hypercapnia, and liver disease leading to disturbances of brain glucose and amino acid metabolism. Studies of acute hyperammonemia on amino acids, pH_i, and high energy phosphates have been reported in rat brain following ammonium acetate infusion (Fitzpatrick *et al.*, 1989b), and acute liver ischemia (Deutz, 1988). Chronic hepatic encephalopathy has been reported in studies of portacaval shunted rats (Fitzpatrick *et al.*, 1989a,b; Behar and Fitzpatrick, 1989; Deutz, 1988) and recently in human patients with end-stage liver cirrhosis (Luyten *et al.*, 1989). Fitzpatrick *et al.* (1989b) studied the effects of an elevated and constant concentration (≈ 500 μM) of blood $[NH_4^+]$ on the content of glutamine, glutamate, lactate, high energy phosphates and pH_i. Shortly after the start of the infusion (10–20 min), the levels of glutamine and lactate increased to 170 ± 25% and 260 ± 60% of control, respectively after 50 min. Glutamate decreased over the same time interval to 80 ± 4% of control. No changes were observed in intracellular pH or high energy phosphates in the ^{31}P spectrum as a result of the ammonium acetate infusion.

Measurements of Flux Using ^{13}C-Labelled Glucose

An important application of ^1H NMR spectroscopy involves the indirect detection of ^{13}C-labelled metabolites following an infusion of ^{13}C-glucose. Referred to as ^1H-observed, ^{13}C-decoupled spectroscopy ($[^{13}C] - {}^1H$), ^{13}C-labelled ^1H resonances are selected on the basis of the $^1H-{}^{13}C$ heteronuclear scaler coupling of the ^1H resonance when the ^{13}C-coupled resonance is selectively irradiated (Rothman *et al.*, 1985). The method is an order of magnitude more sensitive than ^{13}C detection methods, permitting the measurement of the turnover of metabolites having a concentration of a few μmol/g. Cerebral lactate turnover has been measured by $[^{13}C] - {}^1H$ NMR during ischemia in rat (Rothman *et al.*, 1985) and following electroshock seizures in rabbits (Petroff *et al.*, 1989). Following anoxia-induced cardiac arrest during an infusion of $[1-{}^{13}C]$ glucose in the rat, the relative contribution of glucose and glycogen to the total glycolytic production of lactate was determined using [^{13}C] – ^{1}H NMR spectroscopy (Rothman *et al.*, 1985). As the total lactate ([^{12}C-lactate] + [^{13}C-lactate]) concentration increased, the fractional labelling of lactate C3 decreased from 38 ± 8% at the beginning of ischemia to 20 ± 4% after 45 sec (Fig. 8). During complete ischemia the brain is converted into a closed system so that no label can enter or leave through the circulation. It was determined that 34 ± 13% (≈4.5 μmol/g wet wt) of the lactate formed during the first 5 min of ischemia was derived from unlabelled carbon sources. The dilution of the C3 labelled lactate was interpreted as the breakdown of unlabelled glycogen that had occurred during ischemia.

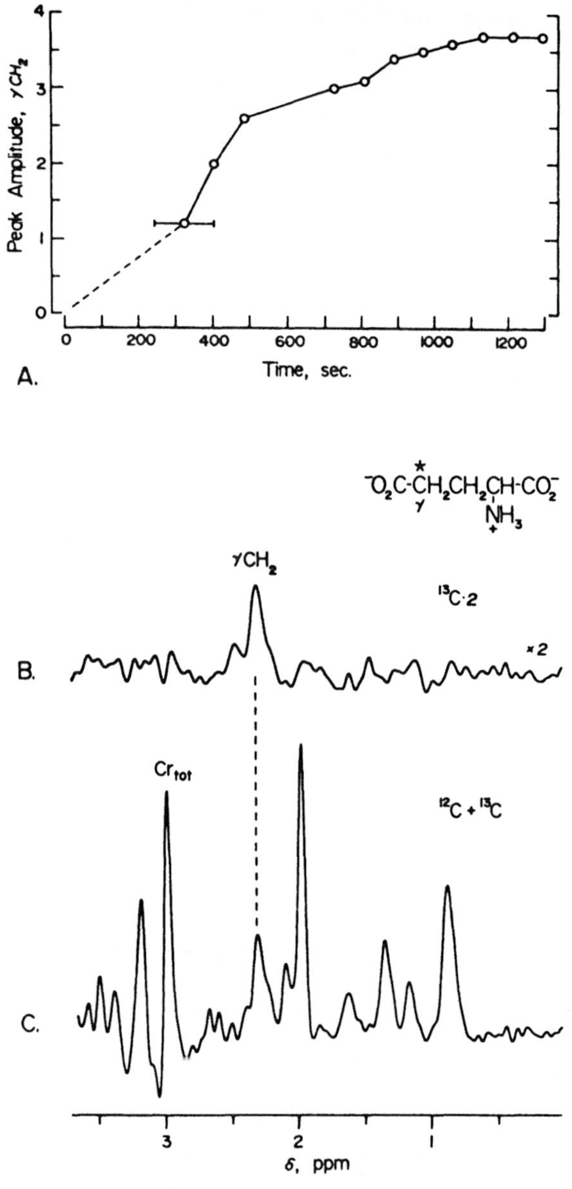

Fig. 9. Timecourse of the ^{13}C enrichment of [4–CH_2] glutamate in the normal rat brain following a programmed infusion of [1–^{13}C] glucose. (A) Timecourse of labelling [4-CH_2] glutamate (γCH_2). All spectra were acquired by the proton-observed, carbon-decoupled method. (B) Difference spectrum containing only ^{13}C-labelled glutamate [4–$^{13}CH_2$]. Vertical scale times 2. (C) Subspectrum containing total ^{12}C + ^{13}C resonance intensity. Cr_{tot}, total creatine. Reproduced from Rothman *et al.* (1985)

Intravenous infusion of [1–^{13}C]glucose leads to rapid labelling of the [4–CH_2] of brain glutamate (Fig. 9). In the normoxic, nitrous oxide anesthetized rat, brain glutamate reaches an isotopic steady-state within 15–20 min and the enrichment follows an exponential timecourse (Rothman *et al.*, 1985; Fitzpatrick *et al.*, 1990). The t $^1/_2$ of labelling of the [4–CH_2] of glutamate was 5.3 min, while the labelling of [3–CH_2], which lagged that of [4–CH_2], had a t$^1/_2$ of ≈26.6 min. These results are explained on the basis of the metabolism of [1–^{13}C]glucose through glycolysis and the TCA cycle, resulting initially in the incorporation of the label in C4 of α-ketoglutarate (α-KG). Aspartate aminotransferase catalyzes a rapid exchange between α-KG and glutamate at a rate that is several times faster than the rate of formation of α-KG or oxaloacetate (Balazs, 1970). Therefore, in the steady-state, the ^{13}C enrichment of [4–CH_2] glutamate is a measure of the cerebral TCA cycle activity *in vivo*. Subsequent turns of the TCA cycle will lead to ^{13}C label incorporation into the C2, C3 and ultimately, the C1 positions of glutamate. The activity of the TCA cycle is determined by the rate of oxygen consumption and this will, in turn, depend on the state of functional activity.

These methods are technically demanding, requiring good hardware stability, precise timing of pulse lengths and extended pulse phase cycling constructions to enhance the signal-to-noise ratio and minimize artifacts. However, the ability to determine substrate utilization and kinetics of selected metabolic pathways offsets the potential difficulties. Following the recent report of the measurement of the timecourse of glutamate labelling in human brain by [1–^{13}C]glucose (Rothman *et al.*, 1989), these techniques should find wide applications in metabolic studies of normal and diseased brain.

References

1. Allen K, Busza AL, Crockard HA, Frackowiack RSJ, Gadian DG, Proctor E, Russell RWR, Williams SR (1988) Acute cerebral ischaemia: concurrent changes in cerebral blood flow, energy metabolites, pH, and lactate measured with hydrogen clearance and ^{31}P and ^{1}H nuclear magnetic resonance spectroscopy. III. changes following ischaemia. J Cereb Blood Flow Metalbol 8: 816–821

2. Arus C, Yen-Chang, Barany M (1985) Proton nuclear magnetic resonance spectra of excised rat brain. Assignment of resonances. Physiol Chem Phys Med NMR 17: 23–33
3. Balazs R (1970) Carbohydrate metabolism. In: Lajtha A (ed) Handbook of neurochemistry, Vol 3. Metabolic reactions in the nervous system. Plenum, New York, pp 1–36
4. Behar KL, Den Hollander JA, Stromski ME, Ogino T, Shulman RG, Petroff OAC, Prichard JW (1983) High-resolution ^1H nuclear magnetic resonance study of cerebral hypoxia *in vivo*. Proc Natl Acad Sci 80: 4945–4948
5. Behar KL, Rothman DL, Shulman RG, Petroff OAC, Prichard JW (1984) Detection of cerebral lactate *in vivo* during hypoxemia by ^1H NMR at relatively low field strenghts (1.9T). Proc Natl Acad Sci 81: 2517–2519
6. Behar KL (1985) Nuclear magnetic resonance of the brain: evaluation of ^1H, ^{31}P, and ^{13}C spectra in normal and pathological states *in vivo*. Thesis, Yale University, New Haven, Connecticut
7. Behar KL, den Hollander JA, Petroff OAC, Hetherington HP, Prichard JW, Shulman RG (1985) Effect of hypoglycemic encephalopathy upon amino acids, high-energy phosphates, and pH_i in the rat brain *in vivo*: detection by sequential ^1H and ^{31}P NMR spectroscopy. J Neurochem 44: 1045–1055
8. Behar KL, Ogino T, Shulman RG (1986) Assignments of resonances in the ^1H spectrum of rat brain *in vitro* and *in situ* by 2D shift correlated and J-resolved spectroscopy. Abstract. 5th Ann Mtng Soc Magn Reson Med 3: 985
9. Behar KL, Rothman DL, Fitzpatrick SM, Hetherington HP, Shulman RG (1987) Combined ^1H and ^{31}P NMR studies of the rat brain *in vivo*: effects of altered intracellular pH on metabolism. In: Cohen SM (ed) Ann New York Acad Sci, Vol 508. Physiological NMR spectroscopy: from isolated cells to man, pp 81–88
10. Behar KL, Fitzpatrick SM (1989) Effects of hypercarbia and porta-caval shunting on amino acids and high energy phosphates of the rat brain: a ^1H and ^{31}P NMR study. In: Butterworth RF, Layrargues GP (eds) Hepatic encephalopathy: pathophysiology and treatment. Humana Press, Clifton, NJ, pp 189–200
11. Behar KL, Rothman DL, Hossmann K-A (1989) NMR investigation of the energy and acid-base homeostasis in the cat brain after prolonged ischemia. J Cereb Blood Flow Metabol 9: 655–665
12. Berkelbach van der Sprenkel JW, Luyten PR, van Rijen PC, Tulleken CAF, den Hollander JA (1988) Cerebral lactate detected by regional proton magnetic resonance spectroscopy in a patient with cerebral infarction. Stroke 19: 1556–1560
13. Bruhn H, Frahm J, Gyngell ML, Merboldt KD, Hänicke W, Sauter R (1989) Cerebral metabolism in man after acute stroke: new observations using localized proton NMR spectroscopy. Abstract 8th Ann Mtng Soc Magn Reson Med 9: 126–131
14. Cady EB, Chu A, Costello AMDL, Delpy DT, Gardiner RM, Hope PL, Reynolds EOR (1987) Brain intracellular pH and metabolism during hypercapnia and hypocapnia in the newborn lamb. J Physiol 382: 1–14
15. Cerdán S, Parrilla R, Santoro P, Rico M (1985) ^1H NMR detection of cerebral myo-inositol. FEBS Lett 187: 167–172
16. Chang L-H, Pereira BM, Weinstein PR, Keniry MA, Murphy-Boesch J, Litt L, James TL (1987) Comparison of lactate concentration determinations in ischemic and hypoxic rat brains by *in vivo* ^1H NMR spectroscopy. Magn Reson Med 4: 575–581
17. Chang L-H, Abiko H, James TL, Mintorovitch J, Weinstein PR (1989) Effects of dichloroacetate on brain lactate levels during complete global ischemia and reperfusion in the rat: an *in vivo* ^1H and ^{31}P NMR study. Abstract 8th Ann Mtng Soc Magn Reson Med 1: 331
18. Corbett RJT, Laptook AR, Nunnally RL, Hassan A, Jackson J (1988) Intracellular pH, lactate, and energy metabolism in neonatal brain during partial ischemia measured *in vivo* by ^{31}P and ^1H nuclear magnetic resonance spectroscopy. J Neurochem 51: 1501–1509
19. Crockard HA, Gadian DG, Frackowiak RSJ, Proctor E, Allen K, Williams SR, Russell RWR (1987) Acute cerebral ischaemia: concurrent changes in cerebral blood flow, energy metabolites, pH, and lactate measured with hydrogen clearance and ^{31}P and ^1H nuclear magnetic resonance spectroscopy. II. changes during ischaemia. J Cereb Blood Flow Metabol 7: 394–402
20. Deutz NEP (1988) Pathophysiological aspects of acute hepatic encephalopathy in the rat. Thesis. Delft University, The Netherlands.
21. Drews LR, Gilboe DD (1973) Glycolysis and the permeation of glucose and lactate in the isolated, perfused dog brain during anoxia and postanoxic recovery. J Biol Chem 248: 2489–2496
22. Duffy TE, Nelson SR, Lowry OH (1972) Cerebral carbohydrate metabolism during acute hypoxia and recovery. J Neurochem 19: 959–977
23. Fan TWM, Higashi RM, Lane AN, Jardetzky O (1986) Combined use of ^1H-NMR and GC-MS for metabolite monitoring and *in vivo* ^1H-NMR assignments.Biochim Biophys Acta 882: 154–167
24. Fitzpatrick SM, Behar KL, Shulman RG (1989a) *In vivo* NMR spectroscopy studies of cerebral metabolism in rats after portal-caval shunting. In: Butterworth RF, Layrargues GP (eds) Hepatic encephalopathy: pathophysiology and treatment. Humana Press, Clifton, NJ, pp 177–187
25. Fitzpatrick SM, Hetherington HP, Behar KL, Shulman RG (1989) Effects of acute hyperammonemia on cerebral amino acid metabolism and pH_i *in vivo*, measured by ^1H and ^{31}P nuclear magnetic resonance. J Neurochem 52: 741–749
26. Fitzpatrick SM, Hetherington HP, Behar KL, Shulman RG (1990) The flux from glucose to glutamate in the rat brain *in vivo* as determined by ^1H-observed, ^{13}C-edited NMR spectroscopy. J Cereb Blood Flow Metabol 10: 170–179
27. Frahm J, Bruhn H, Hyngell ML, Merboldt KD, Hänicke W, Sauter R (1989) Localized high-resolution proton NMR spectroscopy using stimulated echoes: initial applications to human brain *in vivo*. Magn Reson Med 9: 79–93
28. Gadian DG, Frackowiak RSJ, Crockard HA, Proctor E, Allen K, Williams SR, Russell RWR (1987) Acute cerebral ischaemia: concurrent changes in cerebral blood flow, energy metabolites, pH, and lactate measured with hydrogen clearance and ^{31}P and ^1H nuclear magnetic resonance spectroscopy. I. methodology. J Cereb Blood Flow Metabol 7: 199–206
29. Gurdjian ES, Webster JE, Stone WE (1949) Cerebral constituents in relation to blood gases. Am J Physiol 156: 149–157
30. Gurdjian ES, Stone WE, Webster JE (1944) Cerebral metabolism in hypoxia. Arch Neurol Psych 54: 472–477
31. Gyulai L, Schnall M, McLaughlin AC, Leigh JS Jr, Chance B (1987) Simultaneous ^{31}P- and ^1H-nuclear magnetic resonance studies of hypoxia and ischemia in the cat brain. J Cereb Blood Flow Metabol 7: 543–551
32. Hanstock CC, Rothman DL, Howseman A, Lantos G, Novotny EJ, Petroff OAC, Prichard JW, Shulman RG (1989) *In vivo* determination of NAA concentration in the human brain using the proton aspartyl resonance. Abstract 8th Ann Mtng Soc Magn Reson Med 1: 442
33. Hanstock CC, Boisvert DPJ, Bendall MR, Allen PS (1988a) *In vivo* assessment of focal brain lactate alterations with NMR proton spectroscopy. J Cereb Blood Flow Metabol 8: 208–214
34. Hanstock CC, Rothman DL, Jue T, Shulman RG (1988b) Volume-selected proton spectroscopy in the human brain. J Magn Reson 77: 583–588
35. Hanstock CC, Rothman DL, Prichard JW, Jue T, Shulman RG (1988c) Spatially localized 1H NMR spectra of metabolites in the human brain. Proc Natl Acad Sci (USA) 85: 1821–1825

36. Hetherington HP, Avison MJ, Shulman RG (1985) ^1H homonuclear editing of rat brain using semi-selective pulses. Proc Natl Acad Sci (USA) 82: 3115–3118
37. Hope PL, Cady EB, Chu A, Delpy DT, Gardiner RM, Reynolds EOR (1987) Brain metabolism and intracellular pH during ischaemia and hypoxia: an *in vivo* ^{31}P and ^1H nuclear magnetic resonance study in the lamb. J Neurochem 49: 75–82
38. Hope PL, Cady EB, Delpy DT, Ives NK, Gardiner RM, Reynolds EOR (1988) Brain metabolism and intracellular pH during ischaemia: effects of systemic glucose and bicarbonate administration studied by ^{31}P and ^1H nuclear magnetic resonance spectroscopy *in vivo* in the lamb. J Neurochem 50: 1394–1402
39. Lewis LD, Ponten U, Siesjö BK (1973) Arterial acid-base changes in unanesthetized rats in acute hypoxia. Respir Physiol 19: 312–321
40. Ligeti L, Osbakken MD, Subramanian HV, Kovach AGB, Leigh JS Jr, Chance B (1987) ^{31}P and ^1H NMR spectroscopy to study the effects of gallopamil on brain ischemia. Magn Reson Med 4: 441–451
41. Luyten PR, den Hollander JA, Bovée WMMJ, Ross BD, Bosman DK, Chamuleau RAFM (1989) ^{31}P and ^1H NMR spectroscopy of the human brain in chronic hepatic encephalopathy. Abstracts, 8th Ann Mtng Soc Magn Reson Med 1: 375
42. Middlehurst CR, Beilharz GR, Hunt GE, Kuchel PW, Johnson GFS (1984a) Proton nuclear magnetic resonance spectroscopy of rabbit brain homogenate. J Neurochem 42: 878–879
43. Middlehurst CR, King GF, Beilharz GR, Hunt GE, Johnson GFS, Kuchel PW (1984b) Studies of rat brain metabolism using proton nuclear magnetic resonance: spectral assignments and monitoring of prolidase, acetylcholinesterase, and glutaminase. J Neurochem 43: 1561–1567
44. Myers RE, Yamaguchi M (1976) Effects of serum glucose concentration on brain response to circulatory arrest. J Neuropathol Exp Neurol 35: 301
45. Nilsson L, Siesjö BK (1971) The effect of hypoxia upon labile substrates and upon acid-base parameters in the brain. In: Siesjö BK, Sorensen SC (eds) Ion homeostasis of the brain. Munksgaard, Copenhagen, pp 428–436
46. Norberg K, Siesjö BK (1975) Cerebral metabolism in hypoxic hypoxia. I. Pattern of activation of glycolysis: a re-evaluation. Brain Res 86: 31–44
47. Pardridge WM, Connor JD, Crawford IL (1975) Permeability changes in the blood-brain barrier: causes and consequences. CRC Crit Rev Toxicol 3: 159–199
48. Petroff OAC (1988) Biological ^1H NMR spectroscopy. Comp Biochem Physiol 90B: 249–260
49. Petroff OAC, Prichard JW, Behar KL, Alger JR, Shulman RG (1984) *In vivo* phosphorus nuclear magnetic resonance spectroscopy in status epilepticus. Ann Neurol 16: 169–177
50. Petroff OAC, Prichard JW, Behar KL, Rothman DL, Alger JR, Shulman RG (1985) Cerebral metabolism in hyper- and hypocarbia: ^{31}P and ^1H nuclear magnetic resonance studies. Neurology 35: 1681–1688
51. Petroff OAC, Prichard JW, Ogino T, Avison M, Alger JR, Shulman RG (1986) Combined ^1H and ^{31}P nuclear magnetic resonance spectroscopic studies of bicuculline-induced seizures *in vivo*. Ann Neurol 20: 185– 193
52. Petroff OAC, Ogino T, Alger JR (1988) High-resolution proton magnetic resonance spectroscopy of rabbit brain: regional metabolite levels and postmortem changes. J Neurochem 51: 163–171
53. Petroff OAC, Novotny EJ, Avison MJ, Rothman DL, Shulman RG, Prichard JW (1989) Cerebral lactate turnover after electroshock by proton observed, carbon decoupled spectroscopy. Abstract 8th Ann Mtng Soc Magn Reson Med 1: 332
54. Prichard JW, Alger JR, Behar KL, Petroff OAC, Shulman RG (1983) Cerebral metabolic studies *in vivo* by ^{31}P NMR. Proc Natl Acad Sci (USA) 80: 2748–2751
55. Prichard JW, Petroff OAC, Ogio T, Shulman RG (1987) Cerebral lactate elevation by electroshock: A ^1H magnetic resonance study. In: Cohen SM (ed) Physiological NMR spectroscopy: from isolated cells to man. Ann NY Acad Sci 508 p 54–63
56. Remy C, Von Kienlin M, Lotito S, Francois A, Benabid AL, Decorps M (1989) *In vivo* ^1H nmr spectroscopy of an intracerebral glioma in the rat. Magn Reson Med 9: 395–401
57. Richards TL, Keniry MA, Weinstein PR, Pereira BM, Andrews BT, Murphy EJ, James TL (1987) Measurement of lactate accumulation by *in vivo* proton nmr spectroscopy during global cerebral ischemia in rats. Magn Reson Med 5: 353–357
58. Richards T, Budinger TF (1988) NMR imaging and spectroscopy of the mammalian central nervous system after heavy ion radiation. Radiat Res 113: 79–101
59. Rothman DL, Behar KL, Hetherington HP, Shulman RG (1984) Homonuclear ^1H double-resonance difference spectrosocopy of the rat brain *in vivo*. Proc Natl Acad Sci 81: 6330–6334
60. Rothman DL, Behar KL, Hetherington HP, den Hollander JA, Bendall MR, Petroff OAC, Shulman RG (1985) ^1H-observed/^{13}C-decoupled spectroscopic measurements of lactate and glutamate in the rat brain *in vivo*. Proc Natl Acad Sci (USA) 82: 1633–1637
61. Rothman DL, Novotny EJ, Howseman A, Lantos G, Petroff OAC, Hanstock CC, Prichard JW, Shulman GI, Shulman RG (1989) ^1H NMR measurement of 4–^{13}C glutamate turnover in the human brain. Abstract 8th Ann Mtng Soc Magn Reson Med 3: 1060
62. Siesjö BK (1973) Metabolic control of intracellular pH (Editorial). Scand J Clin Lab Invest 32: 97–104
63. Siesjö BK, Messeter K (1971) Factors determining intracellular pH. In: Siesjö BK, Sorensen SC (eds) Ion homeostasis of the brain. Munksgaard, Copenhagen, pp 244–262
64. Schnall MD, Yoshizaki K, Chance B, Leigh JS Jr (1988) Triple nuclear nmr studies of cerebral metabolism during generalized seizure. Magn Reson Med 6: 15–23
65. van Rijen PC, Luyten PR, Berkelback Van Der Sprenkel JW, Kraaier V, van Huffelen AC, Tulleken CAF, den Hollander JA (1989) ^1H and ^{31}P NMR measurement of cerebral lactate, high-energy phosphate levels, and pH in humans during voluntary hyperventilation: associated EEG, capnographic, and doppler findings. Magn Reson Med 10: 182–193
66. Young RSK, Cowan BE, Petroff OAC, Novotny E, Dunham SL, Briggs RW (1987) *In vivo* ^{31}P and *in vitro* ^1H nuclear magnetic resonance study of hypoglycemia during neonatal seizure. Ann Neurol 22. 622–628
67. Young RSK, Chen B, Petroff OAC, Gore JC, Cowan BE, Novotny EJ, Wong M, Zuckerman (1989) The effect of diazepam on neonatal seizure: *in vivo* ^{31}P and ^1H NMR study. Pediatr Res 25: 27–31

Correspondence and Reprints: Kevin L. Behar, Department of Molecular Biophysics and Biochemistry, C-128A Sterling Hall of Medicine, 333 Cedar Street, New Haven, CT 06510, U.S.A.

Acta Neurochir (1993) [Suppl] 57: 21–29

NMR-Spectroscopic Investigation of Cerebral Reanimation After Prolonged Ischemia

K.-A. Hossmann[1], **K. L. Behar**[2], and **D. L. Rothman**[2]

[1] Max-Planck-Institut für Neurologische Forschung, Köln, Federal Republic of Germany and [2] Department of Molecular Biophysics and Biochemistry, Yale University, New Haven, Connecticut, U.S.A.

Abstract

The severity of brain injury following interruption of blood flow depends on a number of ischemic and post-ischemic variables. The most important ischemic variables are the duration of ischemia, the amount of residual blood flow, the type and depth of anesthesia, brain glucose content and temperature. Among the post-ischemic factors the no-reflow phenomenon, edema and a variety of biochemical disturbances are of particular importance. Due to the complex interaction of these factors irreversible brain injury usually occurs after less than 10 min cerebrocirculatory arrest in normothermia. However, the safe ischemia time of the brain can be substantially extended when appropriate therapeutic measures are used to alleviate post-ischemic injury. NMR-spectroscopy is particularly suited for the analysis of this process. Recording of ^{31}P, ^{1}H and ^{19}F spectra allow the continuous non-invasive assessment of such basic parameters as brain energy state, tissue pH, the content of lactate and blood flow (using Freon-23 as an inert tracer). In addition, information is obtained about changes in the content of phosphomonoesters and -diesters, glutamate, glutamine, aspartate and N-acetyl aspartate. These measurements can be combined with *in vivo* electrophysiological and post-mortem biochemical investigations for the further refinement of functional/metabolic monitoring.

We have used this approach to study the potentials of post-ischemic resuscitation after one hour complete ischemia of the normothermic cat brain. The following results were obtained:

1) Cerebrocirculatory arrest caused a suppression of EEG within 15 sec, complete depletion of ATP, and phosphocreatine in less than 10 min, a rise of lactate to about 70% of its maximum value during this interval, and a decrease of pH to between 5.31 and 6.70. After longer ischemia times little further changes occurred.

2) Recirculation of the brain after one hour complete ischemia in normothermia resulted in complete recovery of ATP, CrP, lactate and tissue pH in 30%, in partial recovery of the energy metabolism in 53%, and in no recovery in 17% of cats.

3) Recovery did not depend on pH, lactate or residual blood flow during ischemia but it was critically determined by the speed of ATP resynthesis after ischemia. Complete recovery occurred only when ATP and CrP began to reappear within 5 min and, returned to more than 50% within 20 min after the onset of recirculation; recovery was always incomplete when these intervals were longer.

4) The NMR data correlated with invasive measurements in most but not all animals. In particular, dissociations occurred at low pH and low metabolite levels where NMR either underestimated or overestimated the actual tissue values.

We conclude that recovery of brain metabolism after prolonged ischemia is less sensitive to pH than generally assumed but requires fast blood reperfusion and rapid post-ischemic restoration of energy state. The implications for therapeutic interventions and the monitoring of these interventions by NMR will be discussed.

Keywords: Cerebral ischemia; ^{1}H, ^{31}P-spectroscopy; intracellular pH; energy metabolism.

Introduction

With the advent of *in vivo* NMR spectroscopy non-invasive measurements of metabolites have become possible in living organisms (Prichard, 1986). This allows, for the first time, the continuous evaluation of metabolites during the full length of a pathophysiological process, e.g. before, during and after cerebral ischemia (Naruse *et al.*, 1984; Gadian *et al.*, 1987; Andrews, *et al.*, 1987; Allen *et al.*, 1988). Obviously, this possibility greatly facilitates the analysis of pathological processes under critical conditions because the whole history of the process can be evaluated in individual experiments instead of grouping data from tissue samples that have been harvested from different animals at different time points. This is the reason why we became interested in this technique for the exploration of the limits of cerebral resuscitation after prolonged ischemia. In previous investigations evidence was provided that electrophysiological and even neurological functions may recover after complete cerebrocirculatory arrest in normothermia of 1 h duration (Hossmann and Grosse Ophoff, 1986; Hossmann *et al.*,

1987). However, in about one third of the animals exposed to the same duration of ischemia recovery did not occur, and in another third it was incomplete (Hossmann, 1988). It has been suggested that the outcome of the ischemic impact is mainly determined by post-ischemic factors (Hossmann *et al.*, 1973) but it cannot be excluded that differences in the metabolic response of the brain during ischemia are of equal importance. In particular, the role of a residual level of energy-rich substrates, or of the degree of acidosis developing during ischemia has not been established because the use of conventional techniques for the measurement of these parameters precludes the follow-up of the recovery process.

In the present communication, the results of a recent NMR-spectroscopic investigation (Behar *et al.*, 1989) are correlated with earlier findings obtained in the same model of prolonged cerebrocirculatory arrest (Hossmann *et al.*, 1976; Hossmann, 1988). Our analysis clearly demonstrates that recovery is, in fact, mainly determined by post-ischemic factors, and that the critical period determining the final outcome is the very beginning of post-ischemic recirculation.

Material and Methods

A detailed description of our methods is given in the original publications (Hossmann *et al.*, 1976; Hossmann *et al.*, 1987; Behar *et al.*, 1989). In short, adult normothermic cats were anesthetized with 30 mg/kg pentobarbital, immobilized and placed under artificial ventilation. Body temperature was kept constant throughout the experiment at 37°C, and physiological variables were monitored at short intervals to assure a stable physiological state. Cerebrocirculatory arrest of 1 h duration was produced by intrathoracal occlusion of the innominate and left subclavian arteries which supply both carotid and vertebral arteries. Collateral flow was prevented by ligation of the internal mammary arteries and lowering of the blood pressure to below 80 mmHg. After 1 h ischemia systolic blood pressure was raised to near 200 mmHg in order to restore post-ischemic reperfusion. Brain swelling was alleviated by osmotherapy, and blood acidosis by controlled infusion of buffers.

NMR-spectroscopy was carried out in 7 cats in a 4.7 T ultraconducting magnet with a 30 cm horizontal bore. The animals' head was fixed in a plastic headholder, and spectra were recorded with a 15 mm one-turn surface coil double-tuned to ^{1}H and ^{31}P. ^{31}P spectra were recorded by free induction decay (repetition time 1/s), and water suppressed ^{1}H spectra by semi-selective excitation and pulse editing. Time resolution of ^{31}P spectra was 4 min, and that of ^{1}H spectra 1 min.

In the previous experiments which have been re-analysed for the interpretation of the NMR-data, ischemia was produced by the same method, and post-ischemic recovery was monitored electrophysiologically by recording spontaneous EEG activity, somatically evoked potentials and the pyramidal response following electrical stimulation of the motor cortex. Cerebral blood flow, oxygen consumption and glucose utilisation were measured by a modification of the Kety-Schmidt method, using ^{133}Xenon as the inert flow tracer (Hossmann *et al.*, 1976).

Results and Discussion

Electrophysiological and Hemodynamic Observations

The NMR observations described below are difficult to interpret without knowledge of the electrophysiological and hemodynamic findings. For this reason, previous observations in the same experimental model are briefly reviewed (Hossmann *et al.*, 1976; Hossmann, 1988).

Intrathoracal vascular occlusion results in instantaneous interruption of blood flow and a suppression of spontaneous EEG activity within less than 15 sec after the onset of ischemia. Somatosensory evoked potentials disappear after 2–4 min, and the (electrically) evoked D-wave of the pyramidal response within 8 min. Brain temperature measured 1 mm below the surface of the cortex decreases initially by about 0.2 °C

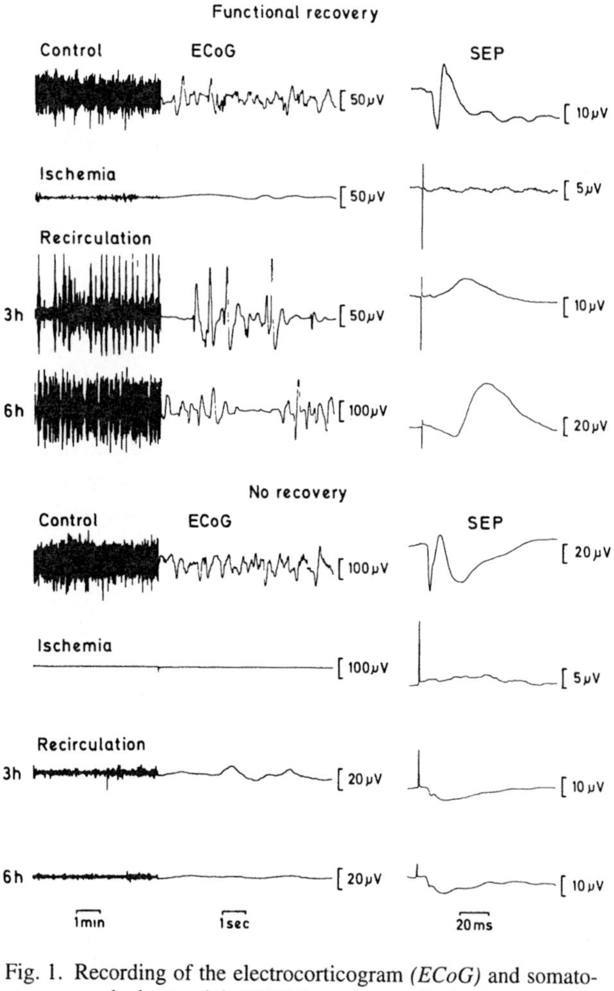

Fig. 1. Recording of the electrocorticogram *(ECoG)* and somatosensory evoked potentials *(SEP)* in two cats submitted to one hour complete cerebrocirculatory arrest in normothermia. Although both animals were treated in the same way, one animal exhibited electrophysiological recovery (above) and the other not (below)

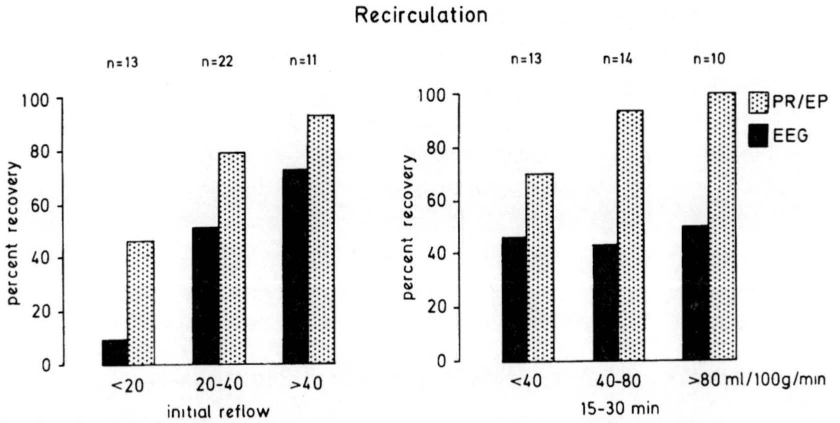

Fig. 2. Effect of post-ischemic recirculation rate on the electrophysiological recovery of cat brain after one hour of global ischemia. Incidence of recovery of the pyramidal response *(PR)* and/or somatosensory potentials *(EP)* is shown by the dotted bars, and incidence of *EEG* recovery by the black bars (n = number of animals in each group). The initial recirculation rate (left) is a sensitive predictor of recovery. Blood flow measured 15–30 min after ischemia (right) does not correlate with recovery (from Hossmann, 1988)

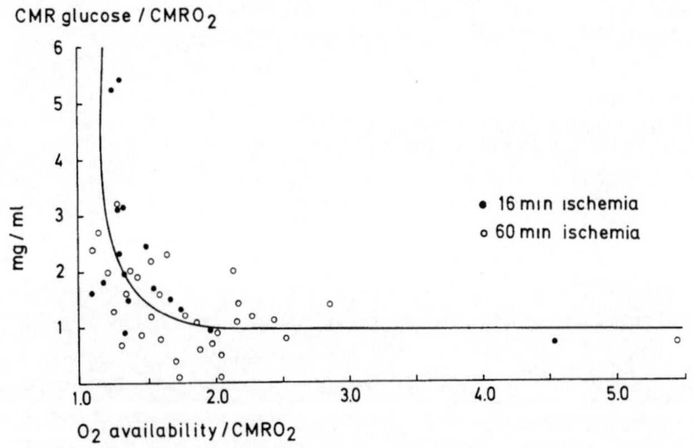

Fig. 3. Relationship between oxidative glucose utilization and the relative oxygen availability during the post-ischemic hypoperfusion phase following 16 min (closed circles) or 60 min (open circles) complete cerebrocirculatory arrest. Stimulation of non-oxidative glucose utilization is expressed by the increase of the glucose/oxygen uptake ratio (ordinate) and a decrease of the relative oxygen availability by the ratio of arterial oxygen supply/oxygen consumption (abscissa). Anaerobic glycolysis is stimulated when relative oxygen availability decreases below 2.0, corresponding to an increase of oxygen extraction by more than 50% (from Hossmann, 1979)

but then declines more slowly and reaches room temperature not before the end of 60 min ischemia.

When blood flow is restored after 1 h ischemia, different electrophysiological recovery patterns are observed (Fig. 1). In 75% of experiments somatosensory evoked potentials and/or the pyramidal response recover. The electrically evoked D-wave of the pyramidal response may appear as early as 8 min after the beginning of recirculation, followed by recovery of the synaptically evoked I-wave of the pyramidal response after 20–40 min, and of somatically evoked potentials after 45–90 min. In 50% also spontaneous EEG activity returns after 2–4 h recirculation. Initially, this activity consists of a burst/suppression pattern, followed after 1–3 h by continuous low voltage slow wave activity. The incidence of recovery of EEG is lower than that of evoked potentials because the electrophysiological recovery process may discontinue before EEG returns. In 25% neither evoked potentials nor EEG return, presumably because of a severe no-reflow.

Measurements of cerebral blood flow after ischemia revealed substantial differences in different animals. The initial recirculation rate after 1 h ischemia – as calculated from the initial post-ischemic clearance of the flow tracer from the pre-ischemically saturated brain – varied over a wide range from less than 20 to more than 150% of control. Fifteen to 30 min later most animals exhibited a transient phase of hyperemia, followed by delayed hypoperfusion to as low as 50% of control after 1–2 h. The incidence of electrophysiological recovery correlated with the initial – but not the later – recirculation rate. As demonstrated in Fig. 2, the incidence of EEG recovery was 8% at an initial reflow rate of less than 20 ml/100 g/min, 45% at a rate between 20 and 40 ml/100 g/min, and 80% when the initial recirculation rate was more than 40 ml/100 g/min. The incidence of recovery of evoked potentials at these flow rates amounted to 46, 73 and 90%, respectively. This relationship was lost when EEG recovery was plotted against blood flow measured 15–30 min after the beginning of recirculation (Fig. 2). Recovery, in consequence, is limited by a flow-dependent process during the initial minutes of recirculation. Or more specifically: recovery requires an initial post-ischemic recirculation rate of about 40 ml/100 g/min; raising blood flow to this or an even higher level after a few minutes of low flow reperfusion does not prevent irreversible injury.

The relationship between functional recovery and delayed hypoperfusion is more complex. We did not observe a direct influence on EEG or evoked potentials but there are indications of an indirect relationship between these parameters. It has been previously shown, that a brief ischemia suppresses the coupling

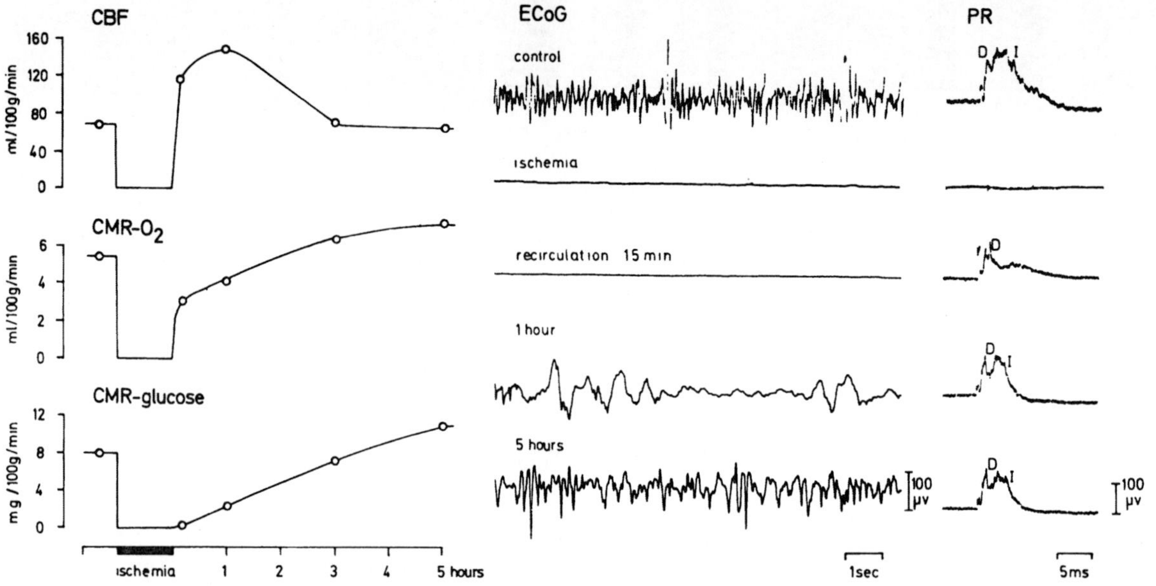

Fig. 4. Recording of the electrocorticogram *(ECoG)*, the pyramidal response *(PR)*, cerebral blood flow *(CBF)*, cerebral oxygen consumption *(CMR-O_2)* and cerebral glucose utilization *(CMR-glucose)* before and after one hour complete cerebrocirculatory arrest of cat. Note the transient recovery of glucose utilization shortly after the onset of post-ischemic recirculation, followed by a more gradual recovery of both glucose and oxygen utilization in parallel with the recovery of EEG activity. Note also uncoupling of cerebral blood flow and metabolism both during the early and later recovery phase (from Hossmann *et al.*, 1976)

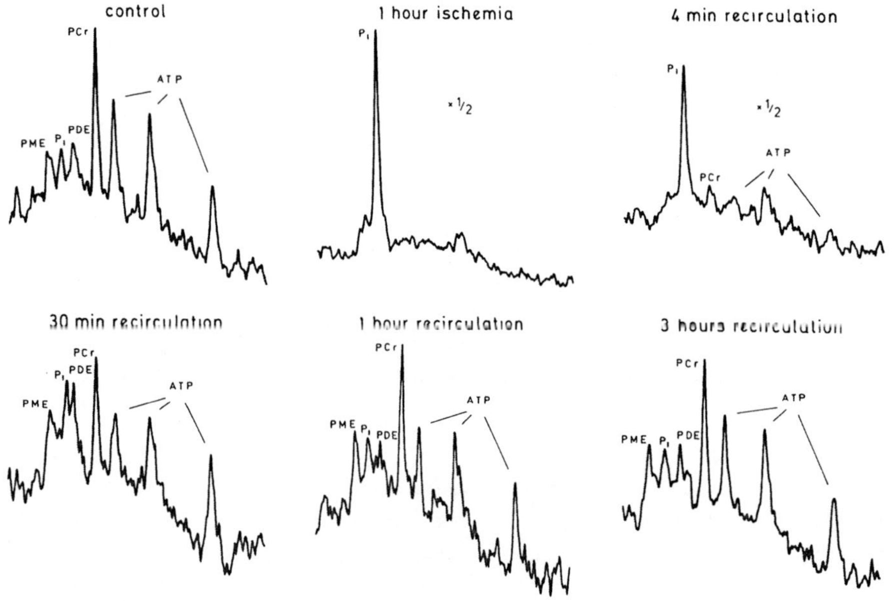

Fig. 5. Recording of phosphorus spectra during and after one hour complete cerebrocirculatory arrest in animal with complete post-ischemic recovery

between metabolism and blood flow for extended periods, i.e. for days or possibly even months after the beginning of recirculation (Dietrich *et al.*, 1986; Schmidt-Kastner *et al.*, 1986a; Ueki *et al.*, 1988). As a consequence, oxygen extraction of the arterial blood greatly varies depending on the individual flow rate, the arterial oxygen content and the cerebral oxygen consumption. Plotting the cerebral glucose/oxygen uptake ratio – which increases with increasing anaerobic metabolism of glucose – against arterial oxygen extraction clearly demonstrates that aerobic metabolism of glucose is impaired when oxygen extraction exceeds 50% (Hossmann, 1979) (Fig. 3).

Shortly after ischemia oxygen requirements of the brain are low (Schmidt-Kastner *et al.*, 1986b) (Fig. 4). However, since oxygen consumption rises when EEG recovers, the metabolic situation may become critical as soon as spontaneous electrical activity returns. Hypoperfusion, in consequence, is a risk factor for secondary hypoxia: not because of an absolute threshold of post-ischemic flow but because of the suppression of the flow-metabolism couple. This may be the reason that not all animals with a return of evoked potentials recovered EEG, and that occasionally secondary suppression of EEG occurred after a few hours despite an apparently normal recovery process. Our NMR data support this hypothesis, as discussed in more detail below.

Nuclear Magnetic Resonance Spectroscopy

Under control conditions fully relaxed *phosphorus spectra* revealed a CrP/ATP_β peak area ratio of 2.24 ± 0.44 (mean ± SD, n = 6) and a P_i/ATP_β ratio of 0.43 ± 0.24 (n = 5). ATP of control animals – measured by *in situ* freezing under the same experimental conditions as in our NMR-study – amounts to 2.1 μmol/g w.w. (Schmidt-Kastner *et al.*, 1986b). Neglecting the contribution of lesser amounts of purine nucleotides to the ATP_β signal, the CrP and P_i concentrations within the sensitive volume of the surface coil are, in consequence, 4.7 and 0.9 μmol/g w.w., respectively. Pre-ischemic intracellular pH (pH_i) amounted to 7.06 ± 0.05 (n = 6). Following vascular occlusion CrP and ATP decreased to noise level between 4 and 8 min of ischemia in 5 out of 7 animals, and after 16 and 18 min in the other two animals (Figs. 5–8). The decrease of CrP and ATP was accompanied by a substantial increase of P_i, and a decline of pH_i to 6.17 ± 0.12 (range 6.05–6.37) after 1 h of ischemia. The maximum of acidosis was attained within 5–17 min of ischemia, with the exception of one animal where pH_i reached its lowest level after 30 min.

Ischemia resulted in a more rapid depletion of CrP than ATP. A plot of the relative amplitudes of the two resonances detected that ATP began to decline when CrP was already reduced to below 50%. A spontaneous recovery of energy-rich phosphates or an increase of pH during the later phases of ischemia was never observed. Complete energy depletion of the brain, in consequence, lasted between 42 and 56 min.

When the brain was recirculated after 1 h ischemia, different degrees of metabolic recovery were observed. In 3 out of 7 animals CrP and ATP was detected within 4 min of recirculation and recovered to a level between 90 and 113% of the pre-ischemic value after 1 h of recirculation (Figs. 5 and 6). In these animals the cerebral content of CrP and ATP was maintained at nearly normal levels throughout the remainder of the 3 h observation period. Intracellular pH recovered more slowly than ATP. Initially, it even transiently further declined but after about 30 min it began to rise and reached control value within 1.5 h of recirculation. This finding is in accordance with observations after shorter periods of ischemia which also

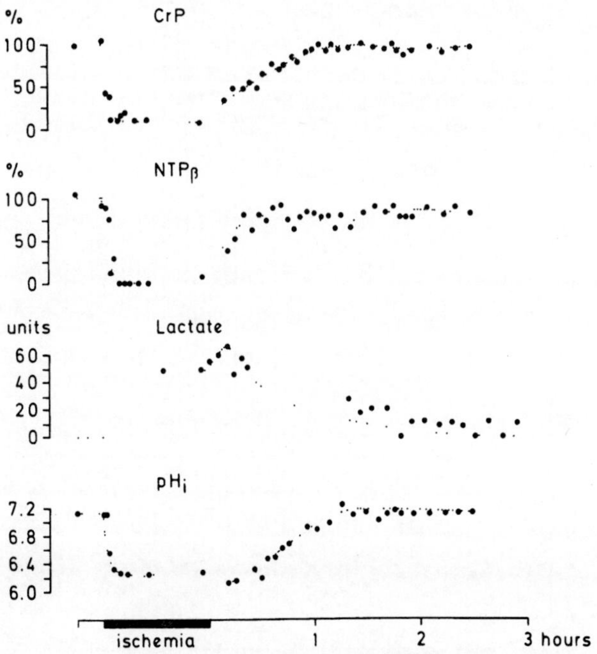

Fig. 6. Timecourse of changes of high energy phosphates, pH_i and lactate during and after complete cerebrocirculatory arrest of one hour. Note the rapid recovery of ATP (measured as the beta peak of nucleotide triphosphate, *NTP_β*), followed by the more slowly recovery of phosphocreatine *(CrP)*, lactate and pH_i (from Behar *et al.*, 1989)

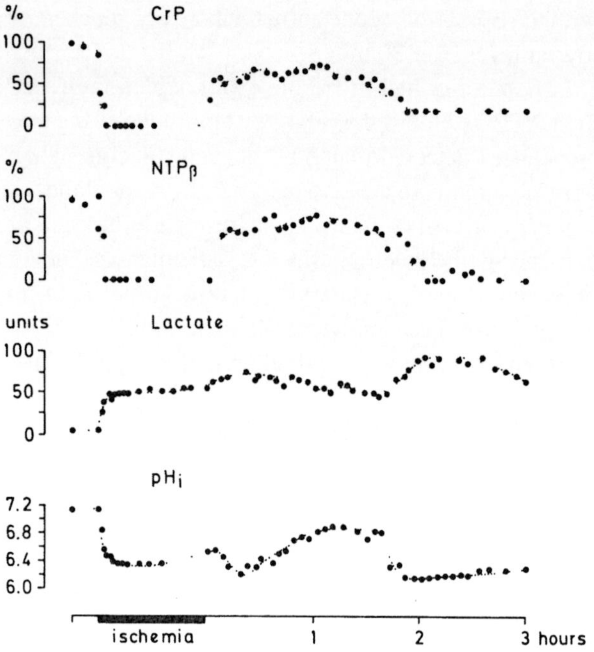

Fig. 7. Timecourse of changes in high energy phosphates, pH_i and lactate during and after one hour complete cerebrocirculatory arrest in an experiment which led to incomplete recovery. After 100 min recirculation pH abruptly declined, lactate further increased and energy-rich phosphates secondarily declined. These changes are indicative of secondary hypoxia resulting from uncoupling between blood flow and oxygen utilization during the phase of post-ischemic hypoperfusion (from Behar *et al.*, 1989)

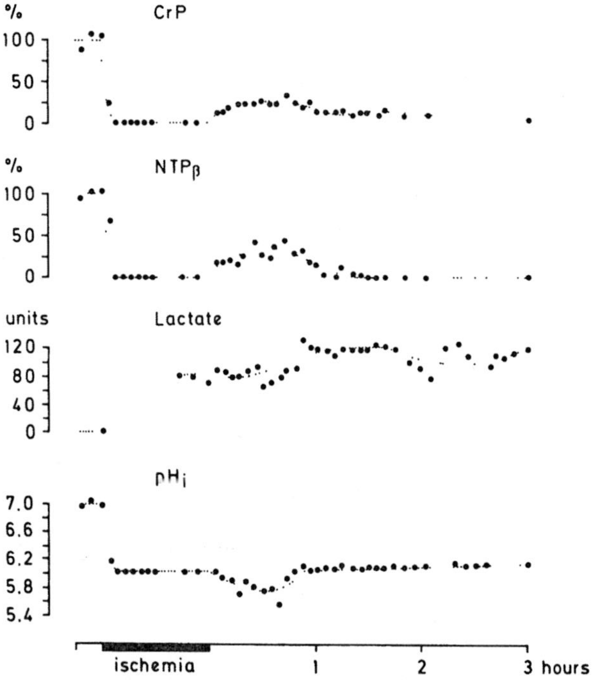

Fig. 8. Timecourse of changes in high energy phosphates, lactate and pH after one hour complete cerebrocirculatory arrest in an experiment with minor post-ischemic recovery. Note the persistence of lactacidosis after ischemia, indicating severe recirculation disturbances (from Behar *et al.*, 1989)

revealed a transient post-ischemic decline of pH (Chopp *et al.*, 1987).

In 4 animals CrP and ATP recovered partially, CrP reaching a maximum of 20–27% and ATP a maximum of 30–80% of the pre-ischemic value (Figs. 7 and 8). Recovery of high energy phosphates was not maintained in these animals due to a spontaneous secondary decline in CrP and ATP after 1–2 h of recirculation. The post-ischemic changes of pH_i were quite variable in these animals, either showing a delayed recovery, no recovery or a secondary decline. The latter occurred abruptly after near-normalization, followed by a secondary decline of ATP and CrP (Fig. 7).

Proton spectra were recorded in all seven cats, but the complete time course (without gaps) from the beginning to the end of the experiment could be measured in only two animals. The data obtained, however, are fully consistent with the phosphorus spectra (Figs. 6–8). Shortly after ischemia lactate sharply increased and leveled off in less than 15–20 min, i.e. at a time when pH had reached a minimum. In animals with metabolic recovery lactate transiently further increased during post-ischemic recirculation, which is in line with the further decline of pH (Fig. 9). Since tissue ATP and creatine phosphate began to replenish during this period, the secondary lactacidosis is clearly the result of post-ischemic stimulation of glycolysis.

This finding is supported by our previous measurements of cerebral oxygen and glucose utilisation (Hossmann *et al.*, 1976). Shortly after the initiation of recirculation glucose but not oxygen was taken up by the brain. After about 20 min glucose uptake fell to the same low level as oxygen uptake but then both glucose and oxygen consumption gradually improved in parallel with the recovery of the EEG.

In animals without post-ischemic normalization of energy metabolism a different pattern was seen (Figs. 7 and 8). In these instances lactate either did not or only partially decline, and it secondarily further increased when delayed post-ischemic hypoxia developed (Fig. 7).

These different post-ischemic recovery patterns can be easily associated with the above described hemodynamic and electrophysiological responses (Hossmann *et al.*, 1976; Hossmann, 1988). The early progressing type of metabolic recovery corresponds to the group of animals with progressing electrophysiological recovery. In fact, the beginning reappearance of the pyramidal response after as little as 8 min is fully consistent with the beginning restoration of energy state during

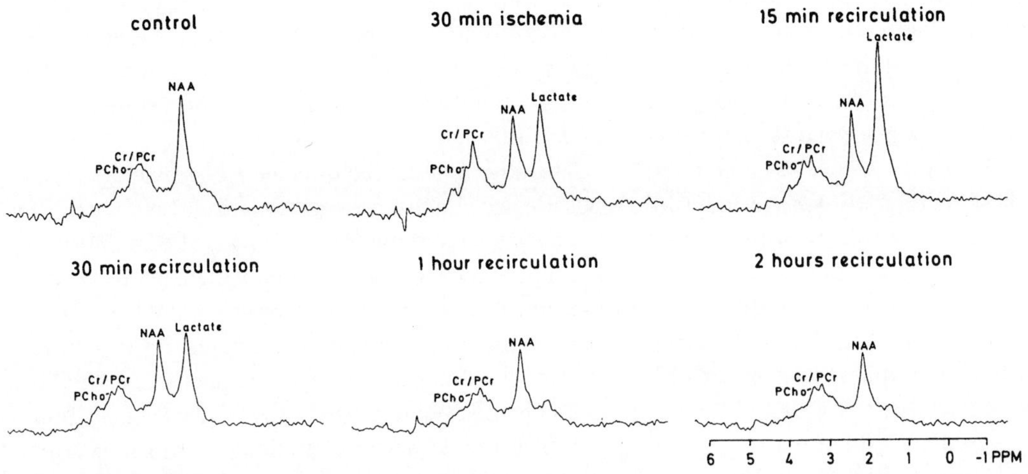

Fig. 9. Water-suppressed proton spectra obtained during and after cerebrocirculatory arrest. Note the increase of lactate during ischemia, followed by a further rise after the beginning of recirculation. With longer recirculation times lactate gradually declined (by courtesy of Höhn-Berlage, Kloiber, Okada, Hossmann)

this time. The much slower normalization of pH and PCr corresponds to the more delayed recovery of spontaneous EEG which previously has been shown to return after ion homeostasis and the extra-/intracellular volume relationship have normalized. Conversely, the animals with failure of metabolic recovery can be associated with the development of no-reflow and the absence of electrophysiological recovery. Of particular interest are the animals with secondary deterioration. It is obvious, that this pattern corresponds to the late decompensation of oxidative respiration, as described above. The NMR findings demonstrate that this disturbance develops abruptly from one minute to the other, and that it is not possible to predict this event from the earlier recovery process. It, therefore, will be difficult to design an appropriate therapeutic approach to prevent this fatal complication.

The continuous recording of NMR during the whole length of the ischemic period provided us with the opportunity to test the possibility that differences of the metabolic impact during ischemia determined the pat-

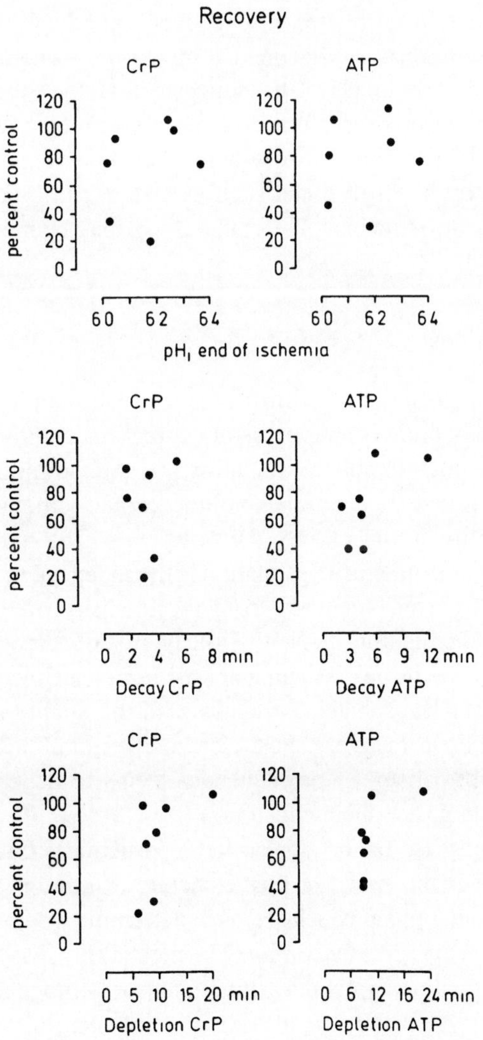

Fig. 10. Recovery of phosphocreatine (*CrP*, left) and *ATP* (right) following one hour complete cerebrocirculatory arrest in cats. The maximum values of metabolites were plotted against end-ischemic pH_i, the initial decay and the depletion time of CrP and ATP. The initial decay was expressed as the time elapsing from vascular occlusion until metabolites declined to 37% of control amplitude. Depletion time is the time elapsing from the onset of ischemia to complete depletion of metabolites. Note the independence of post-ischemic recovery from the degree of acidosis and the speed of depletion of energy metabolites

tern of post-ischemic recovery. The following measurements were carried out: intracellular pH at the end of ischemia, the time constant of the decay of CrP and ATP after the onset of ischemia (defined as the peak amplitude decay to 37% of control), and the depletion time of CrP and ATP. These values were plotted against the maximum of ATP or CrP recovery during post-ischemic recirculation. As shown in Fig. 10 no correlation was found. This clearly demonstrates that the differences in post-ischemic recovery do not depend on differences of the metabolic impact during ischemia. In particular, a major role of ischemic acidosis has to be dismissed, at least within the natural occurring range of pre-ischemic glucose levels which in this study varied between 90 and 258 mg%.

This observation confirms similar findings by Chopp *et al.* (1987), and it is also corroborated by a recent study from our laboratory in which the relationship between blood glucose, cerebral pH and recovery was studied in more detail (Kloiber *et al.*, in preparation). In this investigation end-ischemic pH also failed to correlate with recovery although the incidence of recovery declined at higher pre-ischemic blood glucose levels. This suggests that the well-known deteriorating effects of hyperglycemia are not mediated by a more severe acidosis. Other factors induced by hyperglycemia such as changes in blood viscosity or tissue osmolality, therefore, may be more important.

Conclusions

Our NMR findings confirm our earlier observations that the cat brain is able to restore energy metabolism after one hour complete circulatory arrest in normothermia but that some animals do not recover although they are treated in exactly the same way. The main conclusion from this investigation is the independence of post-ischemic recovery from the intensity of the initial ischemic impact. Neither the degree of acidosis that develops during ischemia nor the initial decline of energy-rich phosphates correlates with the quality of post-ischemic recovery. Post-ischemic restoration of brain metabolism, in consequence, seems to depend mainly on post-ischemic factors.

Two of these factors appear to be particularly critical. One is the speed of post-ischemic restitution of energy metabolism which seems to determine whether recovery is progressive or not. The most likely reason for this relationship is the requirement of homogeneous reperfusion and a rapid restoration of energy state in order to prevent fatal brain swelling because the influx of fluid and electrolytes from the blood into the energy-depleted brain continues as long as energy metabolism and ion homeostasis are not restored (Hossmann, 1988).

The other critical factor is the recovery of spontaneous electrical activity which begins a few hours after the onset of recirculation. This phase bears the risk of secondary hypoxia because the increasing oxygen requirements of the recovering brain are not coupled to a parallel increase of blood flow. During this phase brain metabolism may abruptly decompensate. This has been clearly shown in this experiment (Fig. 7), and it confirms our earlier more indirect conclusions (Hossmann, 1979).

The progression of recovery in those animals in which these secondary complications did not occur, was surprisingly similar to that previously described by other authors after much shorter periods of ischemia (Naruse *et al.*, 1984; Gadian *et al.*, 1987; Chopp *et al.*, 1987; Andrews *et al.*, 1987; Allen *et al.*, 1988). The secondary rise of lactate and the decline of pH after the initiation of ischemia, the peak values of ischemic and post-ischemic acidosis and the final degree of recovery were the same irrespective of the duration of ischemia (see Fig. 2 in Chopp *et al.*, 1987 after 16 min ischemia, and Figs. 5 and 6 of this investigation after 1 hour ischemia). We, therefore, conclude that the undisputed aggravation of ischemic damage with increasing duration of ischemia is due to the increasing risk of post-ischemic hemodynamic disturbances and not to a more severe metabolic injury during ischemia. This provides the opportunity to improve post-ischemic resuscitation by improving post-ischemic hemodynamics. NMR-spectroscopy may prove to be of particular usefulness for this purpose because of its inherent possibility to monitor both metabolism and blood flow during the whole length of the experiment.

References

1. Allen K, Busza AL, Crockard HA, Frackowiak RSJ, Gadian DG, Proctor E, Russell RWR, Williams SR (1988) Acute cerebral ischaemia: concurrent changes in cerebral blood flow, energy metabolites, pH and lactate measured with hydrogen clearance and ^{31}P and ^{1}H nuclear. J Cereb Blood Flow Metabol 8: 816–821
2. Andrews BT, Weinstein PR, Keniry M, Pereira B (1987) Sequential *in vivo* measurement of cerebral intracellular metabolites with phosphorus-31 magnetic resonance spectroscopy during global cerebral ischemia and reperfusion in rats. Neurosurgery 21: 699–708
3. Behar K, Rothman DL, Hossmann K-A (1989) NMR-spectroscopic investigation of the recovery of energy and acid-base

homeostasis in cat brain after prolonged ischemia. J Cereb Blood Flow Metabol 9: 655–665
4. Chopp M, Frinak S, Walton DR, Smith MB, Welch KMA (1987) Intracellular acidosis during and after cerebral ischemia: *in vivo* nuclear magnetic resonance study of hyperglycemia in cats. Stroke 18: 919–923
5. Dietrich WD, Ginsberg MD, Busto R (1986) Effect of transient cerebral ischemia on metabolic activation of a somatosensory circuit. J Cereb Blood Flow Metabol 6: 405–413
6. Gadian DG, Frackowiak RSJ, Crockard HA, Proctor E, Allen K, Williams SR, Ross Russell RW (1987) Acute cerebral ischaemia: concurrent changes in cerebral blood flow, energy metabolites, pH and lactate measured with hydrogen clearance and ^{31}P and ^{1}H nuclear magnetic resonance spectroscopy. J Cereb Blood Flow Metabol 7: 199–206
7. Hossmann K-A (1979) Cerebral dysfunction related to local and global ischemia of the brain. In: Hoffmeister F, Müller C (eds) (1979) Brain function in old age. Springer, Berlin Heidelberg New York
8. Hossmann K-A (1988) Resuscitation potentials after prolonged global cerebral ischemia in cats. Crit Care Med 16: 964–971
9. Hossmann K-A, Grosse Ophoff B (1986) Recovery of monkey brain after prolonged ischemia. J Cereb Blood Flow Metabol 6: 15–21
10. Hossmann K-A, Lechtape-Grüter H, Hossmann V (1973) The role of cerebral blood flow for the recovery of the brain after prolonged ischemia. Z Neurol 204: 281–299
11. Hossmann K-A, Sakaki S, Kimoto K (1976) Cerebral uptake of glucose and oxygen in the cat brain after prolonged ischemia. Stroke 7: 301–305
12. Hossmann K-A, Schmidt-Kastner R, Grosse Ophoff B (1987) Recovery of integrative central nervous function after one hour global cerebro-circulatory arrest in normothermic cat. J Neurol Sci 77: 305–320
13. Naruse S, Horikawa Y, Tanaka C, Hirakawa K, Nishikawa H, Watari H (1984) *In vivo* measurement of energy metabolism and the concomitant monitoring of electroencephalogram in experimental cerebral ischemia. Brain Res 296: 370–372
14. Prichard JW (1986) NMR-spectroscopy of brain metabolism *in vivo*. Ann Rev Neurosci 9: 61–85
15. Schmidt-Kastner R, Grosse Ophoff B, Hossmann K-A (1986a) Delayed recovery of CO_2 reactivity after one hour's complete ischemia of cat brain. J Neurol 233: 367–369
16. Schmidt-Kastner R, Hossmann K-A, Grosse Ophoff B (1986b) Relationship between metabolic recovery and the EEG after prolonged ischemia of cat brain. Stroke 17: 1164–1169
17. Ueki M, Linn F, Hossmann K-A (1988) Functional activation of cerebral blood flow and metabolism before and after global ischemia of rat brain. J Cereb Blood Flow Metabol 8: 486–494

Correspondence and Reprints: K.-A. Hossmann, Max-Planck-Institut für Neurologische Forschung, Gleuler Strasse 50, D-W-5000 Köln 41, Federal Republic of Germany.

Acta Neurochir (1993) [Suppl] 57: 30–34

Magnetite as a Potent Contrast-Enhancing Agent in Magnetic Resonance Imaging to Visualize Blood-Brain Barrier Disruption

J. W. M. Bulte[1], **M. W. A. de Jonge**[1], **R. L. Kamman**[2], **F. Zuiderveen**[3], **T. H. The**[1], **L. de Leij**[1], and **K. G. Go**[3]

Departments of [1] Clinical Immunology, [2]Magnetic Resonance, and [3] Neurosurgery, University of Groningen, The Netherlands

Abstract

Imaging of blood-brain barrier damage by magnetic resonance was currently studied as to the potential of dextran-magnetite particles (DMP) for contrast enhancement. For that purpose, dextran-T10 (average molecular weight: 10.9 Kilodalton) was complexed with magnetite (Fe_3O_4) in ammonia. Experimental testing of the agent was made *in vivo* using Wistar rats with a freezing injury to the brain. DMP was i.v. injected at a dose of 90 μM Fe/kg b.w. followed by 2% Evans blue (0.6 ml). Control animals with trauma were studied without administration of DMP. Histochemical assessments were made to analyze the tissue distribution of DMP in the brain, kidney and liver after fixation in 4% formalin. MR imaging was conducted with 1.5 Tesla field strength with a circular coil 15 min after the freezing insult and administration of DMP. T_1- and T_2-weighted images were obtained using spin echo sequences among others. Regression analyses indicated a 50% reduction of T_1 at a DMP concentration of 48 μM Fe, while for T_2 only 4 μM/Fe(DMP) were sufficient for a 50% reduction. DMP was also accumulating in other organs, particularly in the Kupffer cells of the liver. Administration of DMP led to recognition of the freezing lesion as black area in agreement with macroscopical findings obtained by autopsy. In animals with a freezing lesion without administration of DMP, only T_2-weighted images demonstrated a somewhat higher intensity attributable to the disruption of the blood-brain barrier. The present findings demonstrate the usefulness of DMP for contrast enhancement of lesions following disruption of the blood-brain barrier.

Keywords: Magnetic resonance imaging; contrast enhancement; dextran magnetite particles; blood-brain barrier.

Introduction

Blood-brain barrier (BBB) impairment can be studied in an animal model in which a lesion is induced by applying a local freezing injury. In this model extravasation of macromolecular tracers depends both on the size and charge of these tracers[1–3].

In MR (Magnetic Resonance) imaging of brain lesions without contrast enhancement, the contrast in the picture is due to local differences in both proton density and the nuclear magnetic resonance (NMR) relaxation times T_1 and T_2. For contrast enhancement in MR imaging of bbb-impairment, the contrast agent Gd-DTPA has been used[11,12]. This paramagnetic contrast agent acts by shortening T_1 and T_2 relaxation times and is operative at millimolar concentrations. It has been reported that magnetite particles are superparamagnetic and may result in considerable shortening of T_1 and T_2 already at nanomolar concentrations[8–10]. Until now magnetite has not been applied for contrast-enhancement of bbb-impairment.

The aim of the present study was to investigate if dextran-magnetite particles (DMP) can be used as a sensitive marker in MR imaging to detect bbb-disruption.

Materials and Methods

Preparation of Dextran-Magnetite Particles (DMP)

DMP was prepared essentially as described by Molday and MacKenzie[5] by ammonia-induced precipitation of magnetite (Fe_3O_4) from an aqueous solution containing dextran, and ferrous- and ferric-chlorides. To obtain very small particles we used dextran-T10 (Pharmacia, MW_{av} = 10,9 kD), whereas the precipitation of magnetite was induced by the stepwise addition of ammonia in a MSE ultrasonic power unit operated at about 60 W for 6 min. The particle size of the obtained DMP preparation was analyzed with a transmission electron microscope (Akashi model 002A) at a magnification of 200,000 ×.

NMR measurements of DMP samples were performed at a resonance frequency of 63 MHz and a temperature of 20°C using a Bruker SXP 4–100 high pulse spectrometer. Calculations of T_1 and T_2 of protons in water were made for DMP at several concentrations.

Induction of BBB-Disruption in a Rat Model

2 groups of each 6 Wistar rats with an average body weight of 200 g were anesthetized with i.p. injected urethan (1.5 g/kg) and a freezing injury was induced by placing a rod of dry ice (–85°C) to the dura of the left parietal cortex for 1 minute through a 5-mm trephine hole. Immediately after the induction of the freezing injury 1 ml DMP in 0.15 M NaCl/0.1 M NaAc pH = 6.5 was injected i.v. in the tail vein at a dose of 90 μmol Fe/kg, followed by i.v. injection of 0.6 ml 2% w/v Evans Blue solution. Control rats (2 × 2) were similarly treated, except that no DMP was injected.

Analysis of DMP Distribution

For histochemical detection of localized DMP in tissue sections, the rats of one group (see above) were killed 60 minutes after induction of the freezing injury. Tissue biopsies taken from the brain, the kidney and liver, were fixed in 4% formalin and 6 μm paraffin-embedded tissue sections were made. In these tissue sections DMP was visualized by the Prussian Blue reaction.

For quantitative physicochemical detection of DMP, rats of the other group were killed also 60 minutes after induction of the freezing injury. To clear all endogenous (e.g. porphyrins) and exogenous (DMP) iron present in the vascular system, the animals were killed by cardiac perfusion with 250 ml saline.

Fresh tissue biopsies taken from the freezing lesion, kidney and liver were immediately weighed and dried at 110°C for 24 h. Samples of 5 mg dry tissue were dissolved by addition of a 200 μl mixture containing 56% perchloric acid and 12% nitric acid, followed by incubation at 60°C for 3 h. The samples were subsequently adjusted to 5 ml with distilled water after which iron content was measured with a AAGF (graphite-furnace atomic absorption) spectrophotometer (Perkin Elmer model 3030B).

MR Imaging

In a third group of animals, MR imaging was carried out in a Philips Gyroscan S15 (1.5 T) using the circular coil C4 (ø 8 cm), starting 15 minutes after induction of the freezing injury and i. v. injection of DMP (450 μmol/kg Fe) and 0.6 ml 2% w/v Evans Blue solution. Transversal T_1-weighted images were made using Spin Echo (SE) sequences with a repitition time (TR) = 650 msec and echo time (TE) = 20 msec. T_2-weighted images were made using TR = 2000 msec and TE = 50 msec.

Results

The DMP preparation contained particles in the range of 4 to 12 nm as is shown by electron microscopy (Fig. 1). The paramagnetic properties of the DMP preparation were assessed by NMR-spectroscopy. Figure 2 shows the T_1 and T_2 of water protons measured at various DMP concentrations (expressed as mM iron (Fe)). Linear regression analysis indicates a 50% reduction of T_1 at a DMP concentration of 48 μM Fe, whereas for T_2 a reduction of 50% was calculated at 4 μM Fe DMP.

After i.v. injection of DMP, extravasation of the particles occurred in the freezing lesion. This was demonstrated by Prussian Blue staining of sections

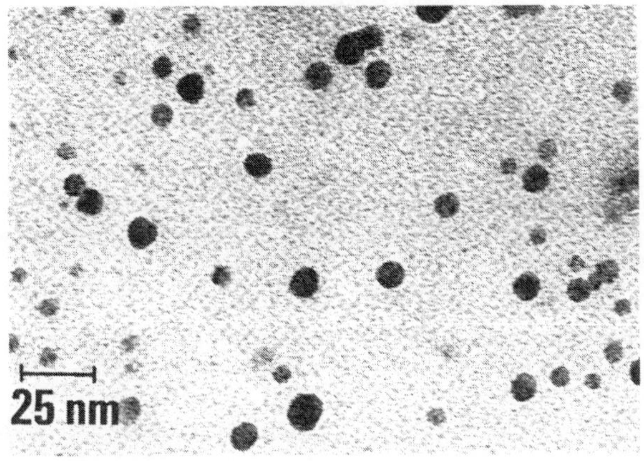

Fig. 1. Transmission electron micrograph of DMP

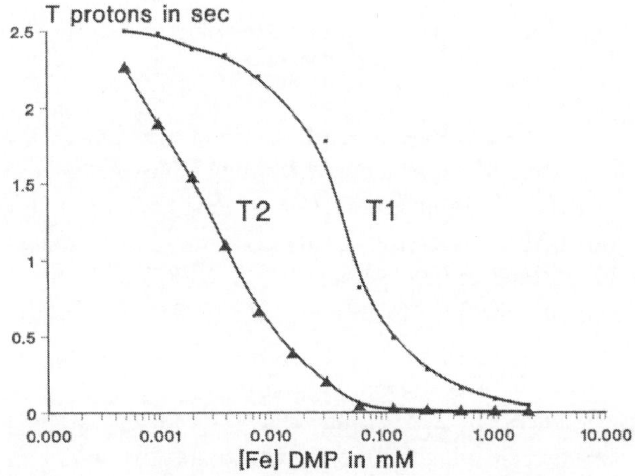

Fig. 2. Proton spin relaxation times T_1 and T_2 in DMP suspensions as a function of DMP concentration (in mM Fe)

made from the brain. Typically, DMP was found clustered in pericytes lining brain capillaries in the freezing lesion. In the remaining part of the brain, where the integrity of the bbb was not disturbed, the particles proved to be retained within the vascular bed, since Prussian Blue staining was negative. When other organs (kidney, liver) were assessed for possible presence of DMP it became clear that prominent trapping of DMP had occurred in liver Kupffer cells. Throughout the entire liver, Kupffer cells were densely packed with these particles. In contrast, in the kidney, a few mesangium cells showed only minimal amounts of DMP.

As shown in Fig. 3 quantification of the iron content of the above tissues by AAGF spectrophotometry showed a slight, but significant cerebral uptake of DMP in biopsies taken from the freezing lesion. In the liver the iron content was found to be considerably increased, whereas in the kidney no uptake of DMP

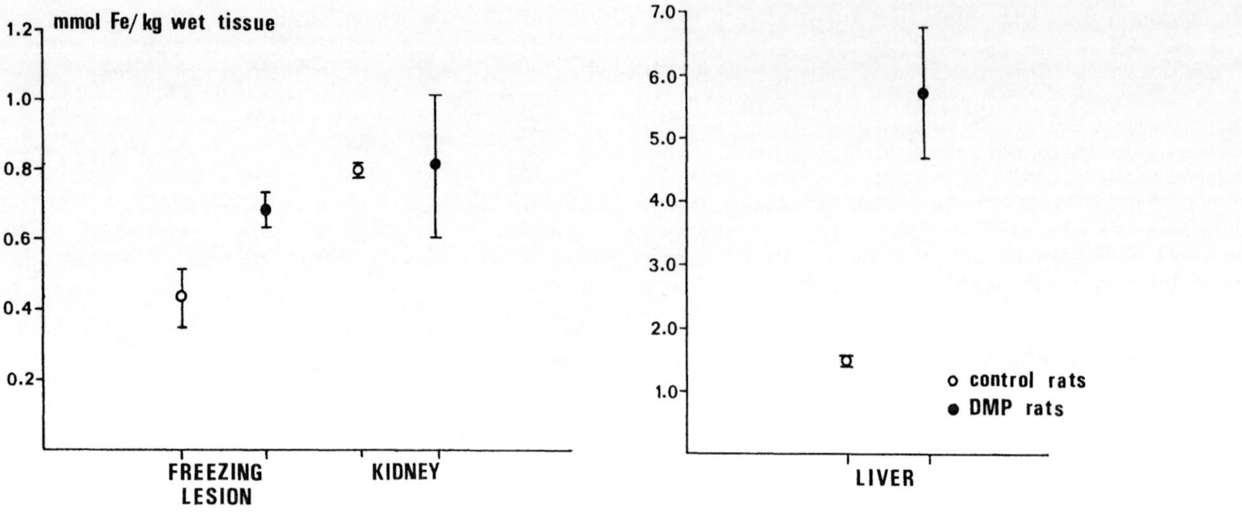

Fig. 3. Iron content (mmol/kg fresh tissue weight) of cerebral freezing lesion, kidney and liver of rats, injected (•), or not injected (○) with DMP; the latter provide the endogenous iron content

was seen. Mean iron content (mmol Fe/kg wet tissue) after DMP administration ($n = 4$) versus control (without DMP administration, $n = 2$) values were respectively 0.67 ± 0.04 versus 0.43 ± 0.08 for the brain region at the freezing lesion, 5.68 ± 0.99 vs. 1.51 ± 0.02 for the liver and 0.81 ± 0.21 vs. 0.80 ± 0.02 for the kidney. This is shown in Fig. 3.

Figures 4a and 4b show a T_1- and T_2-weighted frontal MR image of the rat brain respectively 15 min after i.v. DMP administration. The region containing the freezing lesion is clearly identified as a black spot (see arrows). The exact location of the lesion was confirmed macroscopically following autopsy, since it was found that the distribution of the tracer Evans blue in the brain matched the extent of bbb-disruption as seen in MR imaging. Figs. 4c and 4d show, as a control, T_1- and T_2-weighted MR images, respectively, in which a freezing lesion was analyzed of an animal that had not received DMP. Only for the T_2-weighted image a slightly higher intensity is observed in the edematous tissue induced by the bbb-disruption (see arrow) as compared to normal brain tissue. It can be noted that the black spot in Figs. 4a and 4b is absent in Figs. 4c and 4d.

Discussion

To enable the use of DMP for the study of bbb-disruption, very small, but still superparamagnetic, particles were prepared with molecular sizes between 4 and 12 nm (Fe-core) (see Figs. 1 and 2). These sizes are comparable to those of the proteins IgG (5 nm) and IgM (12 nm). After i.v. injection of a DMP preparation in rats with a local freezing injury in the brain, studies on the distribution of DMP by both histochemical evaluation and AAGF-spectrophotometry showed that the greatest uptake was in the liver Kupffer cells (see Fig. 3). Kupffer cells are known to exhibit phagocytosis of foreign particles. Taking advantage of this property DMP can be used clinically as a (negative) contrast agent for MR imaging of tumor metastases in liver[13–15,17] and spleen[16,18,19].

A new clinical application of DMP is indicated in this report, i.e. the use of DMP as a contrast-enhancing marker for MR imaging to detect disrupted bbb. Since concentrations of 0.004 and 0.048 mM DMP (Fe) induce 50% shortening of T_2 and T_1, respectively, as is shown in Fig. 2, a DMP uptake in the region of the freezing lesion of about 0.24 mM Fe should be enough to induce local reductions of T_1 and T_2 which can then be visualized – as a contrast enhancement in MR imaging. It is known, however, that following BBB-disruption, T_1 and T_2 of the cerebral tissue (without DMP administration) are prolonged due to the increased water content[4,6,7]. By contrast, as is indicated in Figs. 3, 4a and 4b, the extravasation of DMP in a freezing lesion provides considerable shortening of T_1 and T_2. Apparently the prolongation of T_1 and T_2 by the formation of brain edema (see Figs. 4c and 4d) was negligible with respect to the shortening of T_1 and T_2 as induced by DMP (see Figs. 4a and 4b). These results indicate that DMP can be successfully applied as a MR imaging tracer for the detection of bbb-disruption. For this purpose DMP clearance by liver Kupffer cells is

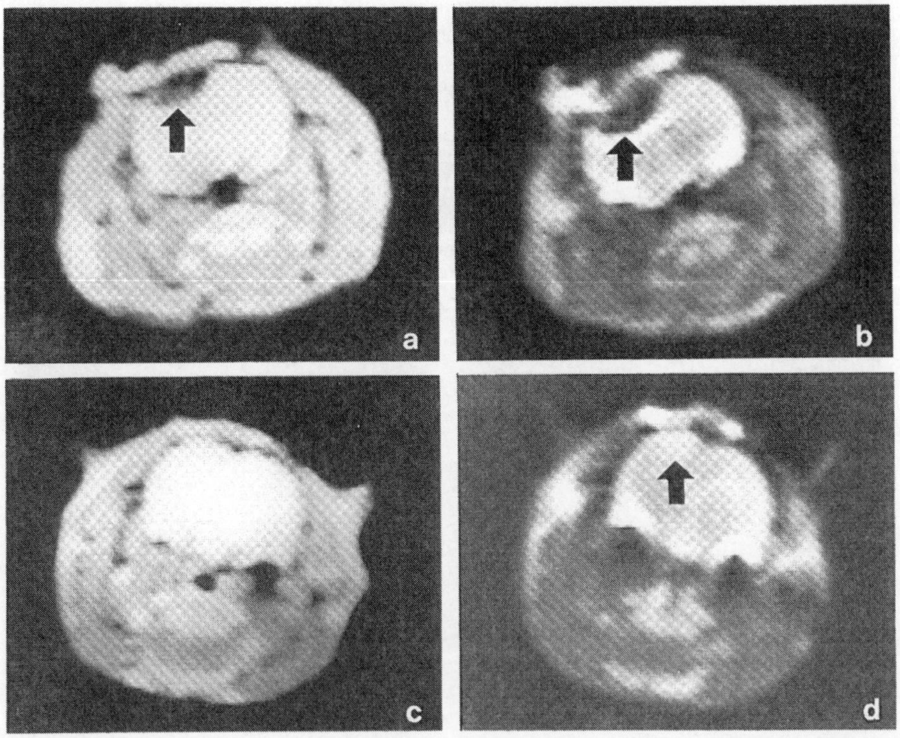

Fig. 4. Frontal MR images (a) T_1-weighted (b) T_2-weighted image of cerebral freezing lesion, 15 min following DMP injection; (c) T_1-weighted (d) T_2-weighted image of cerebral freezing lesion without DMP injection

unfavorable, since it lowers DMP blood concentration. Therefore, although this appears to be only a minor problem, procedures aimed to reduce uptake of DMP by the liver, may further improve the applicability of DMP for the detection of BBB-disruption.

References

1. Go KG, Houthoff HJ, Hartsuiker J, van der Molen-Woldendorp D, Zuiderveen F, Teelken AW (1985) Exudation of plasma protein fractions in vasogenic brain edema. In: Inaba Y, *et al* (eds) Brain edema. Springer, Berlin Heidelberg New York Tokyo, pp 76–87
2. Houthoff HJ, Go KG, Huitema S (1981) The permeability of cerebral capillary endothelium in cold injury: comparison of an endogenous and exogenous protein tracer. In: Cervos-Navarro J, Fritschka E (eds) Cerebral microcirculation and metabolism. Raven, New York, pp 331–336
3. Houthoff HJ, Moretz RC, Rennke HG, Wisniewski HM (1984) The role of molecular charge in the extravasation and clearance of protein tracers in blood-brain barrier impairment and cerebral edema. In: Go KG, Baethmann A (eds) Recent progress in the study and therapy of brain edema. Plenum, pp 67–79
4. Kamman RL, Go KG, Brouwer W, Berendsen HJC (1988) Nuclear magnetic resonance relaxation in experimental brain edema: effects of water concentration, protein concentration, and temperature. Magn Reson Med 6: 265–274
5. Molday RS, MacKenzie D (1982) Immunospecific ferromagnetic iron-dextran reagents for the labelling and magnetic separation of cells. J Immunol Meth 52: 353–367
6. Naruse S, Horikawa Y, Tanaka C, Hirakawa K, Nishikawa H, Yoshizaki K (1982) Proton nuclear magnetic resonance studies on brain edema. J Neurosurg 56: 747–752
7. Naruse S, Horikawa Y, Tanaka C, Hirakawa K, Nishikawa H, Yoshizaki K (1986) Significance of proton relaxation time measurement in brain edema, cerebral infarction and brain tumors. Magn Reson Im 4: 293–304
8. Olsson MBE, Persson BRB, Salford LG, Schröder U (1986) Ferromagnetic particles as contrast agents in T_2 NMR imaging. Magn Reson Med 4: 437–440
9. Renshaw PF, Owen CS, McLaughlin AC, Frey TG, Leigh JS Jr (1986) Ferromagnetic contrast agents: a new approach. Magn Reson Med 3: 217–225
10. Renshaw PF, Owen CS, Evans AE, Leigh JS Jr (1986) Immunospecific NMR contrast agents. Magn Reson Im 4: 351–357
11. Runge VM, Price AC, Wehr CJ, *et al* (1985) Contrast enhanced MRI: evaluation of a canine model of osmotic blood-brain barrier disruption. Invest Radiol 20: 830–844
12. Runge VM, Clanton JA, Price AC, *et al* (1985) The use of Gd DTPA as a perfusion agent and marker of blood-brain barrier disruption. Magn Reson Im 3: 43–55
13. Saini S, Stark DD, Hahn PF, Bousquet J-C, Introcasso J, Wittenberg J, Brady TJ, Ferrucci JT Jr (1987) Ferrite particles: a superparamagnetic MR contrast agent for enhanced detection of liver carcinoma. Radiology 162: 217–222
14. Stark DD, Weissleder R, Elizondo G Hahn PF, Saini S, Todd LE, Wittenberg J, Ferrucci JT (1988) Superparamagnetic iron oxide: clinical application as a contrast agent for MR imaging of the liver. Radiology 168: 297–301
15. Tsang Y-M, Stark DD, Chia-Mei Chen M, Weissleder R, Wittenberg J, Ferrucci JT (1988) Hepatic micrometastases in the rat: ferrite-enhanced MR imaging. Radiology 167: 21–24

16. Weissleder R, Hahn PF, Stark DD, Rummeny E, Saini S, Wittenberg J, Ferrucci JT (1987) MR imaging of splenic metastases: ferrite-enhanced detection in rats. Am J Roentgenol 149: 723–726
17. Weissleder R, Saini S, Stark DD, Wittenberg J, Ferrucci JT (1988) Dual-contrast MR imaging of liver cancer in rats. Am J Roentgenol 150: 561–566
18. Weissleder R, Stark DD, Rummeny EJ, Compton CC, Ferrucci JT (1988) Splenic lymphoma: ferrite-enhanced MR imaging in rats. Radiology 166: 423–430
19. Weissleder R, Hahn PF, Stark DD, Elizondo G, Saini S, Todd LE, Wittenberg J, Ferrucci JT (1988) Superparamagnetic iron oxide: enhanced detection of focal splenic tumors with MR imaging. Radiology 169: 399–403

Correspondence and Reprints: K. G. Go, Department of Neurosurgery, University of Groningen, Academish Ziekenhuis, NL-9700 RB Groningen, The Netherlands.

Acta Neurochir (1993) [Suppl] 57: 35–40

Absent Recruitment of Capillaries in Brain Tissue Recovering from Stroke

A. Gjedde[1] and **H. Kuwabara**[2]

[1] Positron Imaging Laboratories, McConnell Brain Imaging Unit, Montreal and [2] Neurological Institute, Montreal, Quebec, Canada

Abstract

The density of perfused capillaries (d_{CAP}), defined as capillaries that transport glucose, as well as the volume fraction of these capillaries in the vascular bed (f_{CAP}), and the mean transit time of blood through the capillaries (t_{CAP}), were calculated from hemodynamic variables obtained *in vivo* by positron tomography of brains of six patients affected by stroke. Each patient was studied twice, within 38 hrs of the insult, and one week later. 38 ischemic and 38 contralateral mirror regions were compared. The metabolic rate for glucose (CMR_{glc}) was determined on the basis of regional calculations of the lumped constant. No significant change of the lumped constant was observed in any region.

In normal regions, no significant differences of any variables existed between the first and second studies. In the infarct regions of the first study, CMR_{glc} and CMR_{O2} (cerebral metabolic rate for oxygen) were 30–50% of control (deactivation) and CBF (cerebral blood flow), capillary density, and the capillary diffusion capacity for fluorodeoxyglucose (K_1) were similarly reduced, although the oxygen/glucose ratio was only 3.75 in the ischemic regions. While f_{CAP} decreased, t_{CAP} doubled. One week after the first study, blood flow returned to normal in the infarct regions despite continued depression of metabolism. Capillary density and diffusion capacity remained low, indicating absent recruitment of nutrition vessels (perfusion capillaries).

Keywords: PET scanning; capillary perfusion; cerebral metabolic rates; stroke.

Introduction

Different regions of the brain have different capillary densities. In most cases, it is true that the regions of the brain that have the highest rate of metabolism, also enjoy the highest density of capillaries. In the first analysis, this observation suggests that blood flow, metabolism, and the density of capillaries, are functionally coupled[6]. On closer examination, however, this conclusion is not tenable, a priori. Since the variation merely is anatomical, it suggests that the total number of capillaries in any one region is a reflection of the normal average energy requirement of the region, and that individual capillaries have similar properties (radius, length, permeability) in different regions. The regional variation does not address the question of how many capillaries are actually functioning at different times, or how this number varies with the energy requirements of individual regions.

In the present context, we wish to define functioning capillaries as capillaries that permit the transfer of glucose to the tissue. The concept of functioning capillaries does not imply only perfused capillaries. Some capillaries may contain blood that is stationary or moves so slowly that it does not effectively contribute to the nutrition of the tissue. Since the mechanisms of glucose and oxygen transfer are different, it is possible, at least in theory, that capillaries exist which may allow the diffusion of oxygen, yet cannot transport glucose. We will ignore this possibility in the present paper.

In order to prove a coupling between the number of functioning capillaries and the energy requirements of the tissue (recruitment), it is necessary to verify parallel adjustments of flow, density of functioning capillaries, and capillary diffusion capacity when metabolism is subject to change in a region of interest.

Positron emission tomography (PET) allows measurements to be made of several variables, including glucose phosphorylation rate (J_{glc}), fluorodeoxyglucose transfer coefficients (K_1, k_2, and k_3), net uptake of oxygen (J_{O2}), cerebral blood flow (F), and the correction term for the cerebral vascular volume (V_o). However, establishing conditions of activation or deactivation that are sufficiently stable to allow several comparable measurements in the same subject is usually difficult. Alteration of brain function associated with ischemic insults may provide a wide range of flow and metabolism values which are sufficiently stable to allow comparisons in the same subject.

Methods

Six patients affected by recent cerebral stroke were studied. Each patient was examined with positron emission tomography twice. The first study was carried out between 16 and 38 hours after the onset of the insult. and the second study one week later.

PET Scans and Analysis

The patients were scanned in the Therascan 3128, a 3-plane positron tomograph with a resolution of 12 mm (transverse) by 11 mm (axial)[5]. Image reconstruction was performed with correction for random events, detector efficiency variation, attenuation, and scattered events.

In each study, horizontal planes parallel to and 24, 36 and 48 mm above Reid's base line were scanned so that the same region of interest (ROI) from different studies represented the same brain region.

Blood flow, F, was calculated from the brain uptake of radioactive water labeled by inhalation of ^{15}O labeled carbon dioxide, J_{O2}, from the equilibrium distribution of ^{15}O labeled oxygen, and the total vascular volume, V_o, from the equilibrium distribution of ^{15}O labeled carbon monoxide[9,13].

The glucose phosphorylation was calculated from the brain uptake of ^{18}F labeled fluorodeoxyglucose (FDG). A dynamic FDG study consisted of a series of eighteen scans over a period of 35 minutes (ten one-minute scans, five 2-minute scans, and three 5-minute scans) following intravenous administration of FDG. Arterial blood was sampled for analysis of plasma radioactivity.

In the analysis of PET data, regional templates were used. Each template consisted of twenty square regions of interest (ROIs) placed along the cerebral cortex of a PET image. The ROI templates were applied repeatedly to the functional maps of the same subject as well as to the serial images obtained in dynamic studies. The ability to store ROI templates ensured reproducibility of the regional analysis[8].

Determination of Transfer Coefficients and Lumped Constant

The lumped constant of the FDG method was defined as the ratio between the net extraction fractions of FDG and glucose in brain (K^*/K)[#].

Assuming that the ratio between the unidirectional transfer rates of FDG and glucose (τ), and the ratio between the phosphorylation rates (φ), are both constant, and also assuming that the ratio of transport rates from blood to the brain and in the reverse direction ($V_e = K_1/k_2$) has the same value for glucose and the glucose analogs[11], a new formula for the lumped constant can be derived in the following manner:

$$\Lambda = \frac{K^*}{K} = K^* \frac{k^*_2 / \tau) + (k^*_3 / \varphi)}{K^*_1 / \tau)\,(k^*_3 / \varphi)} = \varphi\rho + \tau\,(1 - \rho)\,, \tag{1}$$

where

$$\rho = \frac{k^*_2}{k^*_2 + k^*_3} = 1 - \frac{K^*}{K^*_l}\,, \tag{2}$$

and k_3* is the phosphorylation rate of FDG. The distribution volume (V_e) is,

$$V_e = \frac{K_1}{k_2} = \frac{V_d}{1 + \left(\frac{\tau K^* C_a}{K^*_1 \Lambda K_t}\right)}\,, \tag{3}$$

where K_t is the half-saturation concentration of glucose transport, C_a the arterial plasma glucose concentration, and V_d the brain water content.

Using Eq. (2), the sum of k_2* and k_3* is a function of K* and K_1*:

$$k^*_2 + k^*_3 = \frac{K^*_1}{\rho V_e}\,. \tag{4}$$

Thus, eliminating k_2* and k_3* from the operational equation of the Sokoloff method[15], it is possible to determine the values of K*, K_1* and V_o (blood volume) by a non-linear least-squares method. The values of k_2* and k_3* were calculated as $k_2^* = K_1^* / V_e$, and $k_3^* = K^* / (\rho V_e)$. For the calculation of the rate constants and the lumped constant, we chose the values $\tau = 1.1$, $\varphi = 0.3$, $K_t = 4.8$, and $V_d = 0.77$.

Calculation of Physiological Variables

The capillary permeability-surface area product of fluorodeoxyglucose was calculated from the Bohr equation for capillary exchange[1],

$$\frac{PS}{F} = -\,\alpha \ln\left(\frac{\Delta C_v}{\Delta C_a}\right)\,, \tag{5}$$

where PS is the permeability-surface area product, F the blood flow, α the solubility of the tracer in whole blood relative to that in brain tissue, and ΔC_v and ΔC_a the tracer concentration difference between tissue and blood at the arterial and venous end of the capillary, respectively. If α is unity, and if the tracer clearance is determined initially when tissue concentrations can be assumed to be negligible, as in the case of tracer fluorodeoxyglucose in human whole blood, Eq. (5) can be modified to[7],

$$PS = -F \ln\left(1 - \frac{K^*_1}{F}\right)\,, \tag{6}$$

where K_1* is the initial clearance of the tracer reflecting unidirectional transport. The number of functioning (i.e., glucose-transporting) capillaries, the capillary fraction of the total blood volume in the brain, and the mean capillary transit time of blood were all calculated from the PS product of fluorodeoxyglucose. The capillary surface area (S) is,

$$S = 2\pi r\, D_{CAP}\,, \tag{7}$$

where r is the radius of a capillary, D_{CAP} the total length of capillaries per mm^3 tissue. Since the average length of capillaries is not known with certainty, D_{CAP} is the average number of capillary cross sections per one square millimeter of tissue. Thus,

$$D_{CAP} = \frac{PS}{2\pi r P}\,, \tag{8}$$

where reasonable values of r and P were chosen as 2 µm and 380 nm s^{-1}, respectively. In practice, D_{CAP} has unit of mm^{-2} and PS unit of ml 100 g^{-1} min^{-1}, assuming the specific gravity of brain tissue to be unity. According to this calculation, the value of D_{CAP} is, of course, directly proportional to the value of PS.

The capillary fraction of the total volume of the vascular bed (f_{CAP}) was calculated from the ratio between the capillary blood volume and the total blood volume, measured with labeled carbon monoxide,

[#] Asterisks refer to FDG. Symbols without an asterisk refer to native glucose.

$$f_{CAP} = \frac{V_{CAP}}{V_o} = \left(\frac{r}{2P}\right)\frac{PS}{V_o}, \tag{9}$$

where f_{CAP} is a fraction and V_o has unit of ml 100 g^{-1}.

The mean capillary transit time (t_{CAP}) was calculated from the ratio between the capillary volume and blood flow,

$$t_{CAP} = \frac{V_{CAP}}{F} = \left(\frac{r}{2P}\right)\frac{PS}{F}, \tag{10}$$

where t_{CAP} has unit of seconds (s).

Results

The clinical characteristics of the patients are summarized in Table 1. All but one patient had associated diseases. The patients were hyperglycemic at the time of the positron emission tomography.

We defined "ischemic" regions as those ROIs with an average blood flow less than 20 ml 100 g^{-1} min^{-1} at the time of the first scan. Each patient had four to ten ischemic ROIs and 32 in all were found. These ischemic ROIs were always adjacent to each other and located in the area of the cerebral infarct identified by CT scans. Contralaterally matching ROIs were used as normal controls. On PET images of the second study, the same ROIs were analyzed. Thus, the investigation included a total of 128 ROIs.

Table 1. *Clinical Characteristics and Plasma Glucose Levels*

	Age	Sex	Neurological signs[1]	Associated conditions	Glucose levels[2] (μmol/ml)	
1	65	M	R hemiparesis aphasia	ASHD	5.5	6.0
2	79	M	Anosognosia	HBP	6.0	6.0
3	43	F	R hemiparesis	–	7.8	8.2
4	51	F	L hemiparesis	D.M.	6.6	5.1
5	51	M	R hemiparesis aphasia	ASHD	6.8	6.8
6	57	F	R hemiparesis aphasia	HBP D.M.	9.9	12.4

Abbreviations: *M* = male; *F* = female; *R* = right; *L* = left; *ASHD* = arteriosclerotic heart disease; *HBP* = high blood pressure; *D.M.* = Diabetes mellitus

[1] Neurological signs were those at the time of admission.

[2] Plasma glucose levels at the time of the first PET study (left side column) and at the time of the second study (right side column).

Table 2. *Summary of Variables in Ischemic and Contralateral Matching Regions at the First and the Second Studies*

	1st study (< 38 hours)		2nd study (7 days later)		Elderly controls
Variables	normal	ischemic	normal	ischemic	(mean age 63yrs)
F	49.0 ± 10.5	12.7 ± 1.3*	37.7 ± 1.5	38.5 ± 5.3	36.8 ± 2.8
E_{O2}	0.46 ± 0.04	0.54 ± 0.04*	0.43 ± 0.01	0.16 ± 0.03*	0.44 ± 0.03
V_O	3.9 ± 0.4	3.9 ± 0.6	3.7 ± 0.3	4.4 ± 0.4	4.4 ± 0.5
J_{O2}	187 ± 28	59 ± 9*	143 ± 5	48 ± 7*	144 ± 9
K_1	0.081 ± 0.013	0.041 ± 0.003*	0.065 ± 0.012	0.043 ± 0.007*	0.091 ± 0.002
Λ	0.54 ± 0.02	0.59 ± 0.03	0.54 ± 0.02	0.53 ± 0.02	0.58 ± 0.01
J_{glc}	30.9 ± 4.2	16.1 ± 0.9*	25.4 ± 4.9	16.0 ± 2.0*	29.5 ± 0.57
D_{CAP}	360 ± 58	204 ± 20	294 ± 59	187 ± 32	
f_{CAP}	11.4 ± 0.9	6.8 ± 1.0	9.8 ± 2.0	5.1 ± 0.7	
t_{CAP}	0.6 ± 0.1	1.3 ± 0.2*	0.6 ± 0.1	0.4 ± 0.1*	

Mean values ± standard errors of mean are shown.

Abbreviations and units: F = regional cerebral blood flow (ml hg^{-1} min^{-1}); E_{O2} = oxygen extraction fraction (ratio); V_O = regional cerebral blood volume (ml hg^{-1}); J_{O2} = regional cerebral metabolic rate for oxygen (μmol hg^{-1} min^{-1}); K_1 = unidirectional transfer rate of fluorodeoxyglucose (ml g^{-1} min^{-1}); Λ = lumped constant (ratio); J_{glc} = regional cerebral metabolic rate for glucose (μmol hg^{-1} min^{-1}); D_{CAP} = number of capillary cross sections (mm^{-2}); f_{CAP} = capillary fraction of vascular bed (%); t_{CAP} = mean capillary transit time (s); hg^{-1} = per hundred gram.

* Denotes significant difference from the normal ROIs of the second study ($p < 0.05$).

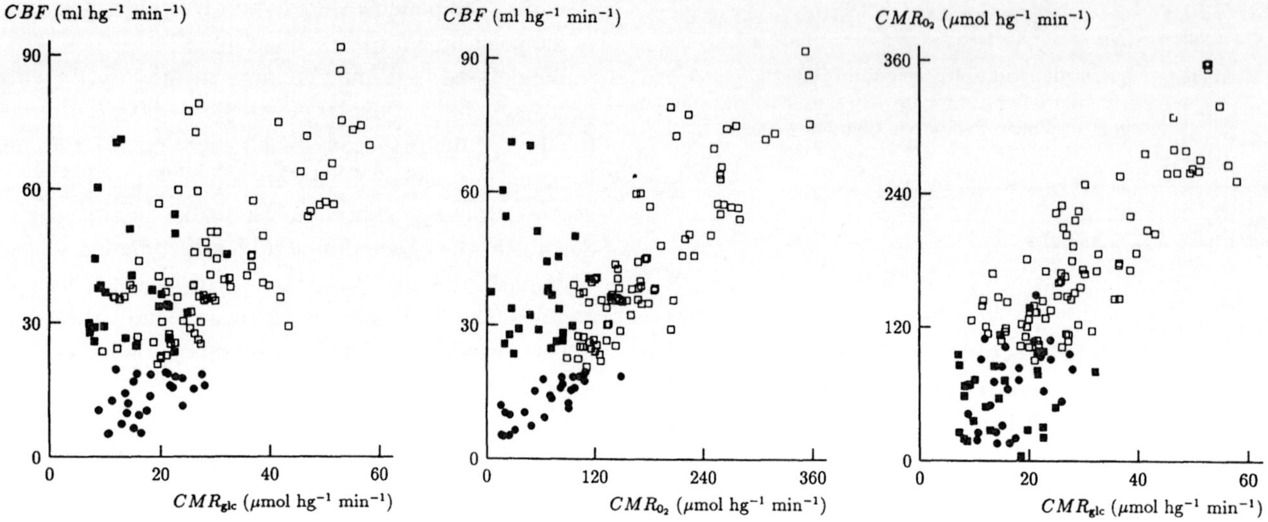

Fig. 1. Relationships between blood flow and metabolic rates for oxygen and glucose. Open squares: contralateral, normal regions. Closed circles: ischemic regions, first study. Closed squares: ischemic regions, second study

Table 2 summarizes the findings of the study. The median values of normal elderly controls (mean age 63 ± 3 years) were also included. Normal ROIs of the first study had higher regional cerebral blood flow and regional oxygen metabolic rate for oxygen than those of the second study, which agreed well with the values of the elderly controls. In the ischemic ROIs, flow and oxygen metabolism were severely depressed and the oxygen extraction fraction (E_{O2}) was slightly elevated at the time of the first study. Seven days later, the E_{O2} decreased significantly below normal and oxygen consumption remained low despite restoration of adequate blood flow.

Normal ROIs of the first study had higher unidirectional uptake of FDG (K_1*) and regional glucose metabolic rate than those of the second study, although K_1* was suppressed in both studies in comparison to older controls. In ischemic ROIs, both K_1* and J_{glc} were markedly reduced and the regionally calculated lumped constant was elevated.

The relationship between cerebral blood flow, oxygen metabolic rate, and glucose utilization rate is shown in Fig. 1. There was a positive correlation between flow and metabolism over a wide range of F (4–124 ml 100 g^{-1} min^{-1}), J_{O2} (75–351 μmol 100 g^{-1} min^{-1}), and J_{glc} (7–52 μmol 100 g^{-1} min^{-1}), except for the ischemic regions one week after the first study.

The average number of capillaries actually transporting FDG (D_{CAP}) was 360 mm^{-2} in the normal ROIs at the first study. The mean value decreased to 294 mm^{-2} seven days later. In the ischemic ROIs at the early stage, the capillary density was 204 mm^{-2}, and it remained low at the time of the second study.

The capillary density was plotted against blood flow and metabolism in Fig. 2. We observed parallel changes of the capillary density with flow and metabolism when metabolism was subject to change.

For the contralateral normal ROIs, the average transit time of blood through the capillaries (t_{CAP}) was 0.6 s in both studies. In the ischemic ROIs in the early stage, t_{CAP} was almost twice as long (1.3 s) but, it decreased below normal seven days later.

Discussion

Studies on six stroke patients in the early stage of a cerebrovascular accident (16–38 hours after the onset), and one week later, provided a wide range of metabolic states. We confirmed that the density of glucose-transporting capillaries, blood flow, and oxygen and glucose metabolic rates change in parallel (recruitment) over this wide range of metabolic states. We found that the coupling between capillary density, metabolism, and flow was lost in ischemic regions at the time of the second study[10].

In the present study, we measured the density of functioning capillaries on the basis of two simple assumptions, i.e. that capillaries are identical, rigid tubes in all regions of the brain cortex, and that changes of the permeability of the capillary endothelium to hexoses

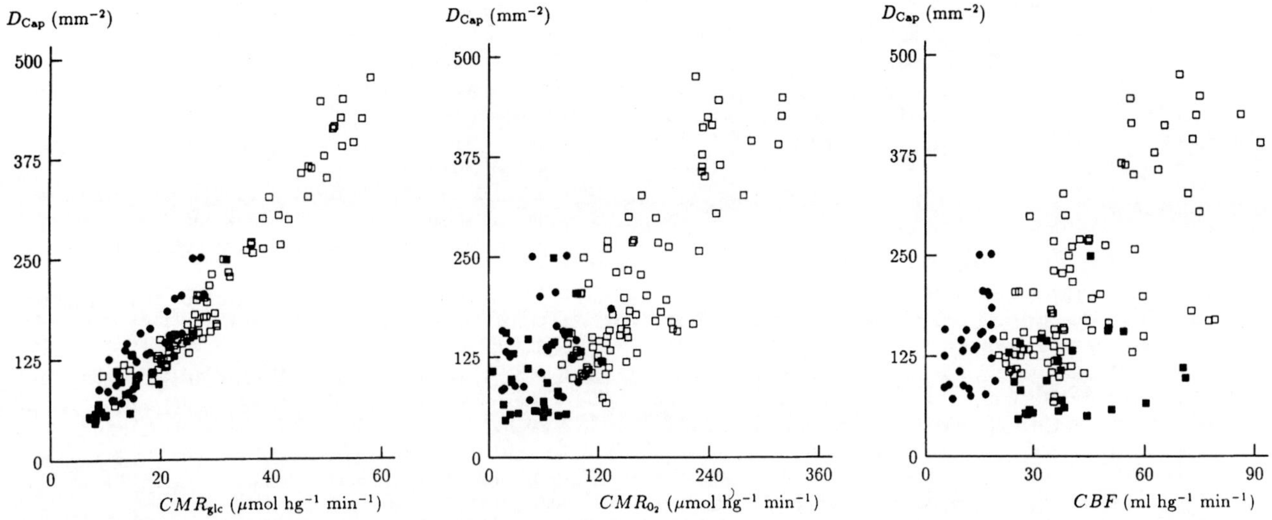

Fig. 2. Relationship between the density of functional capillaries (ordinates) and blood flow and metabolic rates for oxygen and glucose (abscissae). Open squares: contralateral, normal regions. Closed circles: ischemic regions, first study. Closed squares: ischemic regions, second study

like glucose occur only as a result of changes of the capillary surface area available for facilitated diffusion. While specific changes of the endothelium in ischemia, relevant to glucose transport, cannot be ruled out in ischemia, we assumed that such changes did not occur.

The average normal density of cortical capillaries transporting glucose was 294 ± 114 mm^{-2} and ranged from 122 to 675 mm^{-2} in healthy brain tissue. Our results suggested that less than 50% of all capillaries available were actually functioning in the resting state. Reported values of capillary density of the cerebral cortex range from 589 mm^{-2} in rats to 939 mm^{-2} in cats, and 50 to 84% of them were shown to be perfused at a given time[3,4,14]. Since the situation is dynamic, different capillaries may be perfused at different times, even when the average density remains the same. This implies that methods using dyes that require more than ten seconds to circulate may overestimate the density of perfused capillaries[4].

An *in vivo* microscopic study of blood flow in cortical capillaries of cats by Pawlik *et al.* reported that the flow velocity in capillaries ranged from 0.4 to 3.9 mm s^{-1} with a mean value of 1.5 mm s^{-1} [14]. These values correspond to transit times ranging from 0.3 s to 2.0 s with a median of 0.7 s, which agree very well with the mean capillary transit time observed in the present study.

The hypothesis of recruitment is necessitated by the diffusion-limited transfer of oxygen from blood to brain which is close to being rate-limiting for oxygen utilization and thus is responsible for the low oxygen tension in the tissue. Increasing the linear velocity of the blood perfusing the brain does not significantly improve the delivery of oxygen. Hence, decreases of the intercapillary distance are necessary to allow increases of oxygen utilization during activation.

In this study, we have used a new method for the determination of the lumped and rate constants of the three compartment model of the FDG method. This method described the data statistically as well as that conventionally used. In the ischemic regions, the lumped constant was slightly elevated from 0.54 of the normal regions to 0.59. This implies that more tracer FDG, once taken up by brain, was phosphorylated in ischemia than in the normal state, while less glucose was phosphorylated. Although the values of F, E_{O2}, and J_{O2} of the normal regions of the second study were in good agreement with the median values of the older controls, the unidirectional uptake of FDG and the lumped constant were lower. Brooks *et al.* studied the transport of hexose across the blood-brain barrier and reported that hyperglycemic subjects had a lower unidirectional FDG clearance (K_1) (0.055 ml g^{-1} min^{-1}) than normoglycemic subjects[2]. With direct enzymatic measurement of brain glucose content[11], Gjedde and Diemer showed that the lumped constant was lower in hyperglycemic rats. Hence, the explanation for the lower unidirectional clearance and lumped constant in these patients is the elevated plasma glucose level (mean 7.3 ± 2.0 vs. 5.3 ± 0.7 μmol ml^{-1} for older controls).

Acknowledgements

The authors thank the staff of the Positron Imaging Laboratories, including the Cyclotron and Radiochemistry Facility. This study was supported by a grant from the Medical Research Council of Canada (PG-41) and the Quebec Heart Foundation.

References

1. Bohr C (1909) Über die spezifische Tätigkeit der Lungen bei der respiratorischen Gasaufnahme und ihr Verhalten zu der durch die Alveolarwand stattfindenen Gasdiffusion. Skand Arch Physiol 22: 221–280
2. Brooks DJ, Gibbs JSR, Sharp P, Herold S, Turton DR, Luthra SK, Kohner EM, Bloom SR, Jones T (1986) Regional cerebral glucose transport in insulin-dependent diabetic patients studied using [^{11}C]3-O-methyl-D-glucose and positron emission tomography. J Cereb Blood Flow Metabol 6: 240–244
3. Buchweitz E, Weiss H (1986) Alterations in perfused capillary morphometry in awake vs anesthetized brain. Brain Res 377: 105–111
4. Collins RC, Wagman IL, Lymer L, Matter JM (1987) Distribution and recruitment of capillaries in rat brain. J Cereb Blood Flow Metabol 7 [Suppl 1]: S336
5. Cook BE, Evans AC, Fanthome EO, Alarie R, Sendyk AM (1984) Performance figures and images from the Therascan 3128 positron emission tomograph. IEEE Trans Nucl Sci NS-31 (1): 640–644
6. Craigie, EH (1945) The architecture of the cerebral capillary bed. Biol Rev 20: 133–146
7. Crone C (1963) The permeability of capillaries in various organs as determined by the use of the "indicator diffusion" method. Acta Physiol Scand 58: 292–305
8. Evans AC, Beil C, Marrett S, Thompson CJ, Hakim A (1988) Anatomical-functional correlation using an adjustable MRI-based region-of-interest atlas with positron emission tomography. J Cereb Blood Flow Metabol 8: 513–530
9. Frackowiak RSJ, Lenzi G-L, Jones T, Heather JD (1980) Quantitative measurement of regional cerebral blood flow and oxygen metabolism in man using ^{15}O and positron emission tomography. Theory, procedure, and normal values. J Comput Assist Tomogr 4: 727–736
10. Garcia JH, Conger KA, Briggs L (1981) Brain capillaries in acute and subacute regional ischemia. In: Cervós-Navarro J, Fritschka E (eds) Cerebral microcirculation and metabolism. Raven, New York, pp 83–91
11. Gjedde A, Diemer NH (1983) Autographic determination of regional brain glucose content. J Cereb Blood Flow Metabol 3: 303–310
12. Gjedde A, Wienhard K, Heiss W-D, Kloster G, Diemer NH, Herholz K, Pawlik G (1985) Comparative regional analysis of 2-fluorodeoxyglucose and methylglucose up-take in brain of four stroke patients. With special reference to the regional estimation of the lumped constant. J Cereb Blood Flow Metabol 5: 163–178
13. Lammertsma AA, Jones T (1983) The correction for the presence of intravascular ^{15}O in the steady-state technique for the measurement of oxygen extraction in the brain. 1: description of the method. J Cereb Blood Flow Metabol 3: 416–424
14. Pawlik G, Rackl A, Bing RJ (1981) Quantitative capillary tomography and blood flow in the cerebral cortex of cats; an *in vivo* microscopic study. Brain Res 208: 35–58
15. Sokoloff L, Reivich M, Kennedy C, Des Rosiers MH, Patlak CS, Pettigrew KD, Sakurada O, Shinohara M (1977) The [^{14}C] deoxyglucose method for the measurement of local cerebral glucose utilization: theory, procedure and normal values in the conscious and anesthetized albino rat. J Neurochem 28: 897–916

Correspondence and Reprints: A. Gjedde, Neurological Institute, 3801 University Street, Montreal, Quebec, H3A 2B4, Canada.

Acta Neurochir (1993) [Suppl] 57: 41–48

Quantification of Primary and Secondary Lesions in Severe Head Injury

D. I. Graham[1], **J. H. Adams**[1], **D. Doyle**[1], **I. Ford**[1], **T. A. Gennarelli**[3], **A. E. Lawrence**[2], **W. L. Maxwell**[2], and **D. R. McLellan**[2]

Departments of Neuropathology, [1] Statistics and [2] Anatomy, University of Glasgow, Scotland, U.K.., and [3] Division of Neurosurgery, Hospital of the University of Pennsylvania, Philadelphia, U.S.A.

Abstract

The current trend is to classify brain damage due to non-missile head injury as focal or diffuse. Quantitative findings will be discussed and reference will be made to both human and experimental non-human primate.

Focal Injury

Contusions: Based on the concept of an index which takes into account the depth and extent of contusions in various parts of the brain, it has been determined that contusions are most severe in the frontal and temporal lobes; they may be entirely absent in a patient dying as a result of a head injury; there is no correlation between their severity and the nature of the injury; the concept of contrecoup must continue to be questioned; they are more severe in patients who have a fracture of the skull in comparison to those who do not; they are more severe in patients who do not experience a lucid interval than in those who do; and they are less severe in patients with diffuse axonal injury than in those who do not have diffuse axonal injury.

Raised intracranial pressure: It has been established that the pathognomonic feature of a high ICP due to a supratentorial expanding lesion is the presence of a wedge of pressure necrosis in one or both parahippocampal gyri. Pressure necrosis of this type was seen in every patient known to have had an ICP of more than 5.3 kPa (40 mm Hg) during life, and most of the patients with an ICP between 2.7 and 5.3 kPa (20–40 mm Hg), and in no patient with an ICP of less than 2.7 kPa (20 mm Hg).

Diffuse Injury

Axonal injury (DAI): Increasing experience in man has allowed of the identification and grading of 122 cases of DAI in a series of 434 fatal non-missile head injuries – 10 Grade 1, 29 Grade 2 and 83 Grade 3. In our earlier studies of severe DAI, not one of the patients was able to talk after their injury: it is of interest therefore that 17 (14%) of the 122 cases – all with less severe DAI – were known to have talked soon after their injury. Evidence of lesser degrees of DAI were also found in the Penn 2 model of axonal injury.

Ischemic damage: This was found in the brains of 92% of 151 cases of fatal non-missile head injury dying between 1968–72 and in 88% of 112 cases dying between 1981–82. There was no statistical difference in amount of moderately severe and severe ischemic damage in the two groups, 55% and 54% respectively. There was evidence that an increased number of patients with severe ischemic brain damage was admitted in 1981–82 as a result of a change in the admission policy of the Department of Neurosurgery that resulted in an increased detection of intracranial haematomas.

Keywords: Head injury; contusion index; intracranial hypertension; axonal injury; secondary ischemia.

Introduction

In an attempt to provide clinico-pathological correlations, there has been a tendency to emphasise the existence of primary and secondary damage[1]. There is much to be said for this approach in a clinical setting since it helps in the identification of at least potentially preventible or avoidable complications, such as the late recognition of intracranial haematoma[2–6] especially in patients with a head injury who talk and die[7]. There is, however, a trend for clinicians and pathologists to classify brain damage as focal or diffuse[8,9]. The principal focal lesions are cerebral contusions, intracranial haematoma, and the various types of secondary brain damage that can be produced by intracranial expanding lesions. Diffuse brain damage exists in four principal forms comprising diffuse axonal injury, hypoxic brain damage, diffuse brain swelling and diffuse vascular injury.

In the course of the past few years, all of the principal types of brain damage that occur in man as a result of a non-missile head injury have been reproduced in non-human primates using devices referred to as Penn I and Penn II[10,11].

In the course of the past decade there have been major advances, not only in defining, but also in quantifying the types of brain damage that occur in non-missile head injury, and in identifying their pathogenesis. It is the purpose of this paper to emphasise quantitative findings with reference to both human and experimental non-human primate material.

Materials and Method

Human Material

During the 15 year period 1968–1982 post-mortem examinations were undertaken in this Institute on 635 fatal non-missile head injuries. There were 497 (78%) males and 138 (22%) females; the age range was 9 weeks to 89 years and the duration of survival from 1 hour to 14 years 3 months. The majority of injuries were attributable to road traffic accidents (335; 53%), or to falls (221; 35%). There was a fracture of the skull in 478 (75%) of the cases. 206 (32%) of the 635 cases had talked at some time after their injury.

A full autopsy was undertaken in every case and the brains and spinal cords were fixed in 10% formol saline for at least 3 weeks before dissection. The cerebral hemispheres were cut in a standard fashion[12] in the coronal plane, the cerebellum at right angles to the folia, and the brain stem horizontally. Comprehensive histological studies were undertaken in 434 of the 635 cases. Macroscopic and histological abnormalities were recorded on a series of line diagrams of the cerebral hemispheres, the cerebellum and the brain stem. All abnormalities were then transferred to a proforma and the data stored and analysed on the University of Glasgow's mainframe computer.

Experimental Material

The accelerating device (Penn I) consisted of a pneumatic activator which moved the head of anaesthetised subhuman primates via a thrust column and linkage system[13,14]. The motions were limited to the sagittal plane with the head constrained to move in an angular arc of 60° from a resting position of 20° of extension in 6–10 msecs. In the Penn II device the acceleration rate was lower and the head could be moved in the oblique, horizontal and coronal as well as the sagittal planes. The head was again moved through a 60° arc in periods ranging from 11–22 msecs.[11]

Varying degrees of injury were produced in 53 (Penn I) and 26 (Penn II) animals. An experimental trauma severity scale (ETS) was devised to express these degrees of injury quantitatively[15]. In the post-injury phase, unconsciousness was defined as the absence of any eye-opening (spontaneous or in response to stimuli) in the absence of ocular injury or oculomotor palsy. Concussion was defined as coma of less than 15 minutes duration. Mild prolonged traumatic coma (PTC) consisted of unconsciousness lasting from 16 to 119 minutes; moderate PTC was unconsciousness lasting between 2 and 6 hours; and severe PTC was unconsciousness lasting for longer than 6 hours. The animals were sacrificed electively from 2 hours to 8 weeks after injury though many would have died within a few moments of injury without life support, the animals being instantaneously unconscious and remaining so until sacrifice. Under deep anaesthesia they were perfused with formaldehyde/glacial acetic acid/absolute methanol, and the brains were dissected in a manner similar to that used for man. Comprehensive histological studies similar to those described above were undertaken in every case.

Results and Discussion

Focal Injury

Contusions

The pia arachnoid is intact over contusions but torn in lacerations. These have long been considered the hallmark of a head injury[16–18]. Nevertheless, both patients and experimental animals can sustain severe and fatal damage to the brain as a result of a head injury without a trace of conventional contusions.

Contusions occur characteristically at the frontal and temporal poles and on the under-aspects of the frontal and temporal lobes where the brain is in contact with bony protuberances at the base of the skull. In the early stages they are haemorrhagic and swollen: later on they are represented by shrunken brown scars.

In an attempt to assess contusional damage quantitatively and more objectively, the contusion index was established for grading the depth and extent of contusions in various anatomical locations of the brain[19]. The higher the total contusion index (TCI), the greater the severity of contusional damage.

The TCI in the human cases ranged from 0 to 61 (median 17). In 27 cases contusions were absent, while in 28 cases they were minimal. The 43 patients (10% of the series) with the highest TCIs were arbitrarily classified as having severe contusions: all had a TCI of greater than 37.

The contusion index reflects the very characteristic distribution of contusions whatever the site of the original injury. Thus, contusions affect particularly the frontal and temporal lobes, and to a lesser extent the cortex above and below the Sylvian fissures. Contusions of the parietal and occipital lobes and the cerebellum are uncommon unless they are directly related to a fracture of the skull. A precisely similar distribution was seen in the Penn I experiments[19]. It was also found with the Penn I device[20] that frontal contusions occur before temporal contusions: in 12 of the 79 animals with frontal contusions, there were no temporal contusions, while in only 2 of the 17 animals with temporal contusions were there no frontal contusions. With the Penn II device, however, there were no frontal contusions in 6 of the 17 animals with temporal contusions, while there were no temporal contusions in 3 of the 15 animals with frontal contusions.

Using the contusion index it was also possible to show that there is no association between the severity of contusions and any particular type of injury, the median contusion index for road traffic accidents (16), falls (18.5) and crush injuries (19) being similar. There was confirmation, however, that contusions are more severe in patients who have a fracture of the skull in comparison to those who do not, and contusions are more severe in patients who do not experience a lucid interval than in those who do.

In an attempt to shed some light on the relationship between the site of head injury and the distribution of contusions, a special study was made of 183 patients with localising fractures of the skull. A localising fracture was defined as one which was predominantly unilateral in the calvaria and over which there was corresponding scalp injury. Quantitative analysis of these cases produced no evidence to suggest that contusions were more severe in the contralateral than in the ipsilateral hemisphere, while in patients with frontal or occipital fractures contusions were more severe in the frontal lobes. Thus, while contusions can be most severe diametrically opposite the point of impact (that is contrecoup contusions), particularly in relation to falls on the occiput, it does not follow that because the most severe contusions are in a particular region of the brain, the impact occurred in the diametrically opposite part of the skull. The concept of contrecoup must therefore continue to be questioned even though a criticism of this conclusion is that Adams *et al.*[19] failed to exclude fracture contusions when comparing frontal and occipital impacts. According to Dawson *et al.*[18] fracture contusions are not relevant to the coup/contrecoup mechanism.

Correlations were established between length of survival and scores for depth and extent of contusions in 298 patients and 26 subhuman primates with contusions and with survival times less than or equal to 7 days. For the human data no meaningful correlations (all less than 0.15 in absolute value) were identified between length of survival and extent in individual regions or in combination of regions. Similar results were found in subgroup analyses on road traffic accidents and falls. For the subhuman primate data, a number of small correlations were identified but without any consistent pattern (see Table 1).

Correlations were also considered between a total depth and total extent score over all regions investigated (frontal, temporal, occipital, cerebellum and parietal). The computed correlations were 0.51 (total depth) and 0.17 (total extent). The results suggest some slight correlation in the subhuman primate data between length of survival and depth of contusion. However, the small sample size and the lack of consistency in the results in the Table across subregions would suggest that these conclusions should be treated with caution.

It has also been possible to show that contusions are less severe in patients with diffuse axonal injury (mean TCI 13.1) than in those who do not have this type of brain damage (mean TCI 19.8). This pattern of damage was maintained in the experimental series when contusions were readily produced in the Penn I device (mean TCI 4.2) but were less severe in the DAI inducing Penn II device (MTCI 1.3). One difference therefore between the animals injured in the Penn I and II devices is the greater incidence of contusions with Penn I. This can be attributed to several factors. Probably most important is that Penn I delivers acceleration much more rapidly than does Penn II, a condition that maximises damage to blood vessels and to areas of the brain near its surface. If acceleration is delivered to brain tissue more slowly as in Penn II, axons rather than blood vessels are more likely to be damaged, and also damage is less likely to be at the surface of the brain. The Penn I contusions are therefore more frequent because of the more rapidly applied acceleration that was delivered. Another factor is the direction of head movement: all of the Penn I animals were moved from posterior to anterior in the sagittal plane while most of the Penn II animals were subjected to coronal plane motions with the head moving from left to right.

Brain Damage Secondary to Raised Intracranial Pressure

Because some degree of shift and herniation of brain can occur during the period of spatial compensation

Table 1. *Correlation with Survival*

Frontal				Temporal			
left		right		left		right	
depth	extent	depth	extent	depth	extent	depth	extent
0.086	0.090	0.189	–0.171	0.571	0.196	–0.184	–0.274

without intracranial pressure (ICP) becoming significantly high, it may be difficult to establish post mortem if the ICP has been high during life. It has been suggested that the pathognomonic feature of a high ICP due to a supratentorial expanding lesion is the presence of a wedge of pressure necrosis in one or both parahippocampal gyri[21]. Pressure necrosis of this type was seen in every patient known to have had an ICP of more than 5.3 kPa (40 mm Hg) during life, in most of the patients with an ICP between 2.7 and 5.3 kPa (20–40 mm Hg) and in no patient with an ICP of less than 2.7 kPa (20 mm Hg). When the pressure necrosis is haemorrhagic it can be identified macroscopically but if there is no haemorrhage it may only be identifiable microscopically. A past episode of tentorial herniation associated with a high ICP can be recognised by the presence of a small organised wedge of pressure necrosis in the parahippocampal gyrus.

Using the criterion of pressure necrosis in the parahippocampal gyrus as evidence of a high ICP due to a supratentorial expanding lesion, it was established that the ICP had been high in almost three-quarters of the patients in the human series (324 of 434 cases), and in these cases there was an increased incidence of intracranial haematoma, cerebral contusions, brain swelling and hypoxic brain damage[22]. The most common types of brain damage secondary to a high ICP were damage to the brain stem (221, 68%), a supracallosal hernia (80, 25%), and a contralateral peduncular lesion in the mid-brain (54, 17%). Other findings included infarction in the anterior lobe of the pituitary in 45% of cases[23], haemorrhage in one or more oculomotor nerves and infarction within the distribution of the posterior cerebral artery. It is also appreciated that shift, distortion and internal herniation of the brain are the cause of necrosis of structures supplied by the the anterior cerebral, the anterior choroidal and the superior cerebellar arteries. Such an analysis further emphasises the importance and frequency of brain damage secondary to high ICP in non-missile head injury.

Some degree of brain shift and tentorial herniation was apparent in a few of the Penn I animals which were maintained on life support systems for some hours after the appearance of acute subdural haematoma, but secondary damage to the brain stem was not identified. This could be attributed to the short period of survival. In none of the Penn II animals was there evidence of raised intracranial pressure.

Fatal non-missile head injury may occur, however, in the absence of high ICP, and the brain damage in such cases is often not easy to identify[24]. In 110 of the 434 human cases there was no evidence of pressure necrosis in the parahippocampal gyri, but nevertheless there are cases where there was clinical (intracranial pressure monitoring) or radiological evidence[25] that intracranial pressure should have been high during life, yet pressure necrosis was not found in the parahippocampal gyrus after death This was usually because survival was too short to permit structural changes in the parahippocampal gyrus or because the haematoma was evacuated soon after injury, or some other type of decompression procedure was carried out so that there was no time for structural changes to occur in the parahippocampal gyrus. There were 39 such cases in the series, leaving 71 in whom there was no evidence that the ICP had been high in life. The commonest type of brain damage identified in this group of patients was diffuse axonal injury which was present in 41% of the cases. The next most commonly found structural damage was ischaemic brain damage which was either severe or moderate in 25 cases. Another unexpectedly frequent complication of the original head injury was acute bacterial meningitis which was present in 9 cases. In 8 of the cases there was a fracture of the skull and in 7 the meningitis developed as a late complication of the head injury after discharge from the Neurosurgical Unit.

For pathologists performing autopsies in cases of fatal head injury in the absence of intracranial haemorrhage or evidence of a high ICP, appropriate investigations to establish whether the patient has sustained diffuse axonal injury or ischaemic brain damage must be carried out: both of these may be difficult or even impossible to establish without appropriate histological studies.

Diffuse Injury

Diffuse Axonal Injury

Recent clinical studies[9] have emphasised that immediate prolonged unconsciousness unaccompanied by an intracranial mass lesion occurs in almost half of severely head injured patients, and is associated with 35% of all head injury deaths. Yet not much more than a decade ago, there was still considerable controversy as to the nature of the brain damage and its pathogenesis. It had been referred to by several names, viz. shearing injury[26,27], diffuse damage to white matter of immediate impact type[28], diffuse white matter shearing injury[29], and inner cerebral trauma[30]. These authors took the view that the damage to white matter occurred at the moment of injury but other experienced neuropathologists were of the opinion that it was secondary to hypoxic brain damage, to cerebral oedema or to secondary damage to the brain stem resulting from an intracranial expanding lesion[31–33].

In our initial attempts to identify diffuse axonal injury as a distinct clinico-pathological entity[34], it is now clear that we concentrated on the severe end of the spectrum. With increasing experience of fatal head injuries in man it has become clear that there are less severe forms of diffuse axonal injury, and we now wish to define three grades. Indeed, attention had been drawn to mild degrees of diffuse axonal injury by Oppenheimer[35] and Pilz[36].

Evidence of diffuse axonal injury was identified in 122 of the 434 human cases (29%). There were 10 cases of Grade 1 diffuse axonal injury, there being evidence of axonal damage in the white matter of the cerebral hemispheres, including the corpus callosum, and in the brain stem, and occasionally in the cerebellum: this damage can only be identified microscopically. There were 29 cases with Grade 2 diffuse axonal injury, that is there was a focal lesion in the corpus callosum in addition to diffuse axonal damage: the focal lesion was identifiable only microscopically in 11 of these cases. The majority of the cases – 83 – had sustained Grade 3 diffuse axonal injury since there were focal lesions both in the corpus callosum and in the dorsolateral quadrant of the rostral brain stem: in only 49 of these cases were both of the focal lesions apparent macroscopically even in a properly fixed and dissected brain. Thus, of the 122 cases the severity of diffuse axonal injury could only be defined by histological studies in 24, whilst its presence in 31 cases would have been missed unless appropriate histological studies had been undertaken. Grades 2 and 3 can be said to be severe when the focal lesions are apparent macroscopically[37].

In contrast to our previous studies on the most severe type of diffuse axonal injury when all of the patients had been unconscious frorn the moment of injury, 17 of the 122 patients with diffuse axonal injury experienced a lucid interval, that is they were able to "talk" immediately after their injury. Only 2 cases had a total lucid interval, that is they were able to talk clearly and rationally and both of these had the least severe type (Grade 1) of diffuse axonal injury. Of the 15 cases who had a partial lucid interval, not one had the most severe type as shown by the presence of macroscopic focal lesions in the corpus callosum and in the dorsolateral quadrant of the brain stem. Five had Grade 2 diffuse axonal injury, and 6 had Grade 3 diffuse axonal injury as identified microscopically.

A spectrum of brain damage ranging from no injury to instantaneous death was produced with the Penn I device[14,38,39]. Using this device it was possible that most of the types of brain damage that are encountered in non-missile head injury in man could be produced. Even though it was not possible to produce prolonged post-traumatic coma in the absence of an intracranial haematoma, it was possible to produce microscopic Grade 1 lesions as shown by the axonal changes and central chromatolysis[40]. Further evidence in support of this interpretation is provided by the demonstration of degenerating axons in the brain stem stainable by the Nauta and Finckheimer technique in some of the mildly injured animals[41].

Subsequent modification of the device (Penn II) allowed the acceleration to be applied more slowly and the direction of head movement to be controlled in different planes[42–44]. In this model all three grades of diffuse axonal injury have been produced, the severity of axonal injury depending principally on the magnitude and duration of the acceleration and the direction of head motion. With sagittal acceleration, Grade 1 but not Grades 2 and 3 can be produced. Acceleration of the same magnitude in the coronal plane produces more severe diffuse axonal injury, usually Grade 3, while rotation in the horizontal plane produces diffuse axonal injury of intermediate grade similar to that seen in man. As the grade of diffuse axonal injury increases from 1 to 3, there is deeper and more prolonged coma and the residual neurological deficits in survivors are more severe. There is therefore increasing acceptance that mild brain damage is part of a spectrum of mechanically induced diffuse brain injury, the basic pathology of which is axonal damage[45,46]. It is postulated that as mechanical imput increases, brain acceleration causes progressively more shear, tensile and compressive strains to be transmitted to brain substance. At first these strains are insufficient to cause any injury. As they increase, a graded series of clinical syndromes known as cerebral concussion results which, in the milder forms, are probably due to reversible membrane changes and in the severe form to structural abnormalities up to and including axotomy.

Ischaemic Brain Damage

It was not until the 1970's that there was full recognition of the frequency and distribution of ischaemic brain damage in fatal non-missile head injury. In the

West of Scotland this led to increased attention to the recognition and treatment of hypoxia and hypotension at the scene of the accident, during interhospital transfer, in critical care units and to the detection and relief of cerebral compression from traumatic intracranial haematoma[47–49]. Regimens of management and organisation of patient care with broadly similar principles have been employed in many other centres[4,6,50–52]. The aims of a recent study were to compare and contrast the prevalence and patterns of ischaemic brain damage in two groups of fatally head injured patients, the first between 1968–1972, and the second between 1981–1982[53].

Ischaemic damage was found in the brains of 92% of the1968–72 cases and in 88% of the 1981–82 cases. The severity of ischaemic brain damage was graded. It was said to be severe when the lesions were diffuse, multifocal or large within the distributions of arterial territories; moderate when the lesions were limited to the arterial boundary zones, singly or in combination with subtotal infarction in the distribution of the cerebral arteries, of if there were 5 to 10 subcortical lesions; and mild if there were 5 or less small lesions in the brain. The study showed that there was no statistical difference in the amount of moderately severe and severe ischaemic damage in the two groups, 55% and 54% respectively. There was evidence, however, that an increased number of patients with severe ischaemic brain damage was admitted in 1981–82 as a result of a changed admission policy of the Department of Neurosurgery which resulted in an increased detection of intracranial haematomas and to improved results in operated cases[49].

The pathogenesis of ischaemic brain damage in head injury is not yet fully understood, but it has been shown to be significantly more common in patients who have sustained a known clinical episode of hypoxia (hypoxaemia or hypotension) and/or high intracranial pressure during life[47]. The high incidence of damage to the cerebral cortex in the arterial boundary zones in the 1968–72 cases suggested that perfusion failure due to a transient reduction in cerebral blood flow was at least a contributing factor. The institution of measures to maintain adequate cerebral perfusion, including maintenance of adequate airway and blood gases, and the prompt correction of hypotension by replacement of blood loss, should therefore be reflected in a reduced incidence of infarction within the arterial boundary zones. Indeed, this was found to be the case in the 1981–82 series of cases. On the other hand, diffuse ischaemic brain damage is usually a consequence of either cardio-respiratory arrest or status epilepticus. The prevalence of this type of ischaemic damage was higher between 1981–82 than in 1968–72. This perhaps is rather surprising as it would have been expected that modern resuscitative measures would have at least to some extent reduced this type of ischaemic brain damage. On the other hand, the events responsible probably occur immediately after injury before first admission to hospital and even before the arrival of any medical personnel at the scene of the accident. In view of the increase in the number of severely head injured patients who were transferred to the Department of Neurosurgery, it is not surprising that there is an increased incidence of diffuse ischaemic brain damage in those cases dying within the Institute who, prior to the change in admission policy, would have died in accident and emergency departments or primary surgical wards. It is probably for this reason also that there is no apparent change in the overall severity of ischaemic brain damage between the two groups of cases[53].

It is concluded from this study that ischaemic brain damage is still common after severe head injury, and it seems likely that it remains an important cause of mortality and morbidity.

Ischaemic damage comprising multiple well-defined foci of neuronal necrosis throughout the cortex was seen in only 4 of the Penn I animals: however, monitoring showed that systemic hypoxia and hypotension did not occur. Such lesions were not seen in the Penn II animals. Many of the animals, however, had ischaemic damage restricted to CA1 of the hippocampus, particularly in the moderately severely injured animals where CA1 lesions were seen in 55%. The high prevalence of ischaemic lesions in CA1 may be due to a combination of mechanical forces engendered by the injury, and the injury-induced hypoventilation[54].

Response of Cerebral Microvasculature to Brain Injury

There is increasing evidence that there is a direct response by the cerebral microvasculature to head injury[55] and that the vascular response is a function of the severity of the injury. One such response is the development of enthothelial microvilli of the type described in response to an ischaemic episode[56].

The response of microvessels within the white matter of the sub-human primate subjected to the Penn II device was studied by scanning and transmission electron microscopy. It was found that there was a rapid endothelial disruption and swelling of perivascular astrocytes near the sites of petechial haemorrhage. The formation of microvilli in all vessels reached a peak at 6 hours and extended at least 5 mm from the site of haemorrhage. The astrocyte response suggested a partial recovery by 6 hours. Endothelial response was most marked in arterioles and venules, and was maintained for 6 days after injury. These results suggested that there is a bi-phasic response of the cerebrovasculature to brain injury. First, that there was a rapid astrocytic swelling possibly correlated with the transient disruptions of the blood-brain barrier, followed by morphological changes in the endothelium of all vessels which were most marked in arterioles and venules and extended considerable distances throughout the neuropil[57].

Conclusion

Our aims in this paper have been twofold. The first was to demonstrate that most of the principal types of brain damage that occur in man as a result of a non-missile head injury can be expressed quantitatively in fatal cases provided the brain is properly examined post mortem. This in turn allows more precise clinico-pathological correlations. The second was that similar techniques can be applied to brain damage produced experimentally in non-human primates. Analysis of the two sets of results have materially contributed to an increasing understanding of the causes of structural brain damage in head injury and, hopefully, may ultimately contribute to the management, if not the prevention, of such damage.

References

1. Graham DI, Adams JH, Gennarelli TA (1987) Pathology of brain damage in head injury. In: Cooper PR (ed) Head injury. Williams and Wilkins, Baltimore, pp 72–88
2. Becker DP, Miller JD, Greenberg RP, Young HF, Sakalas R (1977) The outcome from severe head injury with early diagnosis and intensive management. J Neurosurg 47: 491–502
3. Seelig JM, Becker DP, Miller JD, Greenberg RP, Ward JD, Choi SC (1981) Traumatic acute subdural haematoma. N Engl J Med 304: 1511–1518
4. Miller JD, Becker DP (1982) Secondary insults to the injured brain. J Roy Coll Surg Edinb 27: 292–298
5. Teasdale G, Galbraith S, Murray L, Ward P, Gentleman D, McKean M (1982) Management of traumatic intracranial haematoma. Brit Med J 285: 1695–1697
6. Klauber MR, Marshall LF, Toole BM, Knowlton SL, Bowers SA (1985) Cause of decline in head injury mortality rate in San Diego County, California. J Neurosurg 62: 528–531
7. Reilly PL, Graham DI, Adams JH, Jennett B (1975) Patients with head injury who talk and die. Lancet 2: 375–377
8. Adams JH (1984) Head injury. In: Adams JH, Corsellis JAN, Duchen LW (eds) Greenfield's neuropathology, 4th Ed. Arnold, London, pp 85–124
9. Gennarelli TA, Spielman GM, Langfitt TW, Gildenberg PL, Harrington T, Jane JA, Marshall LF, Miller JD, Pitts LH (1982) Influence of the type of intracranial lesion on outcome from severe head injury. J Neurosurg 56: 26–36
10. Adams JH, Graham DI, Gennarelli TA (1983) Head injury in man and experimental animals – neuropathology. Acta Neurochir (Wien) [Suppl] 32: 15–30
11. Gennarelli TA (1983) Head injury in man and experimental animals – clinical aspects. Acta Neurochir (Wien) [Suppl] 32: 1–13
12. Adams JH, Murray MF (1982) Atlas of post-mortem techniques. Cambridge University Press, Cambridge
13. Abel J, Gennarelli TA, Segawa H (1978) Incidence and severity of cerebral consussion in the rhesus monkey following sagittal plane angular acceleration. In: Proceedings of 22nd Stapp Car Crash Conference. Society of Automotive Engineers, New York, pp 35–53
14. Gennarelli TA, Adams JH, Graham DI (1981) Acceleration induced head injury in the monkey. I. The model, its mechanical and physiological correlates. Acta Neuropathol (Berl) [Suppl] 7: 23–25
15. Gennarelli TA, Segawa H, Wald U, Czernicki Z, Marsh K, Thompson C (1982) Physiological response to angular acceleration of the head. In: Grossman RG, Gildenberg PL (eds) Head injury: basic and clinical aspects. Raven, New York, pp 129–140
16. Lindenberg R, Freytag E (1960) The mechanism of cerebral contusions. Arch Pathol 69: 440–469
17. Lindenberg R (1971) Trauma of meninges and brain. In: Minckler J (ed) Pathology of the nervous system, Vol 2. McGraw-Hill, New York, pp 1705–1765
18. Dawson SL, Hirsch CS, Lucas FV, Sebek BS (1980) The contrecoup phenomenon: reappraisal of the classic problem. Hum Pathol 11: 155–166
19. Adams JH, Doyle D, Graham DI, Lawrence AE, McLellan DR, Gennarelli TA, Pastuszko M, Sakamoto T (1985) The contusion index: a reappraisal in human and experimental non-missile head injury. Neuropathol Appl Neurobiol 11: 299–308
20. Gennarelli TA, Abel J, Adams JH, Graham DI (1979) Differential tolerance of frontal and temporal lobes to contusion induced by angular acceleration. In: Proceedings of 23rd Stapp Car Crash Conference. Society of Automotive Engineers, New York, pp 561–586
21. Adams JH, Graham DI (1976) The relationship between ventricular fluid pressure and the neuropathology of raised intracranial pressure. Neuropathol Appl Neurobiol 2: 323–332
22. Graham DI, Lawrence AE, Adams JH, Doyle D, McLellan DR (1987) Brain damage in non-missile head injury secondary to high intracranial pressure. Neuropathol Appl Neurobiol 13: 209–217
23. Harper CG, Doyle D, Adams JH, Graham DI (1986) Analysis of abnormalities in pituitary gland in non-missile head injury: study of 100 consecutive cases. J Clin Pathol 39: 769–773
24. Graham DI, Lawrence AE, Adams JH, Doyle D, McLellan DR (1988) Brain damage in fatal non-missile head injury without high intracranial pressure. J Clin Pathol 41: 34–37
25. Teasdale E, Cardoso E, Galbraith S, Teasdale G (1984) CT scan in severe diffuse head injury: physiological and clinical correlations. J Neurol Neurosurg Psychiatry 47: 600–603
26. Strich SJ (1970) Lesions in the cerebral hemispheres after blunt head injury. J Clin Pathol 23 [Suppl 4] 166–171
27. Peerless SJ, Rewcastle NB (1967) Shear injuries of the brain. Can Med Assoc 96: 577–582
28. Adams JH, Mitchell DE, Graham DI, Doyle D (1977) Diffuse brain damage of immediate impact type. Its relationship to "primary brain stem damage" in head injury. Brain 100: 489–502

29. Zimmerman RA, Bilaniuk LT, Gennarelli TA (1978) Computed tomography of shearing injuries of the cerebral white matter. Radiology 127: 393–396
30. Grcevic N (1982) Topography and pathogenic mechanisms of lesions in 'inner cerebral trauma'. Rad Jug Acad Znam Unig Od med Nauke 402: 265–331
31. Jellinger K, Seitelberger G (1970) Protracted post-traumatic encephalopathy: pathology, pathogenesis and clinical implications. J Neurol Sci 10: 51–94
32. Jellinger K (1977) Pathology and pathogenesis of apallic syndrome following closed head injuries. In: Ore GD, Gerstenbrand F, Lücking CH, Peters G, Peters UH (eds) The apallic syndrome. Springer, Berlin Heidelberg New York, pp 88–103
33. Peters G, Rothamund E (1977) Neuropathology of the traumatic apallic syndrome. In: Ore GD, Gerstenbrand F, Lücking CH, Peters G, Peters UH (eds) The apallic syndrome. Springer, Berlin Heidelberg New York, pp 78–87
34. Adams JH, Graham DI, Murray LS, Scott G (1982) Diffuse axonal injury due to non-missile head injury in humans: an analysis of 45 cases. Ann Neurol 12: 557–563
35. Oppenheimer DR (1968) Microscopic lesions in the brain following head injury. J Neurol Neurosurg Psychiatry 31: 299–306
36. Pilz P (1983) Axonal injury in head injury. Acta Neurochir (Wien) [Suppl] 32: 119–123
37. Adams JH, Doyle D, Ford I, Gennarelli TA, Graham DI, McLellan DR (1989) Diffuse axonal injury in head injury. Definition, diagnosis and grading. Histopathology 15: 49–59
38. Adams JH, Graham DI, Gennarelli TA (1981) Acceleration induced head injury in the monkey. II. Neuropathology. Acta Neuropathol (Berl) [Suppl] 7: 26–28
39. Adams JH, Graham DI, Gennarelli TA (1982) Neuropathology of acceleration-induced head injury in the subhuman primate. In: Grossman RG, Gildenberg PL (eds) Head injury: basic and clinical aspects. Raven, New York, pp 141–150
40. Graham DI, Lawrence AE, Adams JH, Doyle D, McLellan D, Gennarelli TA (1989) Pathology of mild head injury. In: Hoff JT (ed) Mild to moderate head injury. Contemporary issues in neurological surgery, Vol 3. Blackwell Scientific, Boston
41. Jane J, Stewart O, Gennarelli TA (1985) Axonal degeneration induced by experimental non-invasive minor head injury. J Neurosurg 62: 96–100
42. Gennarelli TA, Thibault LE, Adams JH, Graham DI, Thompson CJ, Marcincin RP (1982) Diffuse axonal injury and traumatic coma in the primate. Ann Neurol 12: 564–574
43. Gennarelli TA, Adams JH, Graham DI (1986) Diffuse axonal injury – a new conceptual approach to an old problem. In: Baethmann A, Go KG, Unterberg A (eds) Mechanisms of secondary brain damage. Plenum, New York London, pp 15–28
44. Gennarelli TA, Thibault LE, Tomei G, Wiser R, Graham DI, Adams JH (1987) Directional dependence of axonal brain injury due to centroidal and non-centroidal acceleration. 31st Stapp Car Crash Conference. Society of Automotive Engineers, New York, pp 49–53
45. Povlishock JT, Becker DP, Cheng CLY, Vaughan GW (1983) Axonal changes in minor head injury. J Neuropath Exp Neurol 42: 225–242
46. Gennarelli TA (1987) Cerebral concussion and diffuse brain injuries. In: Cooper PR (ed) Head injury, 2nd Ed. Williams and Wilkins, Baltimore, pp 108–124
47. Graham DI, Adams JH, Doyle D (1978) Ischaemic brain damage in fatal non-missile head injuries. J Neurol Sci 39: 213–234
48. Gentleman D, Jennett E (1981) Hazards of inter-hospital transfer of comatose head-injured patients. Lancet 2: 853–855
49. Teasdale G, Galbraith S, Murray L, Ward P, Gentleman D, McKean M (1982) Management of traumatic intracranial haematoma. Brit Med J 285: 1695–1697
50. Becker DP, Miller JD, Ward JD, Greenberg RP, Young HF, Sakalas R (1977) The outcome from severe head injury with early diagnosis and intensive management. J Neurosurg 47: 491–502
51. Bowers SA, Marshall LF (1980) Outcome in 200 consecutive cases of severe head in jury treated in San Diego County: a prospective analysis. Neurosurgery 6: 237–242
52. Bricolo AP, Pasut M (1984) Extradural haematoma: toward zero mortality. A prospective study. Neurosurgery 14: 8–11
53. Graham DI, Ford I, Adams JH, Doyle D, Teasdale GM, Lawrence AE, McLellan DR (1989) Ischaemic brain damage is still common in fatal non-missile head injury. J Neurol Neurosurg Psychiatry 52: 346–350
54. Gennarelli TA, Marcincin RP, Thibault LE, Thompson CJ (1983) Effect of direction of head movement on ICP in experimental head injury. In: Ishii S, Nagai H, Brock M (eds) Intracranial pressure V. Springer, Berlin Heidelberg New York Tokyo, pp 483–486
55. Povlishock JT (1985) The morphologic responses to experimental head injury of varying severity. In: Becker DP, Povlishock JT (eds), Central nervous system trauma status report. National Institute of Neurological and Communicative Diseases and Stroke. National Institutes of Health, U.S.A.
56. Dietrich WD, Busto R, Ginsberg MD (1984) Cerebral endothelial microvilli: formation following global forebrain ischaemia. J Neuropathol Exp Neurol 43: 72–83
57. Maxwell WL, Irvine A, Adams JH, Graham DI, Gennarelli TA (1988) Response of cerebral microvilli to brain injury. J Pathol 155: 327–336

Correspondence and Reprints: D. I. Graham, Department of Neuropathology, Institute of Neurological Science, University of Glasgow, Glasgow G5 14TF, U.K.

Acta Neurochir (1993) [Suppl] 57: 49–52

Traumatic Damage to the Nodal Axolemma: an Early, Secondary Injury

T. A. Gennarelli[1], **R. Tipperman**[1], **W. L. Maxwell**[2], **D. I. Graham**[3], **J. H. Adams**[3], and **A. Irvine**[2]

[1] Division of Neurosurgery, University of Pennsylvania, U.S.A. and the Departments of [2] Anatomy and [3] Neuropathology, University of Glasgow, U.K..

Abstract

Electronmicroscopical investigations were made in a model of optic nerve damage in guinea-pigs on the development of acute axonal damage on an ultrastructural basis. It was expected to obtain thereby further information on mechanisms underlying axonal damage in traumatic brain injury. For that purpose an injury apparatus was employed to deliver defined elongation and/or tensile strains to the optic nerve.

Transmission electronmicrographs were examined of longitudinal and transverse nerve sections throughout its entire length. The most severe abnormalities were identified in the prechiasmatic portion of the nerve. Among others, elongations of the nodes of Ranvier were encountered, swollen axons with accumulation of organelles, and even disrupted axons having a morphology similar to retraction balls. In all instances, abnormal axons were found together with axons having a normal structural appearance. Nodes of Ranvier demonstrated outward dilatations of the nodal axolemma and of the adjacent axoplasm, which are named as nodal blebs. Nodal blebs occurred already 15 min after injury, and were fully developed at 6 or 24 hrs. The blebs had disappeared again after 5–7 days. The axoplasm in the blebs demonstrated considerable disorganization of cytoskeletal elements with an array of amorphous material appearing as granular degeneration.

Taken together, the present experimental model is a useful approach to analyse axonal damage at the ultrastructural level as it may occur in white matter of the central nervous system.

Keywords: Head injury; electron microscopy; axonal damage; nodal blebs.

Introduction

Axonal damage is a frequent finding after traumatic brain injury. Although very severe mechanical insults can result in primary axonal disruption (primary axotomy), secondary axonal degeneration (secondary axotomy) appears to be more common. The events leading to secondary axotomy are, as yet, incompletely understood, but could be of great clinical importance, if therapies could be developed to prevent it. Previous studies have suggested that a site of primary mechanical insult is at the node of Ranvier and that many morphological changes of the node occur after injury. We now describe nodal axolemmal changes and their possible genesis and role in secondary axotomy in a model of isolated axonal injury in the guinea pig optic nerve.

Methods

Preparation

Adult male albino Hartley guinea pigs weighing 700 gm (range 650–750) were deeply anesthetized with intramuscular ketamine 50 mg/kg and xylazine 3 mg/kg, supplemented as necessary. The right eyelids were infiltrated with 1% xylocaine containing epinephrine. After retracting the lids with 4-0 silk sutures, a lateral canthotomy was performed. A 360 degree opening of the conjunctiva was performed with the aid of the operating microscope peripheral to the limbus including the extraocular muscles. The globe was retracted temporally and the optic nerve was dissected free from all attachments. A sling was fashioned from sterile umbilical tape with a 5 mm longitudinal slit cut at its midpoint. The sling was placed over the globe and positioned firmly against the posterior pole of the globe so that the optic nerve ran through the slit. The animal's head was then secured and an elongation of the optic nerve produced.

An injury apparatus was designed to deliver elongation or tensile strain to the optic nerve[6]. The anesthetized animal's head was placed into ear canal pins in the head holder of the apparatus. The two free ends of the umbilical tape sling previously placed over the globe were attached to a cylinder on the apparatus. A pulse generator voltage was set that delivered 5 mm of displacement of the cylinder, sling and optic nerve over a 20 msec period of time.

After the injury was completed, the sling was removed and a lateral tarsorrhaphy was performed with 4-0 silk. Animals were examined at regular times after injury. In all cases, the animals resumed normal feeding and motor behavior. No instances of operative infection or discomfort were observed. All animal procedures conform to current U.S. Public Health Service standards and were approved by the University of Pennsylvania Institutional Animal Care and Use Committee.

Electron Microscopy

At selected intervals after injury, animals were terminally anesthetized with ketamine, xylazine and pentobarbital (30 mg/kg). The previous tarsorrhaphy and canthotomy were reopened and the globe retracted temporally. Both globes were opened on their equators and vitrectomies performed. The retina and orbit were flooded with a fixative of 3% glutaraldehyde and 1% paraformaldehyde in 0.1 M phosphate buffer for 10 minutes followed by transcardiac perfusion with normal saline and then by the same fixative. After fixation was complete, the globes, optic nerves, chiasm and optic tracts were removed en-bloc and stored in fixative at five degrees Centigrade for 24 hours. After fixation, the globes were removed and the optic nerve and chiasm were divided in four equal segments. The specimens were post-fixed in1% osmium tetroxide followed by dehydration in graded alcohols and propylene oxides and were subsequently embedded in epon resin. After the resin hardened, thin sections (< 1 μm) were cut from selected areas and stained with uranyl acetate/lead citrate and examined with a Joel 100S TEM electron microscope.

Results

Optic Nerve Transmission Electron Microscopy (Fig. 1)

Transmission electron microscopy was performed in longitudinal and transverse sections throughout the entire length of the optic nerve from globe to chiasm. There were small numbers of abnormal axons in all areas, however, the most severe abnormalities were identified in the pre-chiasmatic portion of the nerve. The precise TEM morphology depended on the time from injury to sacrifice and included elongated nodes of Ranvier with normal axoplasm, swollen axons with marked accumulation of organelles, and disrupted axons with an appearance identical to retraction balls. In all instances, abnormal axons were seen adjacent to axons with normal morphological appearance. In all nerves, there were also many axons that were very large, but seemed to have normal axoplasmic morphology. Similar changes have been seen in humans and primates with diffuse axonal injury[1,5].

We were particularly struck by abnormalities at nodes of Ranvier that involved outward dilatations of a portion of the nodal axolemma and adjacent axoplasm. We named these phenomena nodal blebs. They were seen as soon as 15 minutes after injury, but were more developed by 6 to 24 hours. By 72 hours they were more difficult to find and were then often associated with accumulations of large numbers of intraaxoplasmic organelles. Blebs were not seen after 5–7 days.

The time course and appearance of the nodal blebs suggested that they developed progressively after injury. At 15 minutes they were small flattened irregularities while later the blebs most commonly were sausage shaped. These aneurysmal shaped blebs had a small neck and elongated to lengths that exceeded the width of the axon. At all points in time the neck of the blebs appeared to arise from one side of the node, at or very near the last terminal loop of myelin, and rarely, if ever, did they extend along the entire length of the node.

The blebs were covered by axolemmal membrane that was continuous with that of the axon and was morphologically no different from the non-bleb axolemma. However, the axoplasm within the blebs had considerably different morphology from the more central axoplasm within the axon.

The axoplasm contained within nodal blebs is characterized by considerable disorganization of its cytoskeletal elements. There are no recognizable neurofilaments or neurotubular components. Instead, the internal structure of the blebs is an array of amorphous materials not unlike granular degeneration. The striking feature of the bleb axoplasm is a sharp border at the bleb neck; there is cytoskeletal disorganization within the bleb and normal or near normal cytoskeletal composition within the axon adjacent to the neck of the bleb. In addition the blebs may contain one or more membrane-bound vesicular elements that could represent abnormal mitochondria or non-specific vesicles.

Discussion

Axonal damage has now been identified as a principal primary response of brain tissue to mechanical insult[1,5,9,11]. Much has been learned about the prevalence and distribution of axonal damage from morphological examination of the brains of humans and experimental animals. However, the whole brain presents several limitations to an in-depth understanding of the events surrounding trauma to axons. The large size of the brain makes detailed histologic examination difficult and comprehensive electron microscopic analysis impossible. In the brain it is impossible to precisely determine the origin, direction of travel and termination of axons that have been injured. Therefore, we developed a model of axonal injury in a more simple biological system by providing rapid elongation (tensile strain) to the optic nerve of the albino guinea pig[6].

Anatomically, the anterior visual system of the albino guinea pig is ideal for experimental studies since the chiasm is 95–98% crossed in the albino but not the pigmented guinea pig[2–4]. Thus, projections from the injured eye can be followed independently throughout the posterior optic pathways,while those from the un-

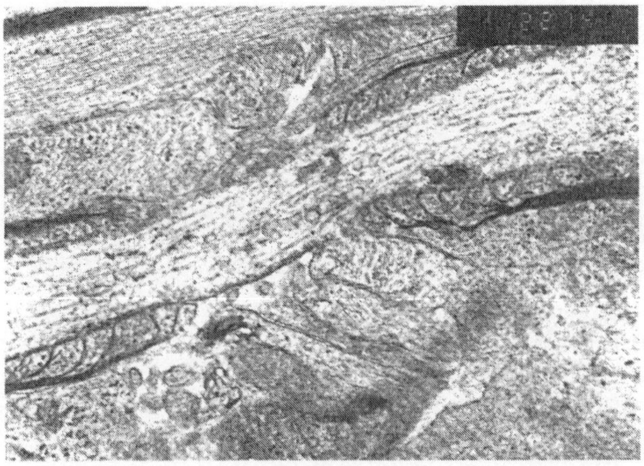

Fig. 1. Transmission electron micrograph of guinea-pig optic nerve subjected to brief displacement trauma administered within 20 msec. The section demonstrates a node of Ranvier (cf. text)

injured contralateral eye can serve as controls in each animal. The anatomy is well described[8] so that injured neurons can be assessed from the cell body in the retina to terminal fields in the superior colliculus or lateral geniculate body and secondary visual neurons can be followed to brainstem or cortex if desired. The relatively large size of the optic nerve (2–3 mm in diameter, 12–20 mm in length) contrasts with the much smaller nerve of the rat, making it easier to dissect and to perform histological evaluation, yet the guinea pig nerve is not so large as to preclude quantification of the number of injured axons in it as might be the case in larger animals (ie. rabbit, cat, dog, or monkey). Also, the retinal ganglion cell of the guinea pig, and its projection, appears to be insignificantly, if at all, different from other mammalian optic systems[10].

Because the mechanics of the optic nerve injury are similar to those of brain injury, the resultant injury to the optic nerve is different from other optic nerve models and is much more like that seen in brain injury. Complete transection or avulsion of the optic nerve produces instantaneous complete axotomy of all optic nerve fibres as well as disruption of all blood vessels in the nerve and exposes the cut ends to the extracellular and extra-neuronal environment. Wallerian degeneration including demyelination occurs in all injured axons, a phenomena that does not occur in human DAI. Crush injury spares some fibres in the nerve but most axons are also transected by this type of injury. Blood vessels are damaged to a variable degree resulting in a great change in the extracellular environment. Instead, human and experimental axonal injury in the brain is more subtle. It is characterized by an admixture of normal, axotomized and damaged (but no axotomized) axons all present in the same microscopic area, adjacent to one another[1,5]. Except in a few small regions where injury is greatest, blood vessels are not transected and the extracellular milieu is not grossly altered. The model produces the same side by side heterogeneity of injured and uninjured axons seen in human axonal injury. This is of great importance because, although axotomized fibers are bound to undergo Wallerian degeneration, the fate of damaged axons is less certain. They may undergo secondary axotomy or, alternatively, they may begin a reparative process and recover[7], both of which are poorly understood phenomena that have great potential clinical significance. If, for example, secondary axotomy could be prevented or if recovery of damaged axons could be promoted by therapeutic intervention, then neurological impairments due to axonal damage could be minimized.

This paper describes changes in the nodal axolemma that follow elongation injury in the optic nerve. These blebs are characterized by an outpouching of normal axolemma within which degraded, disorganized axoplasm is present. The blebs appear to enlarge and become more fully formed between 15 minutes and 72 hours after injury and are seen infrequently thereafter. An attractive hypothesis to explain them would relate the blebs to the progression of events leading to secondary axotomy. However, we cannot, at this time, conclude that axolemmal blebs, in the absence of intra-axonal accumulation of organelles, are precursors of the degeneration or retraction bulbs seen in secondary axotomy. An alternative explanation is that blebs represent regenerative sprouting from the node of Ranvier. However, because of their early appearance and lack of neurotubules, we are not convinced that nodal blebs represent a sprouting phenomenon.

We are attracted to the hypothesis that axolemma blebs are the secondary reflection of a traumatically induced membrane defect that results in an increased intracellular calcium load to the axon. That the calcium load was insufficient to produce widespread dissolution of the cytoskeleton is evidenced by its relative preservation in the axons. It may be, however, that a calcium gradient is established near its point of entry such that calcium dependent neutral proteases (CANP) are activated only near the axolemmal membrane. Degradation by CANP of subsurface proteins that are involved with cell shape (e.g. fodrin or actin) could then lead to axolemmal blebs in the absence of dissolution of neurotubules, neurofilaments or microtubular associated proteins elsewhere in the central portion of the axon. Thus, axolemmal changes after traumatic injury may be independent of the processes involved with secondary axotomy or may represent a form fruste, localized secondary consequent of axonal damage. Further studies are required to determine the specific relationship between nodal axolemmal changes and secondary axotomy.

Acknowledgement

This work was supported by NINCDS program project grant NS-08803-18.

References

1. Adams JH, Graham DI, Murray LS, Scott G (1982) Diffuse axonal injury due to nonmissile head injury in humans: an analysis of 45 cases. Ann Neurol 12: 557–563
2. Creel DJ (1972) Retinogeniculostriate projections in guinea pigs: albino and pigmented strains compared. Exper Neurol 36: 411–425
3. Creel DJ, Dustman RE, Beck EC (1970) Visually evoked response in guinea pigs: strains compared. J Comp Physiol Psych 73: 490– 493
4. Dahl D, Crosby CJ, Bignami A (1981) Filament proteins in rat optic nerves undergoing wallerian degeneration. Exper Neurol 71: 421–430
5. Gennarelli TA, Thibault LE, Adams JH, *et al* (1982) Diffuse axonal injury and traumatic coma in the primate. Ann Neurol 12: 564–574
6. Gennarelli TA, Tipperman R, Maxwell WL, Graham DI, Adams JH, Irvine A: Traumatic damage to the nodal axolemma: an early, secondary injury (in preparation)
7. Gennarelli TA, Adams JH, Graham DI (1986) Diffuse axonal injury. A new conceptual approach to an old problem. In: Baethmann A, Go KG, Unterberg A (eds) Mechanisms of secondary brain damage. Plenum, New York, pp 15–28
8. Grafstein B, Murray M (1969) Transport of protein in goldfish optic nerve during regeneration. Exper Neurol 25: 494–508
9. Hess A (1958) Optic centres and pathways after eye removal in fetal guinea pigs. J Comp Neurol 109: 91–115
10. Jane JA, Stewart D, Gennarelli TA (1985) Axonal degeneration induced by experimental non-invasive minor head injury. J Neurosurg 63: 96–100
11. Levine J, Willard M (1980) The composition and organization of axonally transported proteins in the retinal ganglion cells of the guniea pig. Brain Res 194: 137–154
12. Povlishock JT, Becker DP, Miller JD, Jenkins LW, Dietrich WD (1979) The morphopathologic substrates of concussion. Acta Neuropathol (Berl) 47: 1–11

Correspondence and Reprints: T. A. Gennarelli, Division of Neurosurgery, University of Pennsylvania, 3400 Spruce Street, Philadelphia, PA, U.S.A.

Acta Neurochir (1993) [Suppl] 57: 53–55

Morphometrical Evaluation of Triflusal in Brain Infarction

N. Heye, A. Campos, S. Sampaolo, and **J. Cervos-Navarro**

Institute of Neuropathology, Klinikum Steglitz, Freie Universität Berlin, Berlin, Federal Republic of Germany

Abstract

MCA occlusion in animals is a common model for experimental stroke. In previous studies we have shown that one of the factors, which influence evolution of an infarct is microthrombosis in the area of infarction and in the surrounding brain tissue.

The present study was undertaken for assessment of the number of microthrombi and of the size of brain infarcting in rats treated with the antiaggregatory substance Triflusal.

7 groups of Sprague-Dawley rats, each group consisting of 6 animals, underwent transsphenoidal MCA occlusion. The animals received Triflusal in various amounts from day 2 till day 6. At day 7 animals were decapitated and the brains were fixed in formaldehyde. The brain was dissected at the level of the optic chiasm and embedded in paraffin. Fresh microthrombi were detected py PTAH (Phosphotungstic acid hematoxylin) staining. In each animal the hemisphere with the ischemic lesion as well as the contralateral hemisphere were examined. The area of both hemispheres was calculated by subtraction of the ventricle area from the total brain area of a section. Infarct was defined as the region of necrosis which was sharply demarcated from normal brain. The infarcted area was planimetrically measured to obtain a ratio of infarcted to normal brain.

A correlation between the effect of Triflusal, number of microthrombi and size of the infarcted area could be demonstrated. The pathogenetic role of the microthrombi in the evolution of cerebral infarction as well as the effect of Triflusal in different dosages on the number of microthrombi could be clearly assessed by quantitative morphometry.

Keywords: Cerebral infarction; microthrombosis; antiaggregatory therapy; triflusal.

Introduction

Several studies on experimental middle cerebral artery (MCA) occlusion have shown that after onset of ischemia microthrombi occupy the lumen of microvessels (Sampaolo *et al.*, 1987) and platelets adhere to endothelial cells (Dietrich *et al.*, 1988). A maximum of microthrombus formation was found 7 days after MCA occlusion. Microthrombi and platelet aggregates did occur not only in ischemic brain tissue but also in the contralateral cerebral hemisphere (Sampaolo *et al.*, 1987; Dietrich *et al.*, 1988). This suggests development of a generalized coagulopathy in cerebral ischemia which may offer a target for antiaggregatory treatment. Aim of the present study was to evaluate the influence of triflusal (2-acetoxy-4-trifluoromethylbenzoic acid, Uriach and Cia, S.A., Barcelona, Spain), an antiaggregatory substance, on the number of microthrombi and size of infarction after MCA occlusion.

Material and Methods

The experiments were performed in 29 male Sprague-Dawley rats of 350–400 g b.w. Anesthesia was induced by ketamine given intraperitoneally. Body temperature of the animals was maintained at 37°C by a heating pad. A transsphenoidal approach was used for MCA occlusion. The MCA was occluded by electrocauterization under continuous saline irrigation. Sham operation with only touching the MCA by the coagulation forceps was performed in 6 animals. After surgery the rats received intragastrically triflusal diluted in 1 ml solvent/100 g b.w. Triflusal was given in a dose of 12.5 or 50 mg/100 g b.w. over a period of 3 or 6 days. The rats were decapitated on the 7th day after MCA occlusion. The brain was removed from the cranial vault and placed in chilled formaldehyde (10%) for one week. After fixation the brain was dissected at the level of the optic chiasm, embedded in paraffin and sliced. The histological sections were stained by hematoxylin-eosin, Nissl, elastica van Gieson, and PTAH (phosphotungstic acid hematoxylin). For histological evaluation the specimens of control and experimental animals with and without treatment were examined by two blinded individuals. The sections were observed using a light microscope with orthogonal millimeter scales. One eyepiece had a graticule indicating the center of the microscopic field. By using x- and y-reference points each microthrombus could be exactly localized. The number and distribution of microthrombi of each animal were systematically assessed. The location of microthrombi of both hemispheres was recorded on sectional paper. Infarct morphometry was performed by using the Bioquant system (Bioquant, Bilaney, Nashville, U.S.A.). The total areas of the brain parenchyma, the ventricles, and infarct were measured. The infarct was identified as tissue with sharp demarcation from normal brain parenchyma, proliferation of capillaries and destruction of the normal anatomy. The total area of the brain parenchyma proper was determined after subtracting the ventricle area. The ratios of the infarct/brain parenchyma areas in percent and the number of microthrombi were statistically analyzed. Paired Student's t-test and linear regressions were employed to assess differences between treated and untreated animals.

Results

PTAH which is staining thrombotic material deeply blue was used for identifuation of microthrombi in the histological sections (Fig. 1). The number of microthrombi in the brain parenchyma was highly variable among the different groups (Fig. 2). The largest number of thrombi was found in group C I not receiving triflusal. Group T I receiving 12.5 mg/100 g b.w. for three days had a significantly lower number of microthrombi than group C I ($p < 0.001$). Groups T II and T III which were administered with 50 mg/100 g b.w. of triflusal had also a significantly lower number of microthrombi than the animals of the control group C I. No differences, however, were found between groups T II and T III. In both groups a higher number of microthrombi was present as compared with group T I.

Due to a considerable standard error no significant differences were found in the infarct/brain ratio between control groups without treatment and after administration of trifusal. Significant correlations could neither be established between the number of microthrombi and the infarct/brain ratio, although a respective tendency became recognizable.

Discussion

The smallest number of blood vessels occluded by thrombi was found in the sham-operated control group C II, the highest number in the group with MCA occlusion without treatment (C I). This demonstrates that occlusion of the middle cerebral artery induces microthrombosis of cerebral blood vessels. The different numbers of microthrombi found in experimental groups with treatment and without indicates that triflusal had antithrombotic properties under the prevailing experimental conditions. It should be mentioned that the low-dose regimen of triflusal appeared to be more effective to prevent formation of microthrombi than the high dose. Pharmacological findings are not available at the moment to explain this seemingly paradox effect. Nevertheless, a similar phenomenon has been observed in studies on treatment by acetylsalicylic acid, which may relate with the inhibitory effects of the compound on the cyclooxygenase pathway (de Castellarnau *et al.*, 1988). In a higher dose, acetylsalicylic acid inhibits endothelial cyclooxygenase and, thereby, antagonizes formation of antiaggregatory prostacyclin in endothelial cells.

The current investigations do not provide at a statistically significant level evidence for a relationship between the number of microthrombi and the size of infarction. After MCA occlusion Duverger *et al.* (1985) found a rat strain-specific variability in the size of infarction. The highest variability of infarct volume was found in Sprague-Dawley rats, which also were

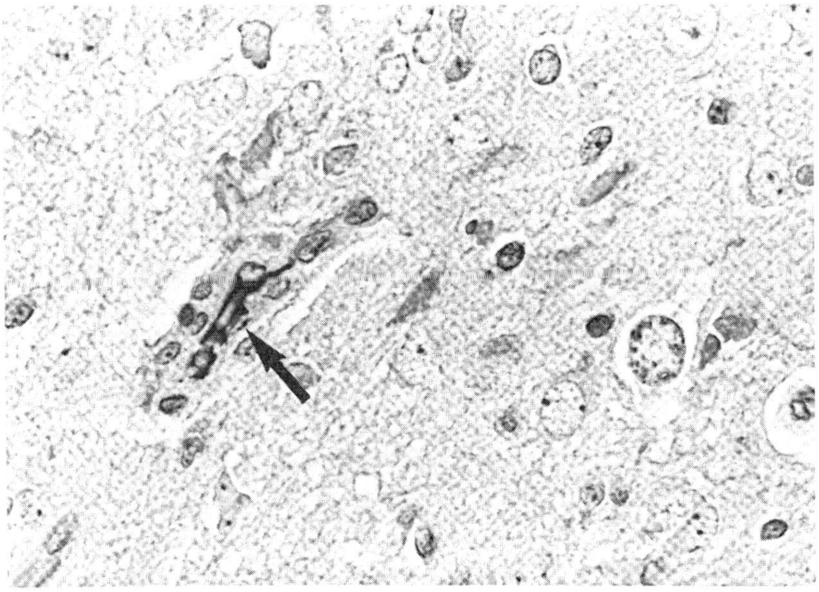

Fig. 1. Occlusion of cerebral blood vessel by PTAH-positive thrombotic material

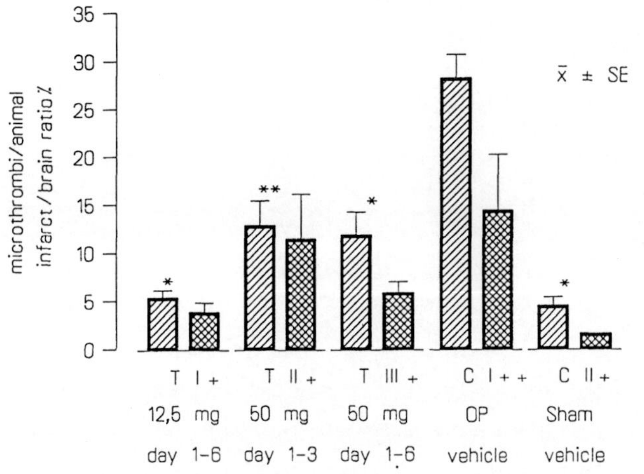

Fig. 2. Number of microthrombi (hatched bars) and ratio (in percent) of infarcted-to-normal brain (cross-hatched bars) in animals with MCA occlusion without (C I) and with treatment (T I, T II, T III) by triflusal. * $p < 0.001$; ** $p < 0.1$; + n = 6; ++ n = 5

used for the present experiments. The authors recommended to use the Fischer 344 strain for ischemia experiments to obtain more reproducible results (Duverger *et al.*, 1985). It is conceivable that strain-specific differences of the vascular anatomy play a role, as occlusion of the MCA by using the transsphenoidal approach could result in a differing severity of ischemic brain damage. This is supported by findings of Benderson *et al.* (1986) on a correlation between the very locus where the MCA is coagulated and the resulting size of infarction. Further, Rubino and Young (1988) have reported an individual variability of the degree of natural preformed anastomoses between the frontal and parietal branches of the middle cerebral artery.

Assessment of the number of microthrombi resulting from vessel occlusion can be utilized as an important parameter to evaluate the therapeutical benefit of antiaggregatory treatment. In a clinical study of Shah *et al.* (1985) an increase of the platelet-specific protein β-thromboglobulin was found within the first week of an ischemic insult in patients with thromboembolic disease. Blood samples drawn later than one week after the event had significantly lower levels. The increase of β-thromboglobulin may indicate enhancement of platelet activation and turnover.

In studies of Sampaolo *et al.* (1987) using light microscopy, microthrombi were found in ischemic brain areas within the first minutes after MCA occlusion. The highest number of microthrombi occurred on day 7 after operation with a nearly symmetrical distribution between both cerebral hemispheres. Moreover, Dietrich *et al.* (1988) found in electronmicroscopical studies an increased adherence of platelets at endothelial cells of cerebral blood vessels. Again, the phenomenon was not restricted to the ischemic hemisphere. Degranulated platelets were also demonstrated in blood vessels of the contralateral brain. The authors obtained similar morphological findings in the brain after ligation of the femoral artery only, when the cerebral blood vessels were not occluded. Based on these observations the authors suggested that development of cerebral microthrombosis is not restricted to local factors, but may be induced by systemic alterations of the coagulatory system. It is concluded that in ischemic brain infarction systemic conditions play a role in coagulation and formation of microthrombi. For examinations of the therapeutical benefit of antithrombotic treatment, it is suggested to assess the number of microthrombi formed under these conditions rather than the dimensions of brain tissue infarction, which is influenced by many variables. Taken together, administration of the platelet inhibitor triflusal in a therapeutical dose range may be beneficial in stroke by reducing formation of microthrombi in ischemic and non-ischemic brain tissue.

References

1. Benderson JB, Pitts LH, Tsuji M, Nishimura MC, Davis RL, Bartowski H (1986) Rat middle cerebral artery occlusion: evaluation of the model and development of a neurological examination. Stroke 17: 472–476
2. Castellarnau de C, Sancho MJ, Vila L, Albors M, Rutllant MLL (1988) Effects and interaction studies of triflusal and other salicylic derivates on cyclooxygenase in rats. Prostaglandins Leukotrienes and Essential Fatty Acids 31: 83–89
3. Dietrich WD, Prado R, Watson BD, Nakayama H (1988) Middle cerebral artery thrombosis: acute blood-brain barrier consequences. J Neuropath Exp Neurol 47: 443–451
4. Duverger D, Lecoffre C, MacKenzie ET (1985) Histological quantification of cerebral infarction following middle cerebral artery occlusion in various rat strains. J Cerebral Blood Flow Metabol 5 [Suppl 1]: 415–416
5. Rubino GJ, Young W (1988) Ischemic cortical lesions after permanent occlusion of individual middle cerebral artery branches in rats. Stroke 19: 870–877
6. Sampaolo S, Cervós-Navarro J, Djouchadar D, Figols J (1987) Clinical and experimental evidence of microthrombosis in cerebral ischemia. In: Hartmann A, Kuschinsky W (eds) Cerebral ischemia and hemorheology. Springer, Berlin Heidelberg New York, pp 386–393
7. Shah AB, Beamer N, Coull BM (1985) Enhanced *in vivo* platelet activation in subtypes of ischemic stroke. Stroke 16: 643–647

Correspondence and Reprints: J. Cervos-Navarro, Institute of Neuropathology, Klinikum Steglitz, Freie Universität Berlin, Hindenburgdamm 30, D-1000 Berlin 45, Federal Republic of Germany.

Acta Neurochir (1993) [Suppl] 57: 57–63

Mediators and Antagonism in Secondary Brain Damage

In vivo and *in vitro* Control of Acid-Base Regulation of Brain Cells During Ischemic and Selective Acidic Exposure

F. Plum

Cerebral Vascular Disease Research Center and The Raymond and Beverly Sackler Foundation, Department of Neurology and Neuroscience, Cornell University Medical College New York, New York, U.S.A.

Abstract

The three-compartment model of brain acid-base regulation postulates that under circumstances of changing function or disease, hydrogen ion concentrations may differ considerably in the interstitial space (ISS), the neurons and the glial cells. During hyperglycemia plus profound ischemia, for example, direct measurements by microelectrodes followed by intracellular HRP staining show that intraglial pH can fall transiently as low as 3.9, although more often the nadir drops to the 4.5–5.5 range. Concurrently, ISS-pH and, by calculation, neuronal pH falls to and remains constant (but not necessarily the same) at pH 6.2. By contrast, during spreading depression, ISS and intraglial pH at first move rapidly and transiently in opposite directions, ISS [H^+] rising, intraglial falling. These two then gradually stabilize, whereas neuronal pH remains substantially more steady and near normal, shifting only minimally from resting baseline levels over several minutes' time. Similar but less pronounced effects follow direct electrical stimulation. The net change represents complex biophysical transmembrane and buffering mechanisms that appear to guard neuronal homeostasis.

Studies carried out on embryonic rat forebrain neurons and glia show that these cells have considerably different vulnerabilities to extracellular acidity depending on the anionic nature of the acid in the bathing medium. In cultures to which HCl was added to the medium, neurons and neuronal processes almost all survived ten minute exposures to pH 3.8, whereas glial cells succumbed after ten minute exposures at pH not lower than 4.2. Both types of cells, however, showed much greater vulnerability to lactic acidification in the media; neither neurons nor glia survived exposure for ten minutes at pH 4.8 or for sixty minutes at the lesser degree of acidity of 5.2. Measurements of intracellular pH in cultured mammalian neurons using the fluorescent dye BCECF demonstrated a rapid fall in intracellular pH from 7.18 to 6.80 when 20 mM lactate was added to pH – clamped extracellular medium held at a pH of 7.35. By contrast, when the pH_o in the medium fell from 7.35 to 6.65 in the presence of 20 mM lactate, pH_i fell to 6.48 and failed to recover unless pH_o was quickly restored to 7.35 and lactate was removed from the medium. The findings indicate that CNS cell membranes are substantially more permeable to lactic acid compared to unorganic acids and that the vulnerability relates to the greater capacity of the former to induce a profound and rapid intracellular acidification.

Keywords: Cerebral acid-base regulation; glial neuronal pH; lactacidosis.

Introduction

This manuscript reviews recent studies from the Cornell Cerebrovascular Research Center dealing with acid-base regulation by the brain in states of severe ischemia as well as during functional alteration. The principal scientists responsible are: William Pulsinelli, who now directs the Research Center; Steven Goldman; Richard Kraig; Maiken Nedergaard; and Carol Petito. Mitchell Chesler contributed creatively as a post doctoral fellow. This paper summarizes their work.

Animal Experiments

Several years ago in association with Carol Petito and others[1–3] we noted that acute, near complete forebrain ischemia, delivered for 20 or 30 minutes, caused different morphological changes in adult rat brain, depending on whether the animal had been given glucose (high carbohydrate (CHO) brain) or was fasting (low CHO brain). Within a matter of hours after ischemia was reversed, electron microsopy revealed massive glial cell swelling and rupture in high CHO brains which then went on to develop extensive bilateral cerebral infarction affecting all tissue elements. By contrast low CHO brains showed only selective neuronal necrosis, sparing the structural integrity of glial cells and other elements. Astrocytes were moderately enlarged in the low CHO brains, and water content rose approximately one percent, but these abnormalities disappeared within a few hours or days. Some of the neuronal necrosis, i.e. in the CA1 zone of the hippocampus, was delayed for several days after the insult[4] by a process that has subsequently been repeatedly confirmed. The severe morphologic differences between the high and low CHO brains led us to speculate that glial cells might regulate hydrogen ion homeostasis differently from neurons[3]. Subsequent experiments in our laboratory have pursued this hypothesis and indicate that during states of altered function or disease, not only can hydrogen ion concentrations in glial cells differ widely from the interstitial space (ISS), but from that of neurons as well. In addition, glial cells *in vivo* appear to acidify selectively in their effort to dampen hydrogen ion shifts under circumstances of severe lactic acidosis during high CHO brain ischemia[5,6]. The mechanisms that underlie these changes, however, remain largely unknown and even controversial. This paper reviews our most recent efforts to clarify the subject.

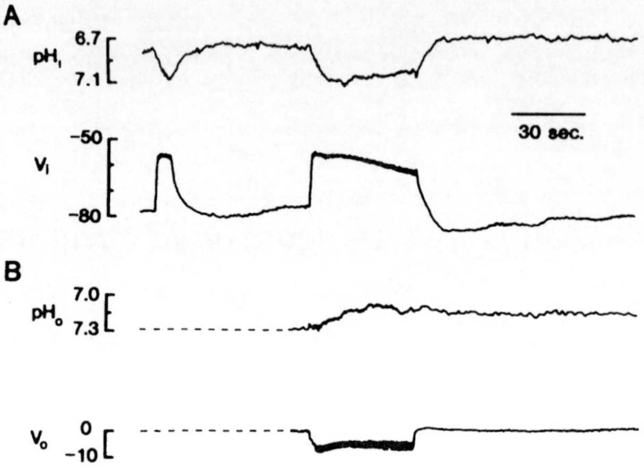

Fig. 2. Intracellular and interstitial responses to cortical stimulation. Stimulus trains (200 µsec pulses at 20 Hz for 0.8 seconds) were delivered once per second. (A) Upper trace shows intracellular space pH (pH_i), bottom trace records intracellular potential (V_i). Stimuli were delivered first for 6 seconds, then for 43 seconds. (B) Upper trace graphs interstitial pH (pH_o), bottom trace shows interstitial potential (V_o). Stimuli were delivered for 41 seconds following withdrawal from cell recorded in (A). Note stimulus trains artifacts on V_i and V_o records. All potentials in millivolts with respect to remote electrodes. Back injection of HRP into similarly recorded cells showed the typical morphology of protoplasmic astrocytes (reprinted from Ref. 10 with permission of the authors and publisher)

Recent experiments by Kraig and Nicholson[7] and Kraig and Chesler[8,9] directly support the conclusion that excess lactic acid accumulates in restricted zones in ischemic, high CHO brains. Fine tipped double barreled pH sensitive electrodes were developed that permitted both recording and labeling of normal and ischemic glial cells. When inserted into cerebral cortex of brains with high carbohydrate stores during severe or complete ischemia, electrode recordings confirmed the expected ISS pH of approximately 6.2, but as the electrode was advanced through the tissue, small pockets of increased acidity were identified[9,10] (Fig. 1). These pockets were subsequently identified as protoplasmic astrocytes by injecting horseradish peroxidase electrophoretically through the recording electrode after completion of the recording[10]. Within the small deeply acidic pockets of high CHO ischemic brains recorded up to one hour following initiation of ischemia, pH values ranged between 4.5 and 5.5 with occasional readings falling as low as 3.9.

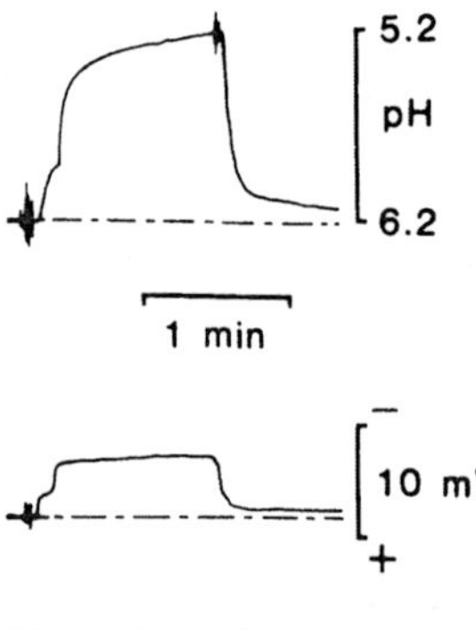

Fig. 1. Intracellular pH changes observed 30 minutes after the onset of complete brain ischemia under hyperglycemic conditions. Fine tipped, H^+ selective microelectrodes could detect areas of brain more acidic than the interstitial space for up to 60 minutes after onset of ischemia (reprinted from Ref. 7 with permission of the authors and publisher)

Subsequently, Chesler and Kraig have compared non-steady state pH changes within rat brain cortical astrocytes and neurons, defined by electrophysiologic criteria, with those found in ISS during functional cortical activation induced by direct electrical stimulation or spreading depression[6,10]. Both these latter measures depolarize neurons, the latter in more extensively coordinated bursts than the former and both increase lactate production upon stimulation. As Fig.2 shows, in astrocytes both short (6 seconds) and long (43 seconds) bursts of cortical electrical stimulation induced oppositely going changes in intracellular pH and ISS pH; glial pH became more alkaline, whereas the ISS

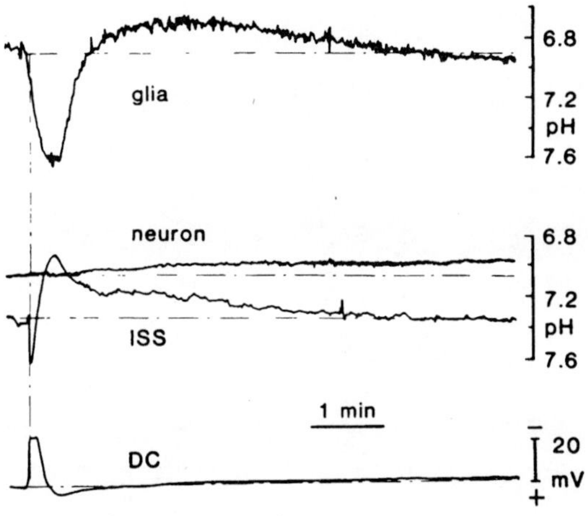

Fig. 3. Intracellular and interstitial space *(ISS)* acid base changes recorded by H^+ sensitive microelectrodes in brain cells from rat frontal cortex during spreading depression *(SD)*. Records shown are typical changes observed during comparable but separate experiments. DC and ISS changes were recorded simultaneously in the ISS. The biphasic waveform of the DC signal corresponds to brain cell depolarization and was used to compare the time course of pH records. Glial (top trace) and ISS pH changes show that the ISS alkaline shift during depolarization, which reached 7.60 in this example, correlated with a more slowly developing glial alkaline transient which reached its peak of 7.61 as ISS pH shifted to become most acid (6.93). Neuronal pH remained unchanged during SD but became somewhat more acidic as the cell repolarized (reprinted from Ref. 8 with permission of the authors and publisher)

became more acidified[10]. Direct measurements comparing intraglial and intraneuronal pH with that of ISS during spreading depression demonstrate heterogenous values in all three spaces[8]. The intraglial transient during the neuronal depolarization-repolarization wave goes alkaline then returns to a normal or slightly acid level (Fig. 3). The ISS pH moves largely oppositely to that in the glial cells, becoming at first transiently alkaline, then shifting to a larger, more acid level before finally returning to baseline. Concurrently, in the diagrammed experiment, the neuron became slightly more acid as the cell repolarized and began spontaneously to discharge. Additional experiments measuring intraglial and tissue PCO_2 suggest that the recurrent neuronal depolarization-repolarization of SD can induce a progressive increase in glial acidity associated with a gradual depletion of bicarbonate stores (see Fig. 5 in Reference 9).

The thermodynamic and physiological forces that underlie these complex and transient intracellular responses to stimulation are presently poorly understood. Both stimulation and spreading depression are known to increase lactate concentrations in the brain with or without immediate declines in ISS pH. Two earlier studies by other investigators[11,12] have reported a pH fall in neurons following depolarizing stimuli, but the parodoxical alkaline shift of glial cells is a new observation and less readily explained. Recorded membrane potential changes during stimulation are insufficient to generate a passive flux of hydrogen ions or bicarbonate to anywhere near the amount required to explain the observed changes. Accordingly, physiologically governed activity appears likely, particularly since comparably large intraglial shifts of potassium, sodium and chloride have been observed during reversal activation[13].

Experiments measuring tissue pH in areas of focal ischemia suggest that strong biological defense mechanisms operate to maintain intracellular brain pH during the early phases of acute brain infarction. Nedergaard *et al.* used a combination of microelectrode and DMO techniques to study interstitial and intracellular acidity during acute focal ischemia in the normoglycemic rat brain[14]. After 60 minutes of ischemia in the middle cerebral artery territory, the non-ischemic hemisphere showed an interstitial pH (pH_o) of 7.24 ± 0.02, and intracellular pH (pH_i) of 7.03 ± 0.02, whereas the interstitial space within the contralateral ischemic zone fell to pH_o 6.44 ± 0.07, and the pH_i fell to 6.82 ± 0.08, ($p<0.01$ Student's t test) the relatively attenuated intracellular pH was taken to reflect an active physiological defense of intracellular homeostasis. After four hours of continuous ischemia the tissue and ISS values in the infarct equilibrated, with pH_o rising to 6.64 and pH_i falling to 6.62. The findings emphasize the continuing capacity for a large proportion of cells in a region of ischemic brain to maintain their physiological buffering mechanisms for at least the first hour of severe ischemia.

If, as Kraig and associates suggest[7,9], protoplasmic astrocytes undergo wider acid-base shifts than occur in neurons or the ISS during stimulation or ischemia, does any evidence indicate that the acidity by itself could harm astrocytes? As noted, the pathologic and electrode recordings suggest this possibility, at least when severe or complete ischemia affects a high CHO brain. Furthermore, our laboratories showed that tiny lactate injections into brain producing ISS pH values below 5.30 for as little as 20 minutes induced local infarction, nearly destroying all cellular elements within the injection site[15]. More recently, the laboratory has employed studies in tissue culture in an effort to answer this question by directly analyzing cellular vulnerability to acidity.

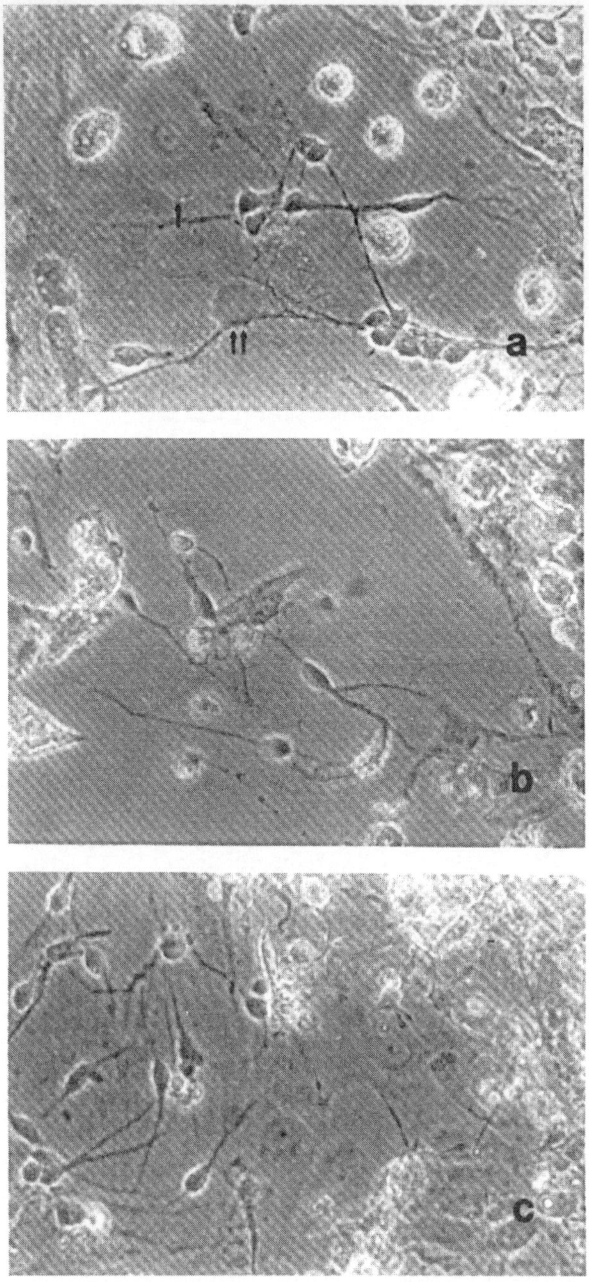

Fig. 4. Neuronal tolerance of HCl-induced extracellular acidosis. Neurons are more tolerant than astrocytes to brief HCl exposure. (a) Shows an 8 day *in vitro* embryonic (E19) forebrain culture, one hour after a ten minute exposure to pH 3.8. Neurons are morphologically indistinguishable from control cultures. However, the underlying astrocytes are undergoing process retraction, with membrane ruffling and substrate denudement (single arrow), as well as intracellular bleb formation (double arrow). Other substrate cells have already rounded up, and are detaching from the fibronectin layer. (b) Shows the same culture three days later. Rounded cells and a denuded substrate are visible, with many surviving neurons. (c) Reveals another area of this culture three days after acid exposure. Surviving neurons are present, as are recolonizing substrate cells of glioblastic morphology (reprinted from Ref. 16 with permission of the authors and publisher)

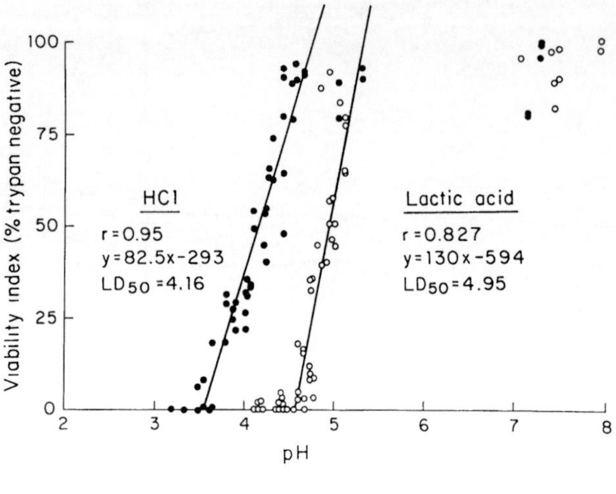

Fig. 5. Determination of thresholds for acid-induced cell death. A total of 129 cultures were exposed for ten minutes each to media titrated with either hydrochloric (n = 43) or lactic (n = 86) acids to the pH levels noted on the abscissa. Cellular viability is indicated on the y-axis, and is defined as the percentage of cells capable of excluding trypan blue 24 hours after acid exposure. Neurons and glia were pooled for this determination. Viability data for cultures exposed to either of the extremes of pH were excluded from the regression line and correlation coefficient determinations, so as to preserve the validity of a linear regression analysis of the data. Thus, the final regression analysis included all data within the pH ranges 4.60–5.20 for lactic acid, and 3.60–4.70 for hydrochloric acid (reprinted from Ref. 16 with permission of the authors and publisher)

Tissue Culture Studies

Goldman *et al.* directly examined the effects of extracellular acid exposure on cultured forebrain neurons and glial cells[16]. Primary cultures were prepared from trypsin dissociated embryonic rat forebrain at 16–17 days gestational age (E16). Cells were placed onto fibronectin substrate and raised in a high protein, high glucose media in a 95% air / 5% CO_2 environment. All measurements were performed at media temperatures of 25°C.

After one week *in vitro*, cultures were exposed for 10 minutes to media acidified with either HCL or lactic acid and brought to pH values ranging between 2.8 and 7.7. Simultaneous control cultures were maintained at pH 7.6 or 7.1. Following exposure, each culture was washed twice in warm Hanks buffered salt solution and returned to fresh media. Cultures were observed and photographed by phase microscopy at hourly intervals over the next two days, using a Leitz inverted microscope. A total of 114 cultures were incubated in HCL titrated media, and 130 in lactic acidified media. Inter-run variability for acid induced death, as measured by the LD_{50} for trypan blue inclusion, was approximately

± 0.1 pH. Criteria for cell death included gross morphologic destruction, trypan blue inclusion, cellular detachment from the substrate and failure of cells to reanchor to new fibronectin layers. Immunocytochemical staining for glial fibrillary acidic protein (GFA) and neurofilament protein (NF) were used to distinguish glial cells from neurons.

Among 114 cultures exposed to HCl for 10 minutes, abnormalities were confined to exposures below pH 4.9. Transient rounding, intracellular granularity and membrane blebbing occurred in both astrocytes and neurons after exposures between pH 4.4 and 4.9, but such groups regained their baseline morphologies by 24 hours after acid exposure. However, most astrocytes exposed for 10 minutes to pH ≤ 4.1 became rounded, detached from their substrate, incorporated trypan blue, and died within hours after exposure (Fig. 4). Neurons proved more resistant to HCl than astrocytes, and survived 10 minute exposures to media of pH ≥ 3.8. Ten minute exposures below pH 3.8 killed essentially all neurons.

Both neurons and glial cells were more vulnerable to lactic than hydrochloric acidosis (Fig. 5). One hundred and thirty forebrain cultures were exposed for 10 minutes to a lactic acidified media titrated between pH 4.0 and 7.6. The threshold for cell death was quantified 24 hours after acid exposure again using trypan blue staining as an endpoint. Fewer than 57% of neurons or glia survived 10 minutes in media titrated below pH 4.6 (Fig. 6), but nearly all tolerated exposures of similar length at pH 5.2 and above. Linear regression of cell viability as a function of pH projected an LD_{50} of pH 4.95, for 10 minute lactic acid exposures. This contrasts with an LD_{50} of 4.1 for HCl exposures of equivalent length (Fig. 5).

These experiments demonstrate that both neurons and glial cells are relatively resistant to severely acidic exposures from strong inorganic acids, but are more susceptible to lesser degrees of environmental acidosis when lactate is the carrier anion of the proton.

Lengthening the duration of exposure to 1 hour intensified the lethal effect of lactic acid. Nearly all neurons and glia died when exposed to media below pH 5.2. In culture, essentially all neurons and glia succumbed following one hour exposures to lactic acid induced pH levels of 5.2, while only 30% died after similar exposures of 10 minutes. These pH values are almost identical to those found to be lethal when injected into the brain *in vivo*.

Several possible explanations could explain the greater effect of lactic acidosis on cellular function,

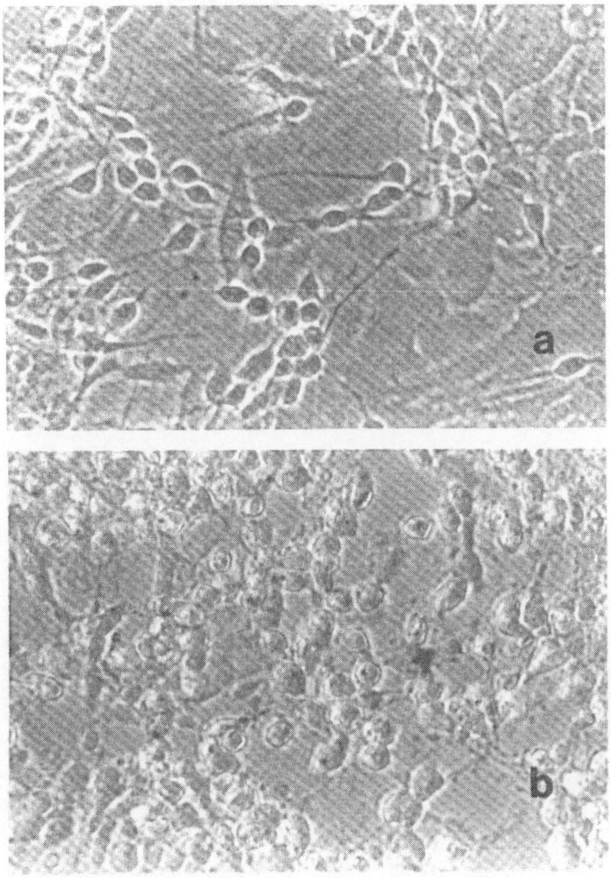

Fig. 6. Heightened cellular vulnerability to extracellular lactic acid. Both neurons and glia are more sensitive to lactic acid exposure than to HCl exposure. Neither neurons nor glia survive lactic acid adjusted pH level of 4.7 or below (ten minute exposures), while both cell types tolerate HCl acidification to pH 4.2. (a) Shows a mixed neuronal-glial population six hours after a ten minute exposure to HCl-titrated media at pH 4.4; this culture is morphologically indistinguishable from control. (b) Displays a similar culture six hours after exposure to lactic acid acidified media a pH 4.4. All cell types in this severely damaged culture have rounded up, many have detached from the fibronectin substrate (reprinted from Ref. 16 with permission of the authors and publisher)

one of which is that the protonated moiety diffuses more readily through the neuronal or glial membranes than do free protons. To examine this question and to compare the relative permeability of glia and neurons to severe degrees of lactic acidosis, the laboratory has turned its attention to studying this problem in single cells in tissue culture.

Nedergaard *et al.* have studied the effect of lactic acid on intracellular pH in cultured neurons and glia, using trypsin dissociated E16 rat forebrain cultures, studied after 7 days in vitro[17,18]. The pH_i was measured using the fluorescence excitation of the dye 2,7-bis-(2-carboxyethyl-5 (and -6) (carboxyfluorescein) (BCE-

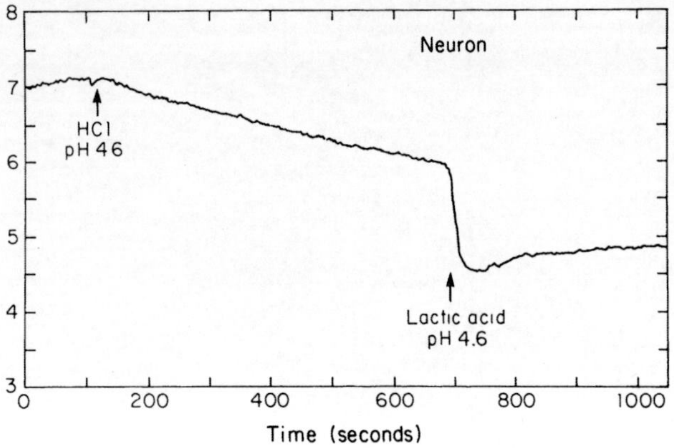

Fig. 7. Intracellular acidification measured by pH-sensitive dye in a single neuron exposed to inorganic (HCl) and organic (lactic) acids. Intracellular acidification to an extracellular hydrogen ion load is relatively slow, the pH (ordinate) falling less than 1.0 unit during a ten minute exposure to HCl at pH 4.6. In contrast, intraneuronal pH falls to ambient levels within a minute of lactic acid exposure at the same extracellular pH

CF; pKa 6.9). The fluorescence intensity ratio of single neurons or glia cells was measured with a PTI Delta scan illumination system coupled to a Leitz Diavert microscope, using excitation wavelengths of 495 and 440 nm, and emission detection of 510 nm and above. Calibrations were performed using the proton ionophore nigericin in a high potassium solution, in order to equalize intra- and extracellular pH. The identities of astrocytes and neurons were confirmed immunocytochemically, using antisera directed against glial fibrillary acidic protein and neurofilaments, respectively.

Experiments to date using these techniques demonstrate that both glial cells and neurons are highly permeable to lactic acid at pH values down to 4.0 and to a much greater degree than to the free protons of highly dissociated HCl solutions of comparable pH (Fig. 7). The tissue culture studies also show that at pH 4.0 astrocytes are more permeable to lactate ion than are neurons, although the rate and degree of exchange of both are rapid, and linearly related to the pH. The modest difference in rate of exchange may reflect the larger surface area of astrocytes *in vitro*, but we have not yet fully investigated this premise.

Summary and Conclusion

Human and animal experimental studies of brain ischemia show greater degrees of tissue lactacidosis and morphological damage affecting brains with high carbohydrate (CHO) stores compared to normal CHO levels. Furthermore, in high CHO ischemic brain, experimental findings suggest that forebrain astrocytes undergo a considerably greater fall in intracellular pH than do either other cellular elements or interstitial fluid. Tissue culture studies, show that astrocytes are more permeable than neurons to the lactate ion, but neurons and glial cells are similarly vulnerable to sustained intracellular pH values below pH 5.2. The tissue culture findings directly support the *in vivo* observations that lactacidosis at pH 5.2 causes pan-necrosis of cerebral tissue. They also are consistent with the hypothesis that selective astrocytic acidosis during high CHO ischemia may represent a first intracellular step that, if prolonged, leads rapidly to astrocytic necrosis and cerebral infarction.

Acknowledgement

Experimental studies described in this review were partly supported by USPHS grant NSO3346-26.

References

1. Petito CK, Babiak T (1982) Early proliferative changes in astrocytes in post-ischemic, non-infarcted rat brain. Ann Neurol 11: 510–518
2. Petito CK, Pulsinelli WA, Jacobson G, Plum F (1982) Edema and vascular permeability in cerebral ischemia: comparison between neuronal damage and infarction. J Neuropath Exp Neurol 41: 423–436
3. Plum F (1983) What causes infarction in ischemic brain. Neurology 33: 222–233
4. Pulsinelli WA, Brierley JB, Plum F (1982) Temporal profile of neuronal damage in a model of transient forebrain ischemia. Ann Neurol 11: 491–498
5. Kraig RP, Pulsinelli WA, Plum F (1985) Hydrogen ion buffering during complete brain ischemia. Brain Research 342: 281–290
6. Kraig RP, Pulsinelli WA, Plum F (1986) Carbonic acid buffer changes during complete brain ischemia. Am J Physiol 250 (Regulatory Physiol 19): R348–R357
7. Kraig RP, Nicholson C (1987) Profound acidosis in presumed glia during ischemia. In: Raichle M, Powers WV (eds) Cerebrovascular diseases. Raven, New York, pp 97–102
8. Kraig RP, Chesler M (1988) Dynamics of volatile buffers in brain cells during spreading depression. In: Somjen G (ed) Cerebral hypoxia and stroke: reversible and irreversible effects and their prevention. Plenum, New York, pp 279–289
9. Kraig RP, Chesler M (1988) Glial acid-base homeostasis in brain ischemia. In: Norenberg M, Hertz L, Schousbou N (eds) The biochemical pathology of astrocytes. Liss, New York, pp 365–376
10. Chesler M, Kraig RP (1987) Intracellular pH of astrocytes increases rapidly with cortical stimulation. Am J Physiol 253 (Regulatory Physiol 22): R666–R670
11. Ahmed Z, Connor JA (1980) Intracellular pH changes induced by calcium influx during electrical activity in molluscan neurons. J Gen Physiol 75: 403–426
12. Enders W, Ballanyi K, Serve G, Grafe P (1986) Excitatory amino acids and intracellular pH in motor neurons of the isolated spinal cord. Neurosci Lett 72: 54–58

13. Ballanyi K, Grafe P, ten Bruggencate G (1987) Ion activities and potassium uptake mechanisms of glial cells in guinea pig olfactory cortex slices. J Physiol (Lond) 382: 159–174
14. Nedergaard M, Kraig RP, Tanabe J, Pulsinelli WA (1988) Focal brain ischemia reverses the normal interstitial/intracellular pH ratio. Neuroscience (Abstr) 14: 1065
15. Kraig RP, Petito CK, Plum F, Pulsinelli WA (1987) Hydrogen ions kill brain at concentrations reached in ischemia. J Cereb Blood Flow Metabol 7: 379–386
16. Goldman S, Pulsinelli WA, Clarke W, Kraig RP, Plum F (1989) The effects of extracellular acidosis on neurons and glia *in vitro*. J Cereb Blood Flow Metabol 9: 471–477
17. Nedergaard M, Goldman SA, Pulsinelli WA (1989) Lactic acid induced intracellular acidification in primary cultures of mammalian brain. J Cereb Blood Flow Metabol 9: S 384
18. Nedergaard M, Goldman SA, Desai S, Pulsinelli WA (1989) Carrier mediated transport of lactic acid in cultured neurons and astrocytes. Neuroscience (Abstr) 15

Correspondence and Reprints: Fred Plum, Cerebral Vascular Disease Center, Raymond and Beverly Sackler Foundation, Department of Neurology and Neuroscience, Cornell University Medical College, New York, NY, 10021, U.S.A.

Acta Neurochir (1993) [Suppl] 57: 64–72

Mediators of Vascular and Parenchymal Mechanisms in Secondary Brain Damage*

M. Wahl[1], **L. Schilling**[1], **A. Unterberg**[2], and **A. Baethmann**[3]

[1] Department Physiology, [2] Department Neurosurgery, and [3] Institute for Surgical Research, Ludwig-Maximilians University, Munich, Federal Republic of Germany

Abstract

Several putative mediators of vasogenic brain edema will be considered with respect to the following criteria: 1) their effect on blood-brain barrier (BBB) permeability, 2) their vasomotor actions which may increase driving forces for transmural bulk flow, 3) their influence on edema formation, 4) their actual tissue concentration in pathological states, and 5) the therapeutic results after specific treatment.

Bradykinin (BK) can induce brain edema by increasing BBB permeability to small solutes and enhancing blood pressure in the microcirculation due to arterial dilatation and venous constriction. Its interstitial concentration is enhanced after experimental trauma. Since kallikrein inhibitors reduce brain swelling all criteria favour BK as a mediator of vasogenic edema.

Arachidonic acid (AA) opens BBB also for large tracers but exerts only small vasomotor effects. The edema formation is associated with an increase of the AA concentration in the interstitial space. However, convincing therapeutic results on inhibition of AA are still lacking. In addition to the formation of vasogenic edema AA has been found to induce cytotoxic edema. From experiments dealing with the vasomotor effects Ellis *et al.* (Am J Physiol 255: H397–H400, 1988) concluded an interaction of BK and AA in brain injury. However, our own results do not favour this hypothesis since we found divergent vasomotor and permeability effects of BK and AA.

Histamine (HA) opens BBB unspecifically and dilates cerebral vessels, mechanisms by which edema formation can be explained. Further requirements, such as formation of histamine in injured brain tissue as well as therapeutical inhibition of edema from mechanical injury by a histamine antagonist, are met supporting a mediator function of the agent in vasogenic brain edema.

Leukotrienes (LT) are potent constrictors of cerebral arteries and their tissue concentrations are increased in pathological states. However, they do not induce extravasation or edema formation. Correspondingly, lipoxygenase inhibitors are not effetive in experimental brain injury.

Free radicals (FR) are released under several pathological conditions but induce only irregular tracer leakage. Formation of edema may be facilitated by some arterial dilatation. Further therapeutic studies are necessary. Moreover, FR have been found to induce cytotoxic edema by similar mechanisms as reported for AA.

Serotonin may increase BBB permeability under certain conditions. However, it appears not to be involved in edema generation since negative results have been described for the other criteria.

Taken together, the requirements of identification of a mediator compound of brain edema are met in the case of bradykinin, arachidonic acid, histamine, and, probably in case of the free radicals. Evidence is less convincing for leukotrienes or serotonin. It should be noted in this context that the above mediator compounds are likely to enhance formation of brain edema not only by their specific pathophysiological properties but also on the basis of mutual activation of a mediator network.

Keywords: Vasogenic brain edema; mediator compounds; blood-brain barrier; vasomotor response.

Introduction

Vasogenic edema is one causal factor of secondary brain damage (Baethmann *et al.*, 1980). A prerequisite for the development of vasogenic edema is the opening of the blood-brain barrier. Evidence is now accumulating that the increase of vascular permeability after ischemia and severe head injury is induced by several mediators released after such insults (Wahl *et al.*, 1988). The mainly discussed mediators are histamine, bradykinin, arachidonic acid, free radicals, leukotrienes, and serotonin. The aim of the present paper is to discuss five criteria for each mediator candidate:

1) the effect on blood-brain barrier (BBB) permeability,

2) the vasomotor action which may increase driving forces for transmural bulk flow,

3) the influence on edema formation,

4) the actual concentration in the tissue or in the interstitial space during pathological states and,

5) the therapeutic results after specific treatment.

* Supported by Deutsche Forschungsgemeinschaft.

Histamine

The effect of histamine on BBB permeability was studied in cats employing cortical superfusion and measurement of tracer extra-vasation by intravital fluorescence microscopy (Schilling *et al.*, 1987a–c). Leakage of Na^+-fluorescein (MW: 376; Stokes' radius: 5.5 Å) started at 10^{-9} M histamine whereas extravasation of FITC (fluorescein isothiocyanate)-dextran 62,000 (Stokes' radius: 60 Å) and 150,000 (Stokes' radius: 87 Å) was first detected at 10^{-8} M and 10^{-6} M histamine, respectively.

Similarly, intravascular administration of histamine has frequently been reported to induce extravasation of various tracers (MW: 103–67,000) into brain tissue (Domer *et al.*, 1983; Dux and Joó, 1982; Gross *et al.*, 1982). Some negative results of others may be explained by methodical reasons or species differences as discussed elsewhere (Wahl *et al.*, 1988). The positive results demonstrate that histamine induces an unspecific opening of the BBB. Furthermore the concentration dependency of the leakage may indicate that an opening of differently sized functional pores is involved in the extravasation of the various tracers. A diameter of about 180 Å can be estimated for the largest pores.

The vasomotor effects of histamine were best studied in cats. Using perivasular microapplication (Edvinsson *et al.*, 1983; Gross *et al.*, 1981; Wahl and Kuschinsky, 1979) or cortical superfusion (Martins *et al.*, 1980; Schilling *et al.*, 1987a–c) histamine dilated pial arteries concentration-dependently. Furthermore, intraparenchymal arteries were dilated by histamine as shown *in situ* (De Ley *et al.*, 1982) and *in vitro* (Dacey and Basset, 1987). Venous diameters increased during cortical superfusion of histamine (Martins *et al.*, 1980; Schilling *et al.*, 1987a–c) but not during perivascular microapplication (Gross et al., 1981).

A mediator role of histamine in formation of brain edema may be deduced from its vasomotor and permeability effects. Indeed, an increase of brain water content and a swelling of astrocytic end-feet was reported by Gross *et al.* (1982) during intracarotid infusion of histamine.

An increase of the histamine concentration in the cortical tissue after cold lesion or stab injury has been recently found by Orr (1988) and Mohanty *et al.* (1989), respectively. However, additional data of the histamine concentration in the interstitial space must still be presented.

Therapeutic results after treatment with histamine antagonists have been reported. A reduction of irradiation induced brain edema was suggested by Csanda (1980) and Joó *et al.* (1976) because the Evans blue extravasation was reduced by application of a H_2 receptor antagonist. Very recently, Mohanty *et al.*(1989) found that the increase of brain water content due to stab injury could be prevented by cimetidine.

As summarized in Table 1 all criteria favour histamine as a mediator of vasogenic brain edema.

Table 1. *Criteria for the Mediator Function of Histamine*

1. Unspecific increase of BBB permeability
2. Dilation of pial arteries and veins may increase driving forces for bulk flow
3. Edema formation
4. Increase of histamine concentration in cortical tissue after stab injury
5. Prevention of stab injury induced increase of brain water content by treatment with cimetidine

The conclusions were mainly drawn from results of Schiling *et al.* (1987a–c) for (1) and (2), Gross *et al.* (1982) for (3), Orr (1988) and Mohanty *et al.* (1989) for (4) and Mohanty *et al.* (1989) for (5).

Bradykinin

Cortical superfusion with bradykinin induced in the cat an extravasation of Na^+ fluorescein from pial vessels at 4×10^{-7} to 4×10^{-6} M which was determined by intravital fluorescence microscopy (Unterberg *et al.*, 1983, 1984; Wahl *et al.*, 1983a, 1985a). A leakage of FITC-albumin (MW: 67,000) or FITC-dextran (MW: 19,400–62,000) was never observed. From this finding a selective increase of BBB permeability by opening of functional pores with a diameter of at least 11 to 15 Å can be calculated. Similarly, intracarotid infusion of bradykinin revealing concentrations of 10^{-6} to 10^{-5} M bradykinin in the sagittal sinus blood induced extravasation of Na^+ fluorescein but not of FITC-dextran (MW: 62,000) as reported by Unterberg *et al.* (1984).

The bradykinin induced leakage is primarily due to an increased permeability of the vessel wall since intra- and extravascular bradykinin caused a comparable extravasation without and with concomitant vasomotor responses, respectively (Unterberg *et al.*, 1984). The same is true for the histamine-evoked leakage (Gross *et al.*, 1982; Schilling *et al.*, 1987a–c). Furthermore, extravascular histamine elicited extravasation of Na^+-fluorescein in a concentration range (10^{-9}–10^{-7} M) in which it did not exert any vasomotor effect (Schilling *et al.*, 1987a–c). Arterial vasodilation which increases the driving force for transmural bulk flow and the vascular surface is per se not sufficient to elicit penetration of tracers. However, there is no doubt that arterial dilation increases leakage if vascular permeability is enhanced (Wahl *et al.*, 1985a).

Bradykinin is a potent dilator of pial arteries when applied by perivascular microapplication (Wahl, 1982; Wahl *et al.*, 1983b; Whalley and Wahl, 1983) or by cortical superfusion (Kamitani *et al.*, 1985a; Kontos *et al.*, 1984; Unterberg *et al.*, 1983, 1984; Wahl *et al.*, 1985a). Similarly, bradykinin dilates intraparenchymal arteries (Dacey *et al.*, 1988).

The increase of vascular permeability and diameter can explain the formation of brain edema found in dogs after ventriculo-cisternal perfusion with artificial cerebrospinal fluid containing 2.5×10^{-6} M bradykinin (Unterberg and Baethmann, 1984).

Bradykinin can be generated in the brain which contains all components of an intraparenchymal kallikrein-kinin system as reviewed elsewhere (Wahl *et al.*, 1986a, 1987). After cold lesion an uptake of plasma kininogen into the brain has been found by Maier-Hauff *et al.* (1984) in cats. From these data bradykinin concentrations of 10^{-7} to 10^{-6} M can be calculated for the interstitial space. That means that after cold lesion the bradykinin concentration raises to values which increase permeability and arterial diameter and induce edema formation.

Further evidence for a mediator role of bradykinin in brain edema is provided by therapeutic results. The cryogenic brain swelling could be significantly reduced by treatment with the kallikrein-inhibitor aprotinin (Unterberg *et al.*, 1986). Similarly, the dilation of cerebral arterioles after concussive brain injury which was accompanied by a seven - to eight - fold increase of brain bradykinin concentration (Ellis *et al.*, 1987) could be diminished by pretreatment with a B_2-kininergic receptor antagonist (Ellis *et al.*, 1988).

As summarized in Table 2 all criteria favour bradykinin as a mediator of vasogenic edema.

Table 2. *Criteria for the Mediator Function of Bradykinin*

1. Increase of BBB permeability only for Na^+-fluorescein (MW 376) but not for FITC-dextran (MW 19400)
2. Arterial dilation may increase driving forces for bulk flow
3. Edema formation
4. Increase of bradykinin concentration in cortical tissue after cold lesion
5. Reduction of cold lesion induced brain swelling after treatment with the kallikrein inhibitor aprotinin

The conculsions were mainly drawn from results of Unterberg *et al.* (1984) for (1) und (2), Unterberg and Baethmann (1984) for (3), Maier-Hauff *et al.* (1984) for (4), and Unterberg *et al.* (1986) for (5).

Table 3. *Criteria for the Mediator Function of Arachidonic Acid*

1. Unspecific opening of BBB for small and large tracers, penetration of polymorphonuclear leukocytes
2. Small vasomotor responses
3. Vasogenic and cytotoxic edema
4. Arachidonic acid concentration in the interstitial space increases up to 2×10^{-4} M after cold lesion
5. Therapeutic results?

The conclusions were mainly drawn from results of Unterberg *et al.* (1987) for (1) and (2), Chan and Fishman (1984) for (3), and Baethmann *et al.* (1989) for (4).

Arachidonic Acid

Arachidonic acid (10^{-5}–10^{-3} M) leads to an unspecific opening of the BBB for small (Na^+-fluorescein) and large (FITC-dextran 62,000) tracers (Unterberg *et al.*, 1985, 1987a; Wahl *et al.*, 1985a, b). This finding is in accordance with investigations using other tracers, e.g. Evans blue-albumin (Chan and Fishman, 1984; Chan *et al.*, 1983; Wakai *et al.*, 1982; Wei *et al.*, 1986). Since pretreatment with indomethacin or BW755C, an inhibitor of cyclooxygenase and lipoxygenase, did not reduce the arachidonic acid-induced barrier opening for Na^+-fluorescein it was concluded that this effect was mediated by the acid itself and not by its metabolites (Unterberg *et al.*, 1987a). The vascular leakage due to arachidonic acid results obviously from an interaction between polymorphonuclear leukocytes and the vascular endothelium. Employing electronmicroscopy sticking of leukocytes and their penetration through endothelial cells was found after cortical superfusion with arachidonic acid (Unterberg *et al.*, 1987a). The conclusion that the lesion of the barrier is due to arachidonate itself is in conflict with findings of Kontos *et al.* (1980) and Black and Hoff (1985). Kontos *et al.* (1980) could prevent the arachidonate-induced lesions of arterioles by indomethacin and Black and Hoff (1985) found that BW755C attenuated the uptake of Evans blue-albumin after intraparenchymal injection of arachidonic acid.

A weak constriction (5–10%) of pial arteries due to 3×10^{-6} to 3×10^{-4} M arachidonate but no significant response to 3×10^{-3} M was reported by Unterberg *et al.* (1987a) for cortical superfusion in cats. In contrast,

Busija and Heistad (1983), Kontos *et al.* (1980), and Wei *et al.* (1980) reported concentration-dependent dilations of pial arteries exposed to arachidonic acid that were explained by generation of free radicals (Kontos *et al.*, 1980). Since arachidonate did also not cause a change of venular diameter the weak vasomotor effects were considered not to participate in the development of vasogenic edema (Unterberg *et al.*, 1987a).

Indeed, arachidonic acid induced cytotoxic and vasogenic edema (Chan *et al.*, 1983; Chan and Fishman, 1984). Recently, Unterberg *et al.* (1987b, 1989a) found that an intraparenchymal infusion-induced brain edema was markedly increased by addition of arachidonic acid to the infusion fluid. Release of arachidonic acid from brain tissue has been found to occur during hypoxia, ischemia, seizure, and head injury (Baethmann *et al.*, 1989; Bazan and Rodriguez de Turco, 1980; Bhakoo *et al.*, 1984; Gardiner *et al.*, 1981; Siesjö, 1981). Baethmann *et al.* (1989) reported that the arachidonic acid concentration in the interstitial space increased up to 2×10^{-4} M after cold lesion, a concentration which is effective to induce tracer extravasation (Unterberg *et al.*, 1987a).

Convincing therapeutic results are still lacking since a complete pharmacological inhibition of the arachidonic acid release is not yet possible. The beneficial effect of steroid therapy on vasogenic edema formation might be explained by an inhibition of phospholipase A_2 and a reduction of arachidonic acid release.

As summarized in Table 3 most of the criteria favour arachidonic acid as a mediator of vasogenic edema. However, convincing therapeutic results must be demonstrated in the future.

Free Radicals

Cortical superfusion with oxygen free radicals caused endothelial lesions and arterial dilation (Wei *et al.*, 1985). In a following study Wei *et al.* (1986) found also a leakage of Evans blue-albumin and of horseradish peroxidase when the feline cortical surface was superfused with arachidonic acid. Since this extravasation could be prevented by superoxide dismutase and catalase the authors concluded the mediation by free radicals which was already proposed by Kontos *et al.* (1980). Similarly, free radicals decreased electrical resistance of frog pial venules indicating an increase of ionic permeability which could be inhibited by superoxide dismutase and catalase (Olesen, 1987). Furthermore, an increased permeability of intraparenchymal vessels was demonstrated by extravasation of intravenously administered Evans blue after injection of a free radical generating system into the brain (Chan *et al.*, 1984, 1987). In contrast to these findings, cortical superfusion with xanthine-oxydase and hypoxanthine caused leakage of Na^+-fluorescein only in 30% of the cats investigated whereas extravasation of FITC-dextran (MW: 62,000) was never observed (Unterberg *et al.*, 1988; Wahl *et al.*, 1986b). Extravasation was increased by an additional hypertonicity of the superfusion fluid which induced also leakage of FITC-dextran (MW: 62,000) in some cases (Unterberg *et al.*, 1988; Wahl *et al.*, 1986b).

Discrepant results concerning the vasomotor effects of free radicals have been reported. Whereas Rosenblum (1983) and Wei *et al.* (1985) found pronounced dilations of pial arterioles, Unterberg *et al.* (1988) and Wahl *et al.* (1986b) found only weak dilation up to 10% when feline cortex was superfused with normotonic hypoxanthine and xanthine-oxydase. These authors found pronounced arteriolar dilation of up to 70% only when undialyzed, hypertonic xanthine-oxydase and hypoxanthine were administered.

Free radicals are generated in the brain tissue during hypertensive insults and after ischemia or trauma (Demopoulos *et al.*, 1979; Kontos, 1985; Kontos and Wei, 1986; Siesjö, 1981; Siesjö *et al.*, 1985; Suzuki and Yagi, 1974). Superoxide radicals can be assayed *in situ* by the reduction of nitroblue tetrazolium to nitroblue formazan. Thus, a sixfold increase of nitroblue formazan brain level was detected in rats 1 h after cold lesion (Chan *et al.*, 1987).

Controversial therapeutic results on treatment of edema with radical scavengers have been reported (Long, 1989). However, Chan *et al.* (1987) reported that the cold lesion induced vasogenic edema in rats was reduced by treatment with liposome-entrapped superoxide dismutase.

Table 4. *Criteria for the Mediator Function of Free Radicals*

1. Increase of BBB permeability
2. Weak arteriolar dilatation up to 10%
3. Edema formation
4. Generation of free radicals in brain tissue after trauma and cold lesion
5. Therapeutic results?

The conclusions were mainly drawn from results of Unterberg *et al.* (1988) for (1) and (2), Chan *et al.* (1984) for (3), and Chan *et al.* (1987) for (4).

As summarized in Table 4 there is some evidence that free radicals are involved in the generation of vasogenic edema. However, further studies are needed to clarify the potential of free radical scavengers for the successful treatment of vasogenic edema.

Interaction Between Bradykinin, Arachidonic Acid and Free Radicals

An interaction between the kallikrein-kinin system, arachidonic acid metabolism and free radicals has been suggested by Ellis *et al.* (1988) to occur after brain injury. These authors hypothesized that brain injury induces either directly or by activation of the kallikrein-kinin system a release of arachidonic acid. Free radicals are generated when arachidonic acid is further metabolized via the cyclooxygenase pathway. Therefore, Ellis *et al.* (1988) concluded that the arterial dilation and endothelial lesions after concussive brain injury are finally mediated by free radicals. However, investigations of Unterberg *et al.*(1984, 1987a, 1988) and Wahl *et al.*(1985a, 1985b, 1986b) do not support the concept that the effects of bradykinin and arachidonic acid are mediated by free radicals. As shown in Table 5 these authors found divergent vasomotor and permeability effects of bradykinin, arachidonic acid and free radicals in their experimental model which are not compatible with hypothesis of Ellis *et al.* (1988).

Leukotrienes

Controversial data on the effect of leukotrienes on BBB permeability have been reported. After intraparenchymal injection of leukotrienes Black (1984) found a brain uptake of intravenously administered Evans blue. However, Unterberg *et al.* (1987b, 1989a) were unable to detect extravasation of Evans blue after intraparenchymal injection of leukotrienes in cats. Correspondingly, employing the cortical superfusion model in cats an opening of BBB permeability of pial vessels for Na^+-fluorescein could not be detected even when leukotriene C_4, D_4 and E_4 were tested up to 2 μM (Unterberg *et al.*, 1987c, 1989a, 1989b; Wahl *et al.*, 1986b). Similarly, Hua *et al.* (1985) found no increased BBB permeability for Evans blue after i.v. infusion of leukotrienes in guinea pigs. However, Olesen and Crone (1986) reported that leukotriene C_4 induced an increase of ionic conductance in frog pial venules.

During cortical superfusion leukotriene C_4, D_4 and E_4 induced constriction up to 25% in feline pial arterioles and venules although they did not affect BBB permeability (Unterberg *et al.*, 1987c, 1989b; Wahl *et al.*, 1986b). In accordance, constriction of pial arteries were found in rabbits, mice and piglets (Kamitani *et al.*, 1985b; Rosenblum, 1985; Busija *et al.*, 1986).

No effect of leukotrienes C_4 and B_4 on cerebral edema formation of cats was detected by Unterberg *et al.* (1987b, 1987c, 1989a). These authors infused artificial cerebrospinal fluid into the frontal white matter of

Table 5. *Comparison of Vasomotor and Permeability Effects of Bradykinin, Arachidonic Acid and Oxygen Radicals*

Bradykinin	dilatation up to 40%[1]	opening of BBB for Na^+-fluorescein (MW 376) in 100%[1]
Arachidonic acid	no change[2]	opening of BBB for FITC-dextran (MW 62,000)[2]
Oxygen radicals	dilatation up to 10%[3]	opening of BBB for Na^+-fluorescein in 30%[3]

The conclusions were mainly drawn from results of Unterberg *et al.* (1984) for [1], Unterberg *et al.* (1987a) for [2], and Unterberg *et al.* (1988) for [3].

Table 6. *Criteria for the Mediator Function of Leukotrienes*

1. Intact BBB without extravasation of Na^+-fluorescein
2. Constriction of arterioles and venules
3. No edema formation
4. Increase of leukotriene concentration in CSF up to 10^{-8} M after concussive brain injury
5. No therapeutic effects with inhibition of cyclooxygenase and lipoxygenase by BW755C after cold lesion

The conclusions were mainly drawn from results of Unterberg *et al.* (1987c, 1989b) for (1) and (2), Unterberg *et al.* (1987b, 1989a) for (3), Moskowitz *et al.* (1984) for (4), and Unterberg *et al.* (1987c, 1989b) for (5).

Table 7. *Criteria for the Mediator Function of Serotonin*

1. Discrepant results on opening of BBB
2. Arteriolar dilatation
3. No edema formation
4. No causal relationship between serotonin release and edema formation after cold lesion
5. No therapeutic results

The conclusions were mainly drawn from results of Westergaard (1977) and Hardebo *et al.* (1981) for (1), Harper and MacKenzie (1977b) for (2), Fenske *et al.* (1976) for (3), and Pappius and Dadoun (1987) for (4).

the right hemisphere to simulate formation and spread of vasogenic edema. Similarly, cerebrospinal fluid containing 15 μM leukotriene C_4 or B_4 was infused into the left hemisphere to test whether the leukotrienes promote edema formation which was determined microgravimetrically in brain slices. The same increase of tissue water content was measured in the samples of both right and left hemisphere. Thus, it was concluded that leukotrienes do not promote edema or act as mediators of brain edema.

Brain concentrations of leukotrienes were enhanced after ischemia, subarachnoid hemorrhage, concussive brain injury (Moskowitz *et al.*, 1984) and in tumor (Black *et al.*, 1986; Kiwak *et al.*, 1985). Thus, one can estimate interstitial leukotriene concentrations in the range of 10^{-8} to 10^{-9} M in these states.

Therapeutic studies of Unterberg *et al.* (1987c, 1989b) demonstrate that treatment and pretreatment with BW755C, an inhibitor of the lipoxyyenase and the cyclooxygenase, did not reduce the cold lesion induced brain swelling.

As summarized in Table 6, leukotrienes do not increase BBB permeability and induce edema formation. Since also negative therapeutic results were obtained the leukotrienes cannot be considered as mediators of brain edema.

Serotonin

Discrepant results on the effect of serotonin on BBB permeability have been published. An opening of the BBB for intravenously given horseradish peroxidase was demonstrated in mice after ventricular application of serotonin (Westergaard, 1975, 1977). In contrast, leakage of Evans blue could not be detected in the rat after an intravenous bolus of up to 6×10^{-7} M serotonin/kg BW (Gabbiani *et al*, 1970; Saria *et al.*, 1983). In accordance, neither intracisternal nor intravenous application of serotonin caused extravasation of ^{14}C-inulin or Evans blue-albumin in the cat (Hardebo *et al.*, 1981) which confirmed studies of Solomon (1974) in the monkey employing ventriculo-cisternal perfusion of serotonin. In contrast, intravascular but not extravascular administration of serotonin increased the electrical conductance of frog pial venules (Olesen, 1985).

The vasomotor effects of serotonin depend on the amount of its release. *In situ*, small amounts of extravascularly administered serotonin dilate small feline pial arteries (Edvinsson *et al.*, 1977; Harper and Mac Kenzie, 1977b). Intravascular application of synthetic diffusible serotonin agonists, or of serotonin with simultaneous inhibition of its degradation, or after osmotic opening of the BBB caused a decrease of cerebral consumption of oxygen and glucose which was accompanied by a decrease of cerebral blood flow (Grome and Harper, 1983, 1985, 1986; Harper and MacKenzie, 1977a). Under these conditions, which may simulate the situation after brain injury with a massive release of serotonin, a drop of blood pressure in the microcirculation can be expected which would counteract edema formation.

A stimulation of cerebral turnover of serotonin has been found in rat brain after cold lesion (Pappius and Dadoun, 1987; Pappius and Wolfe, 1983a, 1983b). However, these authors did not find a causal relationship between release of serotonin and edema formation. They concluded that the reduced glucose consumption but not the edema formation in the lesioned hemisphere was due to serotonin (Pappius, 1986). Correspondingly, Fenske *et al.* (1976) reported that the cold lesion induced edema was not further increased after ventriculo-cisternal perfusion with serotonin.

Convincing therapeutic results of brain edema with antagonists of serotonin are not available. However, it must be mentioned that the heat stress induced extravasation of Evans blue-albumin in rats could be prevented by the serotonin antagonist cyproheptadine (Sharma and Dey, 1986).

As summarized in Table 7 most of the criteria discussed do not favour serotonin as a mediator of vasogenic edema.

Taken together, all criteria favour bradykinin and histamine as mediators of brain edema. A lot of evidence supports the mediator role of arachidonic acid and free radicals. However, convincing therapeutic results are still necessary for these two compounds.

References

1. Baethmann A, Oettinger W, Rothenfußer W, Kempski O, Unterberg A, Geiger R (1980) Brain edema factors: current state with particular reference to plasma constituents and glutamate. In: Cervos-Navarro J, Ferszt R (eds) Brain edema. Adv neurol 28. Raven, New York, pp 171–195
2. Baethmann A, Maier-Hauff K, Schürer L, Lange M, Guggenbichler C, Vogt W, Jacob K, Kempski O (1989) Release of glutamate and of free fatty acids in vasognic brain edema. J Neurosurg 70: 578–591
3. Bazan NG, Rodriguez de Turco EB (1980) Membrane lipids in the pathogenesis of brain edema. Phospholipids and arachidonic acid, the earliest membrane components changed at the onset of ischemia. In: Cervos-Navarro J, Ferszt R (eds) Brain edema. Adv neurol 28. Raven, New York, pp 197–205

4. Bhakoo KK, Crockard HA, Lascelles PT (1984) Regional studies of changes in brain fatty acids following experimental ischaemia and reperfusion in the gerbil. J Neurochem 43: 1025–1031
5. Black KL (1984) Leukotriene C_4 induces vasogenic cerebral edema in rats. Prostagland Leukotr Med 14: 339–340
6. Black KL, Hoff JT (1985) Leukotrienes increase blood-brain barrier permeability following intraparenchymal injections in rats. Ann Neurol 18: 349–351
7. Black KL, Hoff JT, McGillicuddy JE (1986) Increased leukotriene C_4 and vasogenic edema surrounding brain tumors in humans. Ann Neurol 19: 592–595
8. Busija DW, Heistad DD (1983) Effects of indomethacin on cerebral blood flow during hypercapnia in cats. Am J Physiol 244: H519–H524
9. Busija DW, Leffler CW, Beasley G (1986) Effects of leukotrienes C_4, D_4 and E_4 on cerebral arteries on newborn pigs. Pediat Res 20: 973– 976
10. Chan PH, Fishman RA (1984) The role of arachidonic acid in vasogenic brain edema. Federation Proc 43: 210–213
11. Chan PH, Fishman RA, Caronna J, Schmidley JW, Priolieau G, Lee J (1983) Induction of brain edema following intracerebral injection of arachidonic acid. Ann Neurol 13: 625–632
12. Chan PH, Schmidley JW, Fishman RA, Longar SM (1984) Brain injury, edema, and vascular permeability changes induced by oxygen-derived free radicals. Neurology 34: 315–320
13. Chan PH, Longar S, Fishman RA (1987) Protective effects of liposome-entrapped superoxide dismutase on posttraumatic brain edema. Ann Neurol 21: 540–547
14. Csanda E (1980) Radiation brain edema. In: Cervos-Navarro J, Ferszt R (eds) Brain edema. Adv neurol 28. Raven, New York, pp 125–146
15. Dacey RG, Bassett JE (1987) Histaminergic vasodilatation of intra-cerebral arterioles in the rat. J Cereb Blood Flow Metabol 7: 327–331
16. Dacey RG, Bassett JE, Takayasu M (1988) Vasomotor responses of rat intracerebral arterioles to vasoactive intestinal peptide, substance p, neuropeptide Y, and bradykinin. J Cereb Blood Flow Metabol 8: 254–261
17. DeLey G, Weyne J, Demeester G, Leusen I (1982) Response of local blood flow in the caudate nucleus of the cat to intraventricular administration of histamine. Stroke 13: 499–504
18. Demopoulos HB, Flamm ES, Seligman ML, Mitamura JA, Ransohoff J (1979) Membrane perturbations in central nervous system injury: theoretical basis for free radical damage and a review of the experimental data. In: Popp AJ, Bourke RS, Nelson LR, Kimelberg HK (eds) Neural trauma. Raven, New York, pp 63–78
19. Domer FR, Boertje SB, Bing EG, Reddix I (1983) Histamine- and acetylcholine-induced changes in the permeability of the blood-brain barrier of normotensive and spontaneously hypertensive rats. Neuropharmacology 22: 615–619
20. Dux E, Joó F (1982) Effects of histamine on brain capillaries. Fine structural and immunohistochemical studies after intracarotid infusion. Exp Brain Res 47: 252–258
21. Edvinsson L, Hardebo JE, MacKenzie ET, Stewart M (1977) Dual action of serotonin on pial arterioles *in situ* and the effect of propranolol. Blood Vessels 14: 366–371
22. Edvinsson L, Gross PM, Mohamed A (1983) Characterization of histamine receptors in cat cerebral arteries *in vitro* and *in situ*. J Pharmacol Exp Ther 225: 168–175
23. Ellis EF, Heizer ML, Chao J (1987) The kallikrein-kinin system in normal an injured brain. J Cereb Blood Flow Metabol 7 [Suppl 1]: S 631
24. Ellis EF, Holt SA, Wei EP, Kontos HA (1988) Kinins induce abnormal vascular reactivity. Am J Physiol 255: H397–H400
25. Fenske A, Sinterhauf K, Reulen HJ (1976) The role of monoamines in the development of cold-induced edema. In: Pappius HM, Feindl W (eds) Dynamics of brain edema. Springer, Berlin Heidelberg New York, pp 150–154
26. Gabbiani G, Basonnel MC, Majno G (1970) Intra-arterial injections of histamine, serotonin, or bradykinin: a topographic study of vascular leakage. Proc Soc Exp Biol Med 135: 447–452
27. Gardiner M, Nilsson B, Rehncrona S, Siesjö BK (1981) Free fatty acids in the rat brain in moderate and severe hypoxia. J Neurochem 36: 1500–1505
28. Grome JJ, Harper AM (1983) The effects of serotonin on local cerebral blood flow. J Cereb Blood Flow Metabol 3: 71–77
29. Grome JJ, Harper AM (1985) Serotonin depression of local cerebral glucose utilization after monoamine oxidase inhibition. J Cereb Blood Flow Metabol 5: 473–475
30. Grome JJ, Harper AM (1986) Local cerebral glucose utilization following indoleamine- and piperazine containing 5-hydroxytryptamine agonists. J Neurochem 46: 117–124
31. Gross PM, Harper AM, Teasdale GM (1981) Cerebral circulation and histamine: 2. responses of pial veins and arterioles to receptor agonists. J Cereb Blood Flow Metabol 1: 219–225
32. Gross PM, Teasdale GM, Graham DI, Angerson WJ, Harper AM (1982) Intra-arterial histamine increases blood-brain transport in rats. Am J Physiol 243: H307–H317
33. Hardebo JE, Owman Ch, Wiklund L (1981) Influence of neurotransmitter monoamines and neurotoxic analogues on morphologic blood-brain barrier function. In: Cervos-Navarro J, Fritschka E (eds) Cerebral microcirculation and metabolism. Raven, New York, pp 177–180
34. Harper AM, MacKenzie ET (1977a) Cerebral circulatory and metabolic effects of 5-hydroxytryptamine in anaesthetized baboons. J Physiol (London) 271: 721–733
35. Harper AM, MacKenzie ET (1977b) Effects of 5-hydroxytryptamine on pial arteriolar calibre in anaesthetized cats. J Physiol (London) 271: 735–746
36. Hua XY, Dahlen SE, Lundberg JM, Hammarström S, Hedquist P (1985) Leukotrienes C_4, D_4 and E_4 cause widespread and extensive plasma extravasation in the guinea pig. Naunyn-Schmiedebergs Arch Pharmacol 330: 136–141
37. Joó F, Zücs A, Csanda E (1976) Metiamide-treatment of brain edema in animals exposed to ^{90}yttrium irradiation. J Pharm Pharmacol 28: 162– 163
38. Kamitani T, Little MH, Ellis EF (1985a) Evidence for a possible role of the brain kallikrein-kinin system in the modulation of the cerebral circulation. Circ Res 57: 545–552
39. Kamitani T, Little MH, Ellis EF (1985b) Effect of leukotrienes, 12-HETE, histamine, bradykinin, and 5-hydroxytryptamine on *in vivo* rabbit cerebral arteriolar diameter. J Cereb Blood Flow Metabol 5: 554–559
40. Kiwak KJ, Moskowitz MA, Levine L (1985) Leukotriene production in gerbil brain after ischemic insult, subarachnoid hemorrhage, and concussive injury. J Neurosurg 62: 865–869
41. Kontos HA (1985) Oxygen radicals in cerebral vascular injury. Circ Res 57: 508–516
42. Kontos HA, Wei EP (1986) Superoxide production in experimental brain injury. J Neurosurg 64: 803–807
43. Kontos HA, Wei EP, Povlishock JT, Dietrich WD, Magiera CJ, Ellis EP (1980) Cerebral arteriolar damage by arachidonic acid and prostaglandin G_2. Science 209: 1242–1245
44. Kontos HA, Wei EP, Povlishock JT, Christman CW (1984) Oxygen radicals mediate the cerebral arteriolar dilatation from arachidonate and bradykinin in cats. Circ Res 55: 295–303
45. Long DM (ed) (1990) Brain edema, pathogenesis, imaging, and therapy. Adv neurol 52. Raven, New York
46. Maier-Hauff K, Baethmann AJ, Lange M, Schürer L, Unterberg A (1984) The kallikrein-kinin system as mediator in vasogenic brain edema. Part 2: studies on kinin formation in focal and perifocal brain tissue. J Neurosurg 61: 97–106

47. Martins AN, Doyle TF, Wright SJ, Bass BG (1980) Response of cerebral circulation to topical histamine. Stroke 11: 469–476
48. Mohanty S, Dey PK, Sharma HS, Singh S, Chansouria JPN, Olsson Y (1989) Role of histamine in traumatic brain edema. An experimental study in the rat. J Neurolog Sci 90: 87–97
49. Moskowitz MA, Kiwak KJ, Hekimian K, Levine L (1984) Synthesis of compounds with properties of leukotrienes C_4 and D_4 in gerbil brains after ischemia and reperfusion. Science 224: 886–889
50. Olesen SP (1985) A calcium-dependent reversible increase in micro-vessels in frog brain induced by serotonin. J Physiol (London) 361: 103–113
51. Olesen SP (1987) Free oxygen radicals decrease electrical resistance of microvascular endothelium in brain. Acta Physiol Scand 129: 181–188
52. Olesen SP, Crone C (1986) Substances that rapidly augment ionic conductance of endothelium in cerebral venules. Acta Physiol Scand 127: 233–241
53. Orr EL (1988) Cryogenic lesions induce a mast cell-dependent increase in cerebral histamine levels in the mouse. Neurochem Pathol 8: 43–51
54. Pappius H (1986) Mechanisms underlying functional disturbances in traumatized brain. In: Mchedelishvilli GJ, Cervos-Navarro J, Hossmann KA, Klatzo I (eds) Brain edema. A pathogenic analysis. Akademiai Kiado, Budapest, pp 282–286
55. Pappius HM, Wolfe LS (1983a) Involvement of serotonin and catecholamines in functional depression of traumatized brain. J Cereb Blood Flow Metabol 3 [Suppl 1]: S226–S227
56. Pappius HM, Wolfe LS (1983b) Functional disturbances in brain following injury: Search for underlying mechanisms. Neurochem Res 8: 63–72
57. Pappius HM, Dadoun R (1987) Effects of injury on the indoleamines in cerebral cortex. J Neurochem 49: 321–325
58. Rosenblum WI (1983) Effects of free radicals generation on mouse pial arterioles: probable role of hydroxyl radicals. Am J Physiol 245: H139–H142
59. Rosenblum WI (1985) Constricting effect of leukotrienes on cerebral arterioles of mice. Stroke 16: 262–263
60. Saria A, Lundberg JM, Skofitsch G, Lembeck F (1983) Vascular protein leakage in various tissues induced by substance P, capsaicin, bradykinin, serotonin, histamine and by antigen challenge. Naunyn-Schmiedebergs Arch Pharmacol 324: 212–218
61. Schilling L, Ksoll E, Wahl M (1987a) Effects of histamine on vasomotor response and permeability of extraparenchymal cerebral vessels. J Cereb Blood Flow Metabol 7 [Suppl]: S 506
62. Schilling L, Ksoll E, Wahl M (1987b) Increase of pial vessel diameter and permeability during cortical superfusion with histamine. Pflügers Arch 408 [Suppl 1]: R20
63. Schilling L, Ksoll E, Wahl M (1987c) Vasomotor and permeability effects of histamine in cerebral vessels. Int J Microcirc Clin Exp 6: 70
64. Sharma HS, Dey PK (1986) Probable involvement of 5-hydroxytryptamine in increased permeability of blood-brain barrier under heat stress in young rats. Neuropharmacology 25: 161–167
65. Siesjö BK (1981) Cell damage in the brain: a speculative synthesis. J Cereb Blood Flow Metabol 1: 155–185
66. Siesjö BK, Bendek G, Koide T, Westerberg E, Wieloch T (1985) Influence of acidosis on lipid peroxidation in brain tissues *in vitro*. J Cereb Blood Flow Metabol 5: 253–258
67. Solomon LS (1974) Failure of buffered 5-hydroxytryptamine to increase brain capillary permeability to albumin in monkeys. J Neurosurg 40: 717–725
68. Suzuki O, Yagi K (1974) Formation of lipoperoxide in brain edema induced by cold injury. Experientia 30: 248
69. Unterberg A, Baethmann A (1984) The kallikrein-kinin system as mediator on vasogenic brain edema. Part 1: cerebral exposure to bradykinin and plasma. J Neurosurg 61: 87–96
70. Unterberg A, Wahl M, Baethmann A (1983) Effects of bradykinin on cerebrovascular permeability and resistance. J Cereb Blood Flow Metabol 3 [Suppl 1]: S234–S235
71. Unterberg A, Wahl M, Baethmann A (1984) Effects of bradykinin on permeability and diameter of pial vessels *in vivo*. J Cereb Blood Flow Metabol 4: 574–585
72. Unterberg A, Wahl M, Baethmann A (1985) Arachidonic acid induces opening of the blood-brain barrier. Int J Microcirc Clin Exp 4: 302
73. Unterberg A, Dautermann C, Baethmann A, Müller-Esterl W (1986) The kallikrein-kinin system as mediator in vasogenic brain edema. J Neurosurg 64: 269–276
74. Unterberg A, Wahl M, Hammersen F, Baethmann A (1987a) Permeability and vasomotor response of cerebral vessels during exposure to arachidonic acid. Acta Neuropathol 73: 209–219
75. Unterberg A, Baethmann A, Wahl M, Schürer L, Marmarou A (1987b) New aspects in the formation of vasogenic brain edema. In: Baethmann A, Messmer K (eds) Surgical research. Recent concepts and results. Springer, Berlin Heidelberg New York Tokyo, pp 3–8
76. Unterberg A, Schmidt W, Polk T, Wahl M, Ellis E, Marmarou A, Baethmann A (1987c) Evidence against leukotrienes as mediators of brain edema. J Cereb Blood Flow Metabol 7 [Suppl 1]: S625
77. Unterberg A, Wahl M, Baethmann A (1988) Effects of free radicals on permeability and vasomotor response of cerebral vessels. Acta Neuropathol 76: 238–244
78. Unterberg A, Polk T, Ellis E, Marmarou A (1989a) Enhancement of infusion induced brain edema by mediator compounds. In: Long DM (ed) Adv Neurol 52 (1990). Raven, New York, pp 355–358
79. Unterberg A, Schmidt W, Wahl M, Baethmann A (1989b) Role of leukotrienes as mediator compounds in brain edema: In: Long DM (ed) Adv neurol 52 (1990). Raven, New York, pp 211–214
80. Wahl M (1982) The effect of opiate-like substances and bradykinin on cerebrovascular resistance in cats. In: Heistad DD, Marcus ML (eds) Cerebral blood flow. Effects of nerves and neurotransmitters. Elsevier, New York, pp 235–241
81. Wahl M, Kuschinsky W (1976) The dilating effect of histamine on pial arteries of cats and its mediation by H_2-receptors. Circ Res 44: 161–165
82. Wahl M, Unterberg A, Baethmann A (1983a) The effect of bradykinin on the cerebrovascular resistance and blood brain barrier permeability. In: Hossmann KA, Klatzo I (eds) Cerebrovascular transport mechanisms. Acta Neuropathol [Suppl] 8: 132–133
83. Wahl M, Young AR, Edvinsson L, Wagner F (1983b) Effects of bradykinin on pial arteries and arterioles *in vitro* and *in situ*. J Cereb Blood Flow Metabol 3: 231–237
84. Wahl M, Unterberg A, Baethmann A (1985a) Intravital fluorescence microscopy for the study of blood-brain-barrier function. Int J Microcirc Clin Exp 4: 3–18
85. Wahl M, Unterberg A, Baethmann A (1985b) Effects of arachidonic acid on blood-brain barrier function. In: Dietz H, Brock M, Klinger M (eds) Extra- intercranial vascular anastomoses. Microsurgery at the edge of the tentorium. Adv neurosurg 13. Springer, Berlin Heidelberg New York Tokyo, pp 323–325
86. Wahl M, Unterberg A, Whalley ET, Baethmann A, Young AR, Edvinsson L, Wagner FFW (1986a) Cerebrovascular effects of bradykinin. In: Owman C, Hardebo JE (eds) Neural regulation of brain circulation. Elsevier, New York, pp 419–430

87. Wahl M, Unterberg A, Baethmann A (1986b) The effects of free radicals and leukotrienes on blood-brain-barrier function. Int J Microcirc Clin Exp 5: 93
88. Wahl M, Unterberg A, Whalley ET, Baethmann A, Young AR, Edvinsson L, Wagner F (1987) Effect of bradykinin on cerebral hemodynamics and blood-brain-barrier function. In: Edvinsson L, McCulloch J (eds) Peptidergic mechanisms in the cerebral circulation. Horwood, Chichester, pp 166–190
89. Wahl M, Unterberg A, Baethmann A, Schilling L (1988) Mediators of blood-brain-barrier dysfunction and formation of vasogenic brain edema. J Cereb Blood Flow Metabol 8: 621–634
90. Wakai S, Aritake K, Asano T, Takakura K (1982) Selective destruction of the outer leaflet of the capillary endothelial membrane after intracerebral injection of arachidonic acid in the rat. Acta Neuropathol 58: 303–306
91. Wei EP, Ellis EF, Kontos HA (1980) Role of prostaglandins in pial arteriolar response to CO_2 and hypoxia. Am J Physiol 238: H226–H230
92. Wei EP, Christman CW, Kontos HA, Povlishock JT (1985) Effects of oxygen radicals on cerebral arterioles. Am J Physiol 248: H157–H162
93. Wei EP, Ellison MD, Kontos HA, Povlishock JT (1986) O_2 radicals in arachidonate-induced increased blood-brain barrier permeability to proteins. Am J Physiol 251: H693–H699
94. Westergaard E (1975) The effect of serotonin, norepinephrine and cyclic AMP on the blood-brain barrier. J Ultrastruct Res 50: 383
95. Westergaard E (1977) The blood-brain barrier to horseradish peroxidase under normal and experimental conditions. Acta Neuropathol 39: 181–188
96. Whalley ET, Wahl M (1983) Analysis of bradykinin receptor mediating relaxation of cat cerebral arteries *in vivo* and *in vitro*. Naunyn-Schmiedebergs Arch Pharmacol 323: 66–71

Correspondence and Reprints: M. Wahl, Department of Physiology, Ludwig-Maximilians University, Pettenkoferstrasse 12, D-W-8000 München 2, Federal Republic of Germany.

Acta Neurochir (1993) [Suppl] 57: 73–79

Glutamate Receptor Antagonists in Experimental Focal Cerebral Ischaemia

J. McCulloch, E. Ozyurt, C. Kun Park, D. G. Nehls, G. M. Teasdale, and **D. I. Graham**

Wellcome Neuroscience Group, Wellcome Surgical Institute, and Hugh Fraser Neuroscience Laboratories, University of Glasgow, Glasgow, U.K..

Abstract

Excessive activation of the N-methyl-D-aspartate (NMDA) subtype of glutamate receptor has been implicated in the sequence of neurochemical events in cerebral ischaemia that results in irreversible neuronal damage. The effects of the NMDA antagonist MK-801 upon the amount of ischaemic brain damage has been assessed quantitatively in a cat and in a rat model of focal cerebral ischaemia.

In chloralose-anaesthetised cats, focal cerebral ischaemia was produced by permanent occlusion of one middle cerebral artery (MCA) and the animal sacrificed 6 hours later. Pretreatment with the non-competitive NMDA antagonist, MK-801 (5 mg/kg, i.v.) reduced significantly the volume of ischaemic damage in the cerebral cortex by 57% compared to vehicle-treated cats. A similar degree of neuroprotection could be demonstrated in the cat MCA occlusion model if treatment with MK-801 was initiated 2 hours after the induction of ischaemia.

In halothane-anaesthetised rats, focal cerebral ischaemia was produced by permanent MCA occlusion and the animals sacrificed 3 hours later. Pretreatment with MK-801 (0.5 mg/kg, i.v.) reduced the volume of ischaemic damage in the cerebral cortex by 38%; treatment with MK-801 initiated 30 minutes after MCA occlusion was equally effective in reducing cortical damage.

In contrast to calcium entry blockers such as nimodipine in the rat MCA occlusion model, the improved histopathological outcome with MK-801 is not associated with improvement in cerebral tissue perfusion to the ischaemic tissue. The increasing evidence that NMDA receptor antagonists are beneficial in experimental focal cerebral ischaemia is reviewed.

Keywords: Focal cerebral ischemia; infarct formation; NMDA-receptor; MK-801.

Introduction

The goals in the treatment of cerebral ischaemia have traditionally been to prevent the breakdown of the energy producing metabolic processes and to preserve the membrane polarisation of neurones. A wide range of therapeutic strategies have been used in man and experimental animals, with the aim of improving blood flow and substrate delivery to the ischaemic brain, but few have shown convincing clinical benefit. Indeed, there has long been concern that increased blood flow to ischaemic brain tissue may have adverse consequences such as increased cerebral oedema (Hossmann, 1982).

The concept that blockade of excitatory amino acid receptors attenuates the transmembrane ionic fluxes which lead to neuronal death (Rothman and Olney, 1986) provides a therapeutic strategy which does not depend upon improvement in cerebral blood flow. High concentrations of glutamate are neurotoxic (Rothman and Olney, 1986), and in cerebral ischaemia, there is a massive release of glutamate into the extracellular compartment (Benveniste *et al.*, 1984). The administration of glutamate receptor antagonists can markedly reduce the amount of ischaemic brain damage, in the rodent hippocampus (Simon *et al.*, 1984; Gill *et al.*, 1987). However, crucial information is unavailable in 3 areas; firstly, the extent to which glutamate receptor antagonists can protect the cerebral cortex from ischaemia; secondly, data from gyrencephalic species; thirdly, studies in which systemic variables which have an impact on the amount of ischaemic damage (blood pressure, plasma glucose, body temperature) has been monitored throughout the post-ischaemic survival period.

The actions of glutamate in the mammalian central nervous system, are mediated via three pharmacologically distinct subtypes of receptor, the nomenclature of which reflects the preferred agonist at each receptor subtype viz. the quisqualate-, kainate-, or AMPA-, and N-methyl-D-aspartate-(NMDA)-preferring receptor (Greenamyre *et al.*, 1985). Excessive activation of the NMDA receptor has been implicated in the pathophysiology of cerebral ischaemia via promotion of calcium entry into neurones (Rothman and Olney, 1986). Recently a number of agents that are selective NMDA receptor antagonists and that cross the blood-brain barrier have been identified (Wong *et al.* 1986; Davies *et al.*, 1987; Carter *et al.*, 1988). In the present report, the effects of pre-treatment with one such NMDA antagonist, MK-801 (Wong *et al.*, 1986) upon the amount of early ischaemic damage after middle cerebral artery occlusion in cats and rats are described and contrasted with those of other agents in similar models of focal cerebral ischaemia.

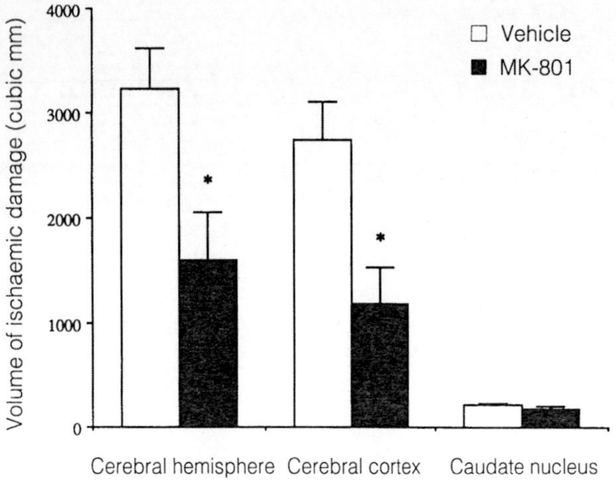

Fig. 1. Focal cerebral ischaemia in the cat: effect of MK-801 pretreatment on the volume of ischaemic brain damage. MK-801 (5 mg/kg) was administered intravenously 30 minutes before MCA occlusion. The animal was sacrificed 6 hours after MCA occlusion. Data are presented as mean + SEM (n = 9 in each group). *$p < 0.01$ (Student's t-test)

Focal Cerebral Ischaemia in the Cat

Methods

The investigations were carried out in 33 cats. The cats were anaesthetised initially with thiopentone (25 mg/kg i.v.) intubated, and connected to a positive pressure ventilator delivering nitrous oxide and oxygen in an open circuit. Polyethylene catheters were inserted into one femoral vein and artery for the administration of drugs and the continuous monitoring of arterial blood pressure, respectively. Anaesthesia was maintained throughout the course of the investigation with chloralose (60 mg/kg, i.v., supplemented as necessary to prevent the return of the corneal reflex). Throughout the experimental period, the animals were maintained normotensive (mean arterial blood pressure (MABP) greater than 80mmHg); normocapnic ($PaCO_2$ close to 32 mmHg by adjusting the stroke volume of the respirator). The cats were maintained normothermic (rectal temperature 37°C) by means of a heating blanket with feedback control via a rectal thermometer. The left middle cerebral artery (MCA) was occluded via a transorbital approach. With microsurgical techniques, the left orbit was exenterated and the optic foramen and optical fissure were enlarged with a dental drill to expose the dura mater overlying the MCA close to its origin. Under the operating microscope the trunk of the MCA and all visible

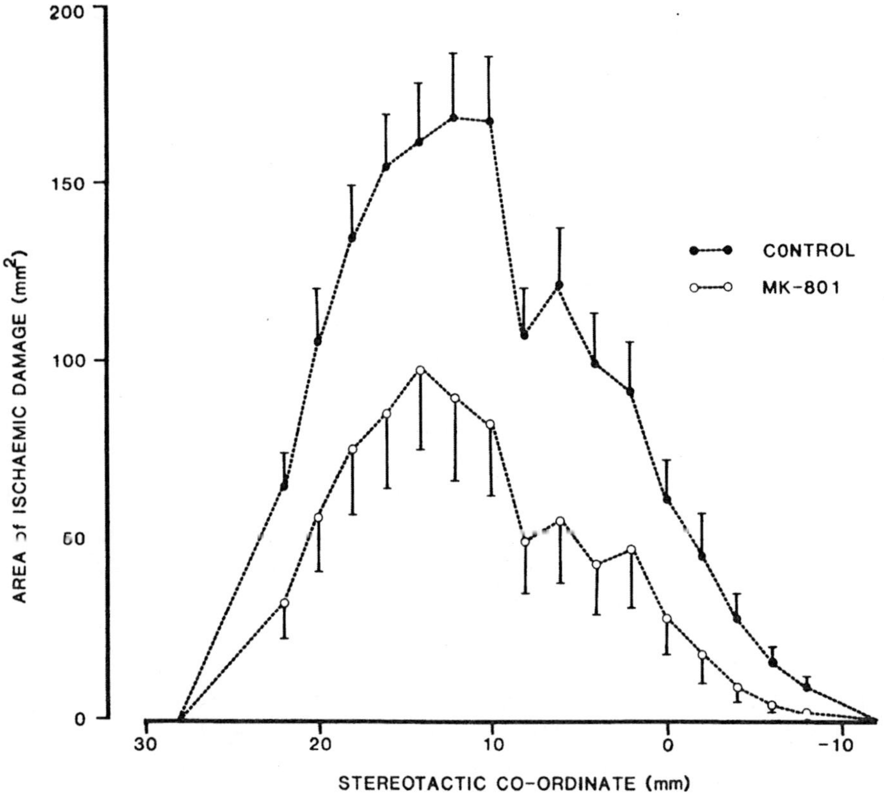

Fig. 2. Focal cerebral ischaemia in the cat: Cerebral cortex. The effect of pretreatment with MK-801 (5 mg/kg i.v.) on the area of ischaemic damage in the cerebral cortex at 16 defined coronal planes. There are significant differences ($p < 0.05$) between vehicle and MK-801-treated groups at all coronal planes examined. Data are presented as mean ± SEM (n = 9 in each group) (from Ozyurt *et al.*, 1988 with permission)

branches of the lenticulostriate arteries were coagulated with a bipolar diathermy and divided with microscissors. Six hours after the occlusion of the middle cerebral artery, the cat was killed by transcardiac perfusion fixation with (40% formaldehyde, glacial acetic acid and methanol [FAM] in a ratio of 1:1:8). After perfusion, the cats were decapitated and the head was stored in fixative at 4°C for at least 12 hours; the brain was then removed; approximately 140 sections were prepared from each forebrain and stained with hemalaun and eosin and by a method combining cresyl violet and Luxol fast blue. Those sections that corresponded most closely to 16 predetermined coronal planes were examined by conventional light microscopy without prior knowledge of the animal's history. Any abnormalities were charted on the diagrams drawn to scale. The area of ischaemic necrosis in the cerebral hemisphere, cerebral cortex, and caudate nucleus was determined from the diagrams at each of the 16 coronal planes using a computer-based image analysis system and the volume of ischaemic tissue in the 3 regions of interest calculated.

Two separate investigations were performed. In the first, the effects of pretreatment with MK-801 (5 mg/kg, i.v., 30 minutes before MCA occlusion were examined) (Ozyurt *et al.*, 1988). In the second series the effects of postocclusion treatment with MK-801 (5 mg/kg, 2 hours after MCA occlusion) were studied (Park *et al.*, 1988b).

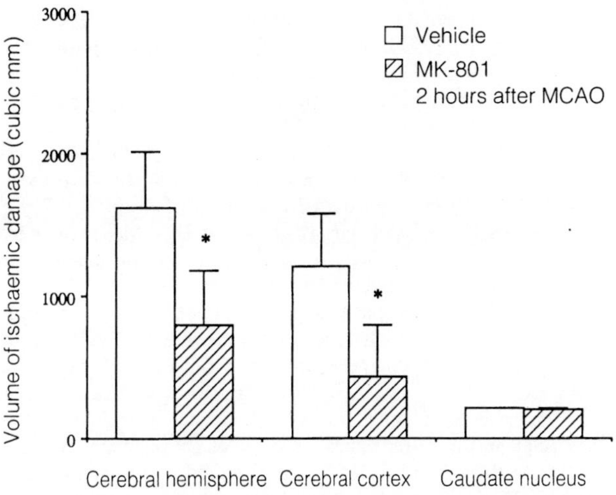

Fig. 3. Focal cerebral ischaemia in the cat: effect of MK-801 treatment after induction of ischaemia, on the volume of ischaemic brain damage. MK–801 (5 mg/kg) was administered 2 hours after MCA occlusion. The animal was sacrificed 6 hours after MCA occlusion. Data are presented as mean + SEM (n = 7 or 8 in each group). *$p < 0.05$ (Mann Whitney U-test)

Ischaemic Brain Damage: Effects of MK-801 in the Cat

All brains were judged to be well perfusion-fixed as evidenced by good neuronal morphology, the absence of intravascular blood, and the lack of cytological artifacts such as "dark cells" or "hydropic cells". Coronal sectioning of the brains revealed the interior of the brains to be well fixed other than in the deeper portions of the left hemisphere where zones of pink discolouration could be seen. Ischaemic damage was observed only within the territory of the occluded MCA, i.e., in the dorsolateral cortex and in the neostriatum. These areas showed the morphological characteristics of early infarction –microvacuolation, shrinkage and triangulation of nucleus and cycoplasm, increased basophilia of cytoplasm, and perineuronal swollen astrocytes. The lesion was sharply delineated with the boundary between the ischaemic area and normal brain being less than 100 µm. Within this narrow boundary, some of the nerve cells appeared normal, whereas others displayed the microscopic features of the ischaemic cell process.

Pretreatment with MK-801 (5 mg/kg) significantly reduced ($p < 0.01$) the volume of ischaemic damage in the cerebral hemisphere (reduced by 50% compared to untreated cats) and cerebral cortex (reduced by 57%) which resulted from occlusion of the MCA (Fig. 1). In the cerebral hemisphere and, in cerebral cortex, MK-801 reduced significantly the area of ischaemic damage in every coronal plane studied. The magnitude of the reductions in the ischaemic areas with MK-801 was similar in the core of MCA territory (stereotactic plane, anterior 12 mm) and at the extremities of MCA territory (Fig. 2). Pretreatment with MK-801 did not significantly modify the volume of ischaemic damage in the caudate nucleus or area of damage in any coronal plane (Fig. 1).

Postocclusion treatment with MK-801 (5 mg/kg) initiated 2 hours after MCA occlusion was similarly effective in reducing the volume of ischaemic damage in the cerebral hemisphere (damage reduced by 51%) and cerebral cortex (damage reduced by 64%). Ischaemic damage in the caudate nucleus was minimally influenced by postocclusion treatment with MK-801 (Fig. 3).

The dramatic effects of MK-801 can be readily established from the number of cats with small ischaemic lesions (e.g. lesions less than 5% of the hemisphere volume). In the control groups from the 2 separate series, 3 of the 17 cats had small ischaemic lesions whereas in the MK-801-treated groups, both pretreatment and postocclusion treatment groups, 10 of the 16 cats had small ischaemic lesions.

Focal Cerebral Ischaemia in the Rat

Methods

The neuropathological investigations with MK-801 were carried out in 20 male Sprague Dawley rats, anaesthetised initially with 2%

halothane and thereafter ventilated mechanically with nitrous oxide-oxygen (70%/30%) containing 0.5 to 1% halothane. A tracheostomy was performed and polyethylene catheters were inserted into one femoral artery and one femoral vein to allow the continuous monitoring of blood pressure, the repeated sampling of arterial blood, and the administration of drugs. Body temperature was monitored by a rectal thermometer, and the animals were maintained normothermic by external heating. Normocapnia and adequate arterial oxygenation ($PaO_2 > 90$ mm Hg) was maintained throughout the experiment. All animals underwent subtemporal craniectomy and exposure of the main trunk of the left MCA. The exposed artery was occluded with microbipolar coagulation from its origin to the point where it crosses the inferior cerebral vein. Three hours after MCA occlusion, the rats were perfusion fixed with FAM and areas of early cerebral infarction were delineated at eight preselected coronal levels in a manner analagous to that described above for the cat. The effects of pretreatment and post-occlusion treated with MK-801 were studied. In the pretreatment group, MK-801 (0.5 mg/kg) was administered intravenously 30 minutes before MCA occlusion, and in the postocclusion treatment group, MK-801 (0.5 mg/kg) was administered 30 minutes after occlusion (Park *et al.*, 1988a).

The effects of nimodipine (1 $\mu g\ kg^{-1}\ min^{-1}$) upon the amount of ischaemic brain damage after MCA occlusion were studied in an essentially similar manner in rats. In the first series, the effect of nimodipine infusion initiated 30 minutes before MCA occlusion and maintained until sacrifice were studied in 8 drug-treated and 8 vehicle-treated animals. In the second series, treatment was initiated 5 mins. after MCA occlusion and maintained until sacrifice in 5 drug-treated and 5 vehicle-treated rats (Mohamed *et al.*, 1985; Gotoh *et al.*, 1986).

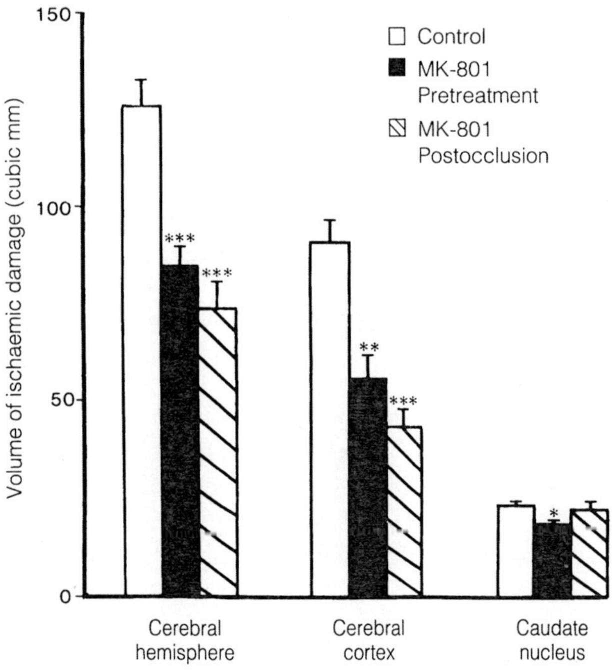

Fig. 4. Focal cerebral ischaemia in the rat: effect of MK-801 upon the volume of ischaemic brain damage. MK-801 (0.5 mg/kg i.v.) was administered either 30 mins. before or 30 mins. after MCA occlusion. The animal was sacrificed 3 hours after MCA occlusion. Data are presented as mean + SEM (n = 6–8 in each group). *$p < 0.05$, **$p < 0.01$, ***$p < 0.001$ (Student's t-test with Bonferroni correction) (from Park *et al.*, 1988b, with permission)

Ischaemic Brain Damage: Effects of MK-801 in the Rat

Pretreatment with MK-801 (0.5 mg/kg) 30 minutes before the occlusion of the MCA significantly reduced the volume of ischaemic brain damage in the cerebral cortex and hemisphere (volumes of ischaemic tissue reduced 38% and 33% compared with untreated controls, respectively; Fig. 4). There was a smaller, though significant, reduction (18% compared with untreated controls) in the volume of ischaemic damage in the caudate nucleus in animals treated with MK-801 prior to MCA occlusion. The areas of ischaemic damage in the cerebral cortex and hemisphere, particularly in the more caudal coronal planes, in the animals with MK-801 pretreatment were smaller than those in the untreated control group.

The administration of MK-801 (0.5 mg/kg) 30 minutes after the occlusion of the MCA significantly reduced the volume of ischaemic brain damage in the cerebral cortex and hemisphere (52% and 41%, respectively) from that observed in untreated control animals. Ischaemic damage in the caudate nucleus was minimally influenced by the administration of MK-801 after MCA occlusion.

Ischaemic Brain Damage: Effect of Nimodipine in the Rat

The effects of the calcium entry blocker nimodipine have been tested in the same model of focal cerebral ischaemia. The nimodipine treatment schedule (continuous intravenous infusion 1 μg/kg/min) had been selected for maximum cerebrovascular effects with minimal hypotension (Mohamed *et al.*, 1984). Pretreatment-with nimodipine (infusion initiated 30 minutes before MCA occlusion) effected a reduction in the volume of ischaemic brain damage in the cerebral hemisphere of 34% which just attained statistical significance ($p < 0.05$ one tailed t-test). In contrast, when the nimodipine infusion was initiated 5 minutes after MCA occlusion, it effected a non-significant increase (15%) in the volume of ischaemic tissue.

Cerebral Blood Flow After MCA Occlusion in the Rat: Effects of MK-801 and Nimodipine

The outcome after an ischaemic insult may be improved either by ensuring that CBF does not fall below a critical threshold and/or by increasing the tolerance of neuronal cells to ischaemia. Nimodipine and MK-

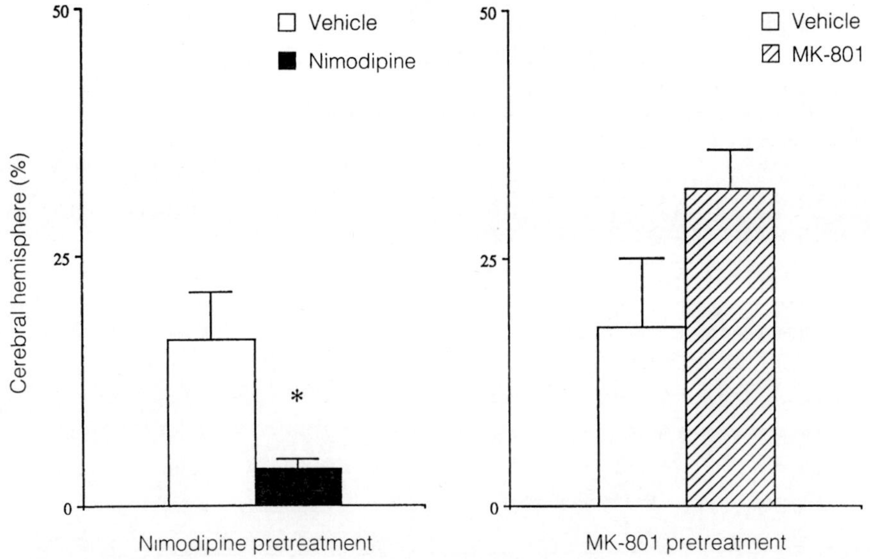

Fig. 5. Anti-ischaemic drugs and cerebral blood flow after MCA occlusion in the rat. Data are expressed as the proportion of the cerebral hemisphere with very low blood flow (i.e. CBF less than 30 ml 100 g^{-1} min^{-1}). Analysis was carried out in autoradiograms at coronal plane ant. 4.4 mm. This mid thalamic plane is outside the densely ischaemic core and corresponds to a level where both agents are neuroprotective. Nimodipine (1 μg kg^{-1} min^{-1}) markedly reduces the amount of tissue with low CBF after MCA occlusion (*$p < 0.05$). Data of Gotoh *et al.*, 1986. MK-801 (0.5 mg/kg) has no significant effect on the amount of tissue with low CBF after MCA occlusion although it is markedly neuroprotective at this anatomical locus

801 can each markedly alter CBF in non-ischaemic tissue via a direct action on the vasculature (e.g. nimodipine) (Mohamed *et al.*, 1985) or indirectly via alterations in cerebral oxidative metabolism (e.g. MK-801) (Nehls *et al.*, 1988). This section, however, will focus exclusively on the effects of nimodipine and MK-801 in the hemisphere ipsilateral to the occluded MCA. Local CBF was assessed with ^{14}C-iodoantipyrine autoradiography, with densitometric analysis of the autoradiograms performed both at neuroanatomically defined loci and with frequency distribution analysis at selected coronal planes.

In brain regions in which blood flow is derived normally from the occluded MCA (e.g. neocortex), pretreatment with MK-801 had no effect on the low levels of CBF which are observed after MCA occlusion, either in the core or in the periphery of MCA territory (e.g. parietal cortex, vehicle, 32 ± 7 ml 100 g^{-1} min^{-1}; MK-801 35 ± 9 ml 100 g^{-1} min^{-1}). In contrast with nimodipine pretreatment, higher levels of CBF relative to vehicle controls were observed in regions in which blood flow is normally derived from the occluded MCA (e.g. parietal cortex vehicle 43 ± 5 ml 100 g^{-1} min^{-1}; nimodipine 80 ± 6 ml 100 g^{-1} min^{-1} $p < 0.05$). The effects of nimodipine were more marked towards the periphery of MCA territory, where the tissue saving effects of nimodipine are noted. The improvement in CBF with nimodipine could be readily demonstrated from the amount of cerebral tissue with low CBF (i.e. CBF less than 30 ml 100 g^{-1} min^{-1}) (Fig. 5). Thus, with nimodipine, its anti-ischaemic effects are associated with improvement in blood flow to the margins of the ischaemic tissue, whereas with MK-801, its anti-ischaemic actions are not correlated with any improvement in blood flow to the ischaemic tissue.

Discussion

Over the years, many therapeutic strategies have been proposed for cerebral ischaemia but few have ultimately found a place in the clinical arena. Only now are convincing data emerging for the use of calcium entry blockers in the treatment of cerebral ischaemia, albeit in highly specific clinical circumstances (Petruk *et al.*, 1988; Pickard *et al.*, 1989). The clinical efficacy of calcium antagonists in carefully conducted, large clinical trials is in contrast to absence of a consistent pattern of response of these agents in the preclinical studies. There was controversy, not just concerning the efficacy of calcium antagonists in various models of experimental ischaemia but also, even where benefit was demonstrable, the extent to which improved CBF was responsible (Mohamed *et al.*, 1985; Gotoh *et al.*, 1986; Steen *et al.*, 1983; Barnett *et al.*, 1986; Sakabe *et al.*, 1986; Berger *et al.*, 1988 among many others).

Table 1. *Glutamate Receptor Antagonists in Focal Cerebral Ischaemia*

Species	Model	Survival	Agent	Neuroprotection	References
Cat	MCA occlusion	6 hours	MK-801	yes	Ozyurt *et al.*, 1988 Park *et al.*, 1988a
Cat	MCA occlusion	4 days	SL-82-715 Ifenprodil	yes	Gotti *et al.*, 1988
Cat	MCA occlusion + reperfusion	6 hours	MK-801	yes	Hamar *et al.*, 1988
Rat	MCA occlusion	1 day	Kynurenic acid	yes	Germano *et al.*, 1987
Rat	MCA occlusion	3 hours	MK-801	yes	Park *et al.*, 1988b
Rat	MCA occlusion	48 hours	MK-801	yes	Park, 1988
Rat	MCA occlusion	5 days	MK-801	yes	Coyle, 1989
Rat	MCA occlusion	2 days	TCP	yes	Gotti *et al.*, 1988
Rat	MCA occlusion	2 days	SL-82-715 Ifenprodil	yes	Gotti *et al.*, 1988
Rabbit	ACA, IC occlusion + reperfusion	5 hours	Dextromethorphan, Dextrorphan	yes	Steinberg *et al.*, 1988

MCA Middle cerebral artery; *ACA* anterior cerebral artery; *IC* internal carotid artery.

For glutamate receptor antagonists, an extensive body of data is becoming available in which the efficacy of these drugs in reducing the amount of histopathological damage in experimental models of focal ischaemia has been clearly demonstrated (Table 1). Neuroprotection which is observed with glutamate receptor antagonists is robust; it is not dependent upon the species used; or the agent studied; or the treatment schedule; or upon the survival period whether acute (hours) or chronic (days); or even the group conducting the research. For no other genre of compound has such a consistent view of efficacy emerged from investigations in animal models of focal cerebral ischaemia. In models of global cerebral ischaemia, the evidence of neuroprotection with NMDA antagonists is less consistent with both positive and negative reports available (Gill *et al.*, 1987; Lanier *et al.*, 1988 inter alia). However, while global ischaemia may provide a model of cardiac arrest, it is models of focal cerebral ischaemia which are the most pertinent for the understanding of stroke in man. In conclusion, the present review highlights the potential of NMDA receptor antagonists in the treatment of cerebral ischaemia in man, particularly as a considerable amount of vulnerable tissue can be saved even when the treatment with NMDA receptor antagonists is initiated several hours after the initial ischaemic insult.

Acknowledgements

These investigations were supported by the Wellcome Trust.

References

1. Barnett GH, Bose B, Little JR, Jones SC, Friel HT (1986) Effects of nimodipine on acute focal cerebral ischemia. Stroke 17: 884–890
2. Benveniste H, Drejer J, Schousboe A, Diemer NH (1984) Elevation of the extracellular concentrations of glutamate and aspartate in rat hippocampus during transient cerebral ischemia monitored by intracerebral microdialysis. J Neurochem 43: 1369–1374
3. Berger L, Hakim AM (1988) Calcium channel blockers correct acidosis in ischemic rat brain without altering cerebral blood flow. Stroke 19: 1257–1261
4. Carter C, Benavides J, Legendre P, Vincent JD, Noel F, Thuret F, Lloyd KG, Arbilla S, Zivkovic B, MacKenzie ET, Scatton B, Langer SZ (1988) Ifenprodil and SL 82.0715 as cerebral anti-ischaemic agents II Evidence for NMDA receptor antagonist properties. J Pharm Expt Ther 247: 1222–1232
5. Coyle P (1989) Altered cerebral collaterals and protection from infarction. In: Hartmann A, Kuschinsky W (eds) Cerebral ischemia and calcium metabolism. Springer, Berlin Heidelberg New York Tokyo, pp 69–78
6. Davies J, Evans RH, Herrling PL, Jones AW, Olverman HJ, Pook P, Watkins JC (1986) CPP, a new potent and selective NMDA antagonist Depression of central neuron resposes, affinity for [^{3}H] D-AP5 binding sites on brain membranes and anticonvulsant activity. Brain Res 382: 169–173
7. Germano IM, Pitts LH, Meldrum BS, Bartkowski HM, Simon RP (1987) Kynurenate inhibition of cell excitation decreases stroke size and deficits, Ann Neurol 22: 730–734
8. Gill R, Foster AC, Woodruff GN (1987) Systemic administration of MK-801 protects against ischemia-induced hippocampal neurodegeneration in the gerbil. J Neurosci 7: 3343–3349
9. Gotoh O, Mohamed AA, McCulloch J, Graham DI, Harper AM, Teasdale GM (1986) Nimodipine and the haemodynamic and histopathological consequences of middle cerebral artery occlusion in the rat. J Cereb Blood Flow Metabol 6: 321–331
10. Gotti B, Duverger D, Bertin J, Carter C, Dupoint R, Frost J, Gaudilliere B, MacKenzie ET, Rousseau J, Scatton B, Wick A (1988). Ifenprodil and SL 82.0715 as cerebral anti-ischaemic

agents I. Evidence for efficacy in models of focal cerebral ischaemia. J Pharm Exp Ther 247: 1211–1221
11. Greenamyre JT, Olson JMM, Penney JB (Jr), Young AB (1985) Autoradiographic characterization of N-methyl-D-Aspartate-Quisqualate- and Kainate-Sensitive glutamate binding sites. J Pharmacol Exp Ther 233: 254–263
12. Hamar J, Reivich M, Greenberg JH, Shearmen GT (1988). Tissue blood flow and metabolic changes of the brain in ischaemia and reperfusion: effect of a glutamate antagonist (MK-801) treatment. Abstract presented to 11th Annual Conference on Shock. Fontana, Wisconsin 5–8 June, 1988
13. Hossmann KA (1982) Treatment of experimental cerebral ischaemia. J Cereb Blood Flow Metabol 2: 275–297
14. Lanier WL, Perkins WJ, Ruud B, Milde JH, Michenfelder JD (1988) Effect of the excitatory amino acid antagonist MK-801 on neurologic function following complete cerebral ischemia in primates. Anesthesiology 69: A846
15. Mohamed AA, Gotoh O, Graham DI, Osborne KA, McCulloch J, Mendelow AD, Teasdale GM, Harper AM (1985) Effect of pretreatment with the calcium antagonist nimodipine on local cerebral blood flow and histopathology after middle cerebral artery occlusion. Ann Neurol 18: 705–711
16. Nehls DG, Park CK, McCulloch J (1989) Cerebral circulatory effects of dizocilpine (MK-801) in the rat. J Cereb Blood Flow Metabol 9 [Suppl I]: S376
17. Ozyurt E, Graham DI, Woodruff GN, McCulloch J (1988) Protective effect of the glutamate antagonist, MK-801 in focal cerebral ischemia in the cat. J Cereb Blood Flow Metabol 8: 138–143
18. Park CK, Nehls DG, Graham DI, Teasdale GM, McCulloch J (1988a) The glutamate antagonist MK-801 reduces focal ischemic brain damage in the rat. Ann Neurol 24: 543–551
19. Park CK, Nehls DG, Graham DI, Teasdale GM, McCulloch J (1988b) Focal cerebral ischaemia in the cat: treatment with the glutamate antagonist MK-801 after induction of ischaemia. J Cereb Blood Flow Metabol 8: 757–762
20. Petruk KC, West M, Mohr G, Weir BKA, Benoit BG, Gentili F, Disney LB, Khan MI, Grace M, Holness RO, Karwon MS, Ford RM, Cameron GS, Tucker WS, Purves GB, Miller JDR, Hunter KM, Richard MT, Durity FA, Chan R, Clein LJ, Maroun FB, Gordon A (1988) Nimodipine treatment in poor-grade aneurysm patients J Neurosurg 68: 505–517
21. Pickard JD, Murray GD, Illingworth R, Shaw MDM, Teasdale GM, Foy PM, Humphrey PRD, Lang DA, Nelson R, Richards P, Sinar J, Bailey S, Skene A (1989) Effect of oral nimodipine on cerebral infarction and outcome after subarachnoid haemorrhage: British aneurysm nimodipine trial. BMJ 298: 636–642
22. Sakabe T, Nagai I, Ishikawa T, Takeshita H, Masuda T, Matsumoto M, Tateishi A (1986) Nicardipine increases cerebral blood flow but does not improve neurologic recovery in a canine model of complete cerebral ischemia. J Cereb Blood Flow Metabol 6: 684–690
23. Rothman SM, Olney JW (1986) Glutamate and the pathophysiology of hypoxic-ischemic brain damage. Ann Neurol 19: 105–111
24. Simon RP, Swan JH, Griffiths T, Meldrum BS (1984) Blockade of N-methyl-D-aspartate receptors may protect against ischemic damage in the brain. Science 226: 850–852
25. Steen PA, Newberg LA, Milde JH, Michenfelder JD (1983) Nimodipine improves cerebral blood flow and neurologic recovery after complete cerebral ischemia in the dog. J Cereb Blood Flow Metabol 3: 38–43
26. Steinberg GK, Saleh J, Kunis D (1988) Delayed treatment with dextromethorphan and dextrorphan reduces cerebral damage after transient focal ischemia. Neurosci Lett 89: 193–197
27. Wong EHF, Kemp JA, Priestley T, Knight AR, Woodruff GN, Iversen LL (1986) The anticonvulsant MK-801 is a potent N-methyl-D-aspartate antagonist. Proc Natl Acad Sci USA 83: 7104–7108

Correspondence and Reprints: James McCulloch, Wellcome Neuroscience Group, Wellcome Surgical Institute, and Hugh Fraser Neuroscience Laboratories, University of Glasgow, Garscube Estate, Bearsden Road, Glasgow G61 1QH, U.K..

Acta Neurochir (1993) [Suppl] 57: 80–88

Cerebral Protection by Adenosine

P. Schubert and **G. W. Kreutzberg**

Max Planck Institute for Psychiatry, Department of Neuromorphology, Martinsried, Federal Republic of Germany

Abstract

Delayed selective nerve cell death as seen in the hippocampal CA1 area of gerbils after transient forebrain ischemia goes along with neuronal hyperactivity and an early demonstrable accumulation of calcium in circumscribed groups of nerve cells. Application of NMDA receptor-blockers such as MK 801 prevents neuronal damage. This suggests the involvement of NMDA receptors which are operated by glutamate and known to mediate a special Ca^{2+} influx required also for establishing sustained enhancement of synaptic efficacy. Thus, the excessive postischemic accumulation of calcium, thought to be instrumental in the generation of nerve cell death, seems to result from turning on a dangerous, but primarily physiological mechanism which ran out of control.

We studied the endogenous control mechanisms by which the firing pattern of nerve cells and the initiation of NMDA receptor-mediated neuronal Ca^{2+} influx are controlled focusing in particular on the role of adenosine. This nucleoside is released from nerve- and glial cells in larger amounts after ischemia. It counteracts at increased extracellular concentration the generation of burst discharges, an effect which is ascribed to a modulation of the dendritic membrane properties. Removal of a possible action of endogenous adenosine by receptor antagonists such as theophylline was found to enhance postischemic nerve cell death. This together with other reported experimental evidences points to a protective action of this nucleoside in the brain. The presumed mechanisms by which this effect is achieved were studied in a rat hippocampal slice using ion selective electrodes. Here, we investigated in particular the influence of adenosine on:

- regulation of the synaptic input strength,
- initiation of NMDA receptor-mediated neuronal Ca^{2+} influx,
- experimental seizure generation,
- possibility of a pharmacological interference with postischemic nerve cell death by drugs related to adenosine action.

Keywords: Forebrain ischemia; hippocampus; delayed neuronal death; adenosine.

Nerve cell damage, as seen after ischemia, goes along with a massive release of excitatory transmitters like glutamate and an increased membrane depolarization (see Diemer *et al.*, this volume). This is accompanied by an excessive loading of nerve cells with calcium which is thought to be instrumental in nerve cell death. However, it is still a matter of discussion whether this is a primary or a secondary event. It has been argued that an increase of intracellular calcium may damage the cells by an enhanced breakdown of phospholipids leading to the formation of free radicals. A calcium-mediated disruption of elements of the cytoskeleton, a damage of mitochondria and an increase of protease activity are considered as well as a pathological gene activation. Another factor which is thought to be relevant for postischemic nerve cell damage is a pathological hyperactivity of neurons, the generation of seizures which are usually observed in the postischemic period (for a review see Meldrum *et al.*, 1982). The present paper will deal with endogeneous mechanisms by which nerve cell activity and neuronal Ca^{2+} entry are controlled and here, in particular, with the neuroprotection exerted by the nucleoside adenosine.

The experimental model used to study postischemic nerve cell damage was the ischemia-induced delayed and selective loss of pyramidal neurons in the hippocampal CA1 area of a gerbil (Kirino 1982). Transient forebrain ischemia was produced by occluding both carotid arteries, usually for a 10 min period. Nerve cell loss was evaluated by microscopic inspection of Nissl stained cross sections of the hippocampus and was quantified by measuring changes in the extinction of the CA1 area by means of densitometry (Fig. 1).

Postischemic delayed nerve cell death was accompanied by a heavy accumulation of calcium in the CA1 area as could be demonstrated histochemically even at the light microscopical level (DeLeo *et al.*, 1987). The calcium accumulation seen at 4 days after the ischemic period represents certainly a final state at a time point when CA1 neurons were severely damaged and had lost their Nissl-stainability. When evaluating this finding, one has to consider that the histochemical visualization of intraneuronal calcium does not allow to discriminate an early transient intracelluar rise of calcium which may initiate the pathological cascade of events from a consecutive calcium loading which results from cell damage.

Under physiological conditions, the cell possesses powerful defense mechanisms which can handle even a massive pathological increase of intracellular calcium. This is shown in the following experiments performed by E. Dux in our department. He investigated the intracellular sequestration of calcium in brain cells of a gerbil following an excessive calcium entrance in the early postischemic phase (Dux *et al.*, 1987). Using an electron microscopic-histochemical technique he found as early as 5 min after a 5 min lasting forebrain ischemia a massive accumulation of calcium in cellular mitochondria. At this stage, the mitochondria were still intact, their structure was well preserved and they had apparently still working mechanisms to get rid of such a calcium overload. Accordingly, the calcium loading was still transient and seen only within 5 and 15 min after the end of the ischemic period whereas the calcium content was back to the control level after 1 hour. This phenomenon was not restricted to hippocampal neurons. It was also seen in other neurons and glial cells of the cortex. However, in the CA1 neurons which are about to die, intracellular calcium raised again and this second increase was sustained. It was accompanied by a structural damage of the mitochondria and may reflect a secondary and more persistent alteration of those mechanisms which control the intracellular calcium level. Since the histochemical technique is rather insensitive and requires a critical duration of the intracellular calcium rise to be recognized, a primary and pathogenetically relevant rise of intracellular calcium may escape detection. The rational of the following experiments was, therefore, to investigate the preceding mechanisms which are known to regulate the intracellular calcium level and which, if altered, may initiate the process of nerve cell damage.

Three of the most obvious regulatory mechanisms are: 1. an increased mobilization of calcium from intracellular stores, 2. an impairment of the physiological extrusion of calcium from the cell into the extracellular space and 3. an increased influx of calcium ions through voltage- and receptor-operated membrane ion channels. So far, most information has been accumulated concerning the control of the activity-evoked neuronal Ca^{2+} influx. Here, the increased Ca^{2+} entrance going along with the activation of NMDA receptors has attracted the particular interest because of its physiological and presumed pathological relevance (see Nicoll *et al.*, 1989).

Blockade of NMDA receptors prevents calcium loading and postischemic nerve cell death in the CA1 area (Fig. 2). If a gerbil has been treated with an NMDA receptor antagonist such as MK-801 shortly before or within hours after a 10 min transient forebrain ischemia, no calcium accumulation can be demonstrated histochemically; the CA1 neurons survive as

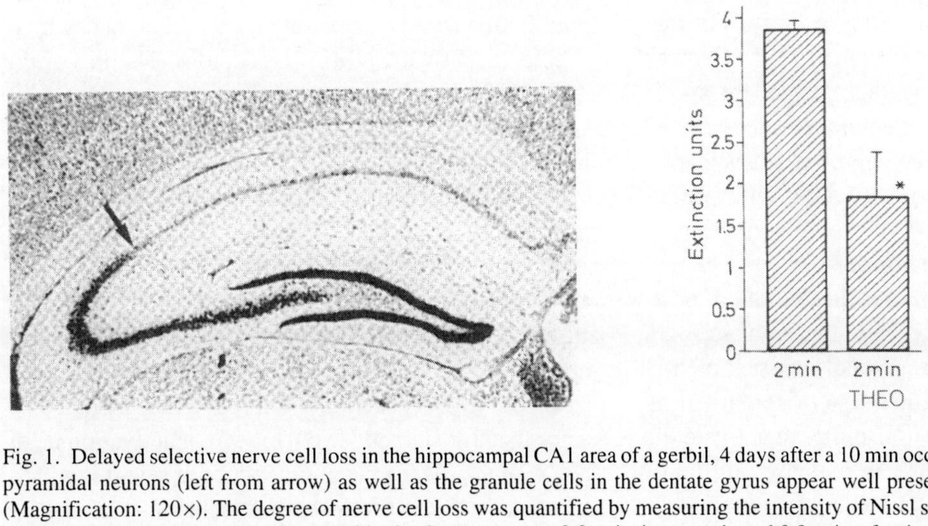

Fig. 1. Delayed selective nerve cell loss in the hippocampal CA1 area of a gerbil, 4 days after a 10 min occlusion of the carotid arteries. CA3 pyramidal neurons (left from arrow) as well as the granule cells in the dentate gyrus appear well preserved in the Nissl-stained section. (Magnification: 120×). The degree of nerve cell loss was quantified by measuring the intensity of Nissl staining by means of densitometry. The average extinction values measured in the CA1 area were 3.8 units in controls and 0.8 units after ischemia. If the ischemic period was restricted to 2 min, no significant damage of CA1 neurons was observed in untreated gerbils. But there was a significant decrease of Nissl staining (average extinction value* 1.8) in animals which had been injected with the adenosine receptor antagonist theophylline *(THEO)* (32 mg/kg i.p.). From: DeLeo *et al.*, 1988a

well as those in the CA3 area which is reflected by a preservation of Nissl-staining 4 days after ischemia. This protection by MK-801 has been shown by several groups (see McCulloch, this volume).

The point I want to make here is that treatment with this NMDA receptor blocker did usually not prevent in our experiments all of the ischemia-induced nerve cell damage. There was still a loss of nerve cells in a circumscribed region in the transition zone between CA3 and CA1 and also in the subiculum (Schubert, in preparation). Histochemical staining for calcium revealed a heavy accumulation of calcium in these nerve cells located in the transition zone (Fig. 2). Calcium loading in this area seems to precede that of the CA1 neurons. Thus, there are apparently different mechanisms responsible for the intracellular rise of calcium: those which have to be ascribed to an activation of NMDA receptors and those which can not be prevented by NMDA receptor antagonists. Different mechanisms seem to be operative in different regions revealing a different timing. The pathological events initiated in the one area may trigger those in the other, but this needs further clarification.

The finding that treatment with MK-801 limits postischemic nerve cell death not only in the hippocampus but also in other brain regions (see McCulloch, this volume) indicates a general pathogenetic significance of a pathological activation of NMDA receptors. The fact that these receptors are dangerous could explain why they are operated like a high security lock which needs not only one but two keys to be opened. When the transmitter binds to the NMDA receptor site, the ion channel opens. But Ca^{2+} is still not allowed to pass through, because the receptor is blocked by physiological Mg^{2+} concentrations in the range of 1 mM (Nowak *et al.*, 1984). This block is voltage-dependent and is removed when the nerve cell membrane is depolarized by a critical amount. Thus, the degree of postsynaptic depolarization is a second critical factor which determines the amount of NMDA receptor activation in addition to the release of glutamate.

Under physiological conditions, the amount of postsynaptic depolarization depends on the input strength and on the input frequency. If both are sufficiently large, the required amount of membrane depolarization will be obtained to allow the transmitter to activate the NMDA receptor. This initiates an increased influx of Ca^{2+} ions through receptor-operated and through additionally opened voltage-gated Ca^{2+} channels. The NMDA receptor-evoked Ca^{2+} influx is necessary, for example, to generate long term potentiation (LTP) which is thought to be a correlate of learning. We know that experimental generation of LTP requires a critical input frequency to be generated, i.e. a frequency which is able to activate NMDA receptors (Nicoll *et al.*, 1988). If the input pattern does not fulfill the required conditions, these dangereous receptors are not brought into play.

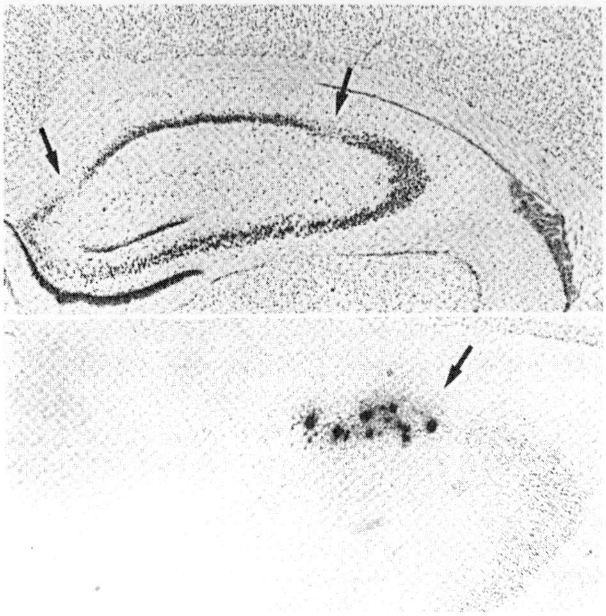

Fig. 2. Preservation of Nissl staining in the CA1 area by MK 801 4 days after 10 min forebrain ischemia. The drug (4 mg/kg i.p.) was injected 2 hours after the ischemic period. Treatment with this NMDA receptor antagonist did not prevent delayed nerve cell loss in the subiculum and in the transition zone between CA1 and CA3 (arrows). Nissl stained section; magnification: 80×. Below: A heavy accumulation of intraneuronal calcium can be demonstrated histochemically in the neurons of the transition zone (arrow) already at 2 days after 10 min ischemia, regardless of whether or not the gerbils had been treated with MK 801. The principle of the histochemical method used is that Ca^{2+} forms a complex with a chelating agent (glyoxal bis-(2-hydroxy-anil) and produces an insoluble red chromophore (Kashiwa and Atkinson 1963). Magnification: 200×. From: DeLeo *et al.*, 1987

The physiologically controlled activation of NMDA receptors may run out of control under pathological conditions. For example, an excessive release of excitatory transmitters, as initiated by ischemia, will lead to an exaggerated depolarization of the postsynaptic membrane and favor a pronounced and sustained activation of NMDA receptors. There is experimental evidence that a sustained NMDA receptor activation is, in fact, responsible for the maintenance of seizures which are usually observed in the postischemic period (see e.g. Mody *et al.*, 1984). Thus, ischemia initiates apparently a vicious cycle: increased depolarization leads to enhanced activation of NMDA receptors which maintains the generation of seizures and this in turn increases depolarization (Fig. 5). It is conceivable that this event is accompanied by an excessive and continuous neuronal Ca^{2+} influx through both, voltage-dependent and receptor-operated ion channels.

The fact that we are dealing here with a primarily physiological mechanism also implies that there are physiological control mechanisms which regulate these dangerous processes. One of these control mechanisms in which we are particularly interested, is the *neuromodulation by adenosine.*

This nucleoside tends to limit the amount of evoked postsynaptic depolarization and influences not only physiological phenomena like LTP (Schubert 1988a) but also pathological nerve cell damage. It has been shown first by the Meldrum group that adenosine agonists like phenyl-isopropyladenosine antagonize the development of ischemia-induced nerve cell death (Evans *et al.*, 1987). Some protection is also exerted by endogeneous adenosine. If the action of endogeneous adenosine is removed, the probability is increased that a short lasting ischemia will cause nerve cell damage (Rudolphi *et al.,* 1987). This could again be demonstrated in the gerbil model. If the ischemic period was short, lasting only for 2 min, we did not find any measurable damage after 4 days. The extinction value was the same as in controls where adenosine was allowed to act (Fig. 1). However, in those animals in which a possible adenosine action was prevented by treatment with the receptor antagonist theophylline, a marked nerve cell loss was seen in the CA1 area (Fig. 1; DeLeo *et al.,* 1988a). This indicates that some neuronal protection is provided by endogeneous adenosine.

A similar protection was achieved by propentofylline (Hoechst, Werk Albert, Wiesbaden) which is also a xanthine derivative and belongs to the same family as theophylline. Therefore, it might be related to adenosine action and this is the reason why we studied the effect of this drug on postischemic nerve cell death. When we injected propentofylline, 10 mg/kg, shortly before a 10 min bilateral carotid occlusion, hippocampal nerve cell death was largely prevented. In untreated controls, the extinction value indicating the loss of Nissl staining was reduced to 20% of the normal value. After pretreatment with propentofylline the extinction was only slightly less than measured in normal animals (DeLeo *et al.*, 1987). The protective effect of this drug was not restricted to the hippocampus; frequently observed postischemic damage of thalamic nuclei was also prevented. It is presumably of particular clinical importance that some protection was still achieved with propentofylline even when administered 1–2 hours after the end of the ischemic period (DeLeo *et al.*, 1988b). We do not know how this drug acts. The only known mechanism of propentofylline is to inhibit the reuptake of adenosine from the extracellular space back into intracellular compartments (Fredholm, personal communication). The drug should therefore help to keep the extracellular adenosine concentration at an elevated level. This leads to the question: *How is the extracellular concentration of adenosine in the brain regulated?*

Similar to a classical neurotransmitter, adenosine is released from axon terminals in an activity-related manner (Schubert *et al.*, 1976). Stimulation of central nerve cell fibers leads to a release of ATP co-packed in synaptic vesicles with the transmitter (see e.g. Zimmermann and Whittaker 1974). The required cascade of ektoenzymes is available which allows the extracellular formation of adenosine from the ATP released. One of the crucial enzymes is the 5′-nucleotidase which splits AMP to adenosine. This enzyme shows a selective distribution in the brain, particularly in the hippocampus and has been extensively studied in our department by EM histochemistry (Kreutzberg *et al.*, 1978). It is usually not associated with neuronal membranes, but with the membranes of astrocytes. Astrocytic processes revealing 5′-nucleotidase activity were often found to engulf synaptic complexes. Thus, the enzyme is in an optimal position to form adenosine from nucleotides which were released from the axon terminal.

On the other hand, adenosine is also formed intracellularly not only from nerve- but also from glial cells (Lewin and Bleck 1979). Since the nucleoside is membrane-permeable it will be released in amounts which are determined by the intracellular breakdown of ATP. Thus, the adenosine appearing in the extracellular space is a rather unique molecular signal which carries two informations:

1. information about the degree of nerve cell activation, and
2. information about the state of the local cellular metabolism.

What are the Mechanisms by which Adenosine Modulates Neuronal Activity Limiting the Amount of Postsynaptic Activation?

One well established effect is the depression of transmitter release (for a review see e.g. Fredholm and Hedqvist, 1980). Here, the release of excitatory transmitters like glutamate seems to be stronger influenced than that of inhibitory transmitters like GABA. Presynaptic receptors are required to mediate the adenosine effect on transmitter release. In order to localize the responsible A1 receptors in the brain, we investigated the distribution of an radioactively labelled A1 adenosine agonist, ^{3}H-cyclohexyl-adenosine (^{3}H-CHA), by means of autoradiography. Light microscopical evaluation revealed a highly selective distribution of A1 binding sites with marked differences between brain areas and even between different parts of such a homogeneously organized structure as the hippocampus. Labelling was most intensive in the CA1 area (Lee *et al.*, 1985). Lesioning of the major input systems originating in the contralateral hippocampus or in the entorhinal cortex, decreased the intensity of labelling in the hippocampal terminal fields of the degenerating axons by 25% respectively by 30% (Schubert *et al.*, 1985). The depletion of binding sites going along with the loss of presynaptic terminals varied strongly among the different input systems. For example, lesioning of the mossy fibers which carry the main input from the dentate gyrus to the CA3 pyramidal neurons was not accompanied by a significant decrease of labelling. This indicates that transmitter release in this system is not or only to a small extent modulated by adenosine, in contrast to the entorhinal and the commissural input. A synaptic localization of A1 binding sites was also verified by EM autoradiography. Interestingly, A1 receptors are not exclusively associated with synapses. The majority of CHA binding sites was found to be extrasynaptical, associated with the dendritic membrane of the pyramidal neurons (Tetzlaff *et al.*, 1987).

If there are synaptic and extrasynaptic A1 receptors, we also have to expect synaptic and extrasynaptic adenosine actions. Both will add in determining the amount of postsynaptic depolarization elicited by a given input. The following experiments are focussed on the

New aspect of a non-synaptic modulation by adenosine, which is presumably exerted by such postsynaptic and extrasynaptic A1 receptors. Since an excessive calcium loading may be of major importance for the development of nerve cell death, we investigated whether adenosine modulates the amount of postsynaptic activation and the activity-evoked neuronal Ca^{2+} influx. We tested this in a hippocampal slice preparation by using Ca^{2+}-sensitive electrodes. This allowed us to measure the evoked neuronal Ca^{2+} influx which results from a repetitive synaptic activation of the pyramidal neurons.

Measurements were performed simultaneously with two Ca^{2+}-sensitive electrodes (Fig. 3). One was located in the synaptic area where the stimulated afferent fibers terminate on the dendrites of the pyramidal neurons; the other one was placed in the soma layer of these neurons. When we stimulated the pyramidal neurons directly by antidromic repetitive stimulation, action potentials were generated in the soma which leads to the opening of voltage-gated Ca^{2+} channels in the soma membrane. This allows Ca^{2+} ions to flow from the extracellular space into the nerve cell bodies which is reflected by a transient decrease of the extracellualar Ca^{2+} concentration. This is what we measured. Usually, pyramidal neurons were activated synaptically by orthodromic stimulation of the afferent fibers with a 20 Hz/10 sec stimulus train. This led to a marked decrease of the extracellular Ca^{2+} concentration in the synaptic area reflecting the influx of Ca^{2+} ions into pre- and postsynaptic elements. If the afferent stimulus intensity was too low to activate the postsynaptic neurons, we only recorded a synaptic Ca^{2+} signal but there was no Ca^{2+} influx into the somas. If the stimulus intensity was increased to get efficient synaptic transmission, action potentials were generated and a marked Ca^{2+} signal was found in the soma layer. Thus, we can use the Ca^{2+} signal in the soma layer to monitor the amount of postsynaptic activation which is achieved by a given repetitive input.

Using this approach, we tested whether the strength of synaptic transmission and evoked neuronal Ca^{2+} influx are influenced by adenosine. Here, we were particularly interested in the action of endogenous adenosine concentrations (in the range of 1 μM; Zetterstrom *et al.*, 1984). We therefore tested the effect of removing a possible adenosine action by adding an adenosine receptor antagonist to the superfusion medium. In these experiments we chose 8-cyclopentyl-1,3-dipropylxanthine (DPCPX) which is a highly selective antagonist of the A1 receptor (Lohse *et al.*, 1987).

Evoked Ca^{2+} signals were markedly enhanced in the presence of DPCPX (Fig. 3; Schubert, 1988b). A stimulus intensity which was too weak to generate action potentials in the presence of adenosine action, now elicited action potentials already early during the train, revealing the phenomenon of frequency potentiation. Even if the stimulus intensity was reduced to a level that the amount of synaptic activation was smaller than before, action potentials together with a marked somatic Ca^{2+} influx were still generated when the action of adenosine was blocked.

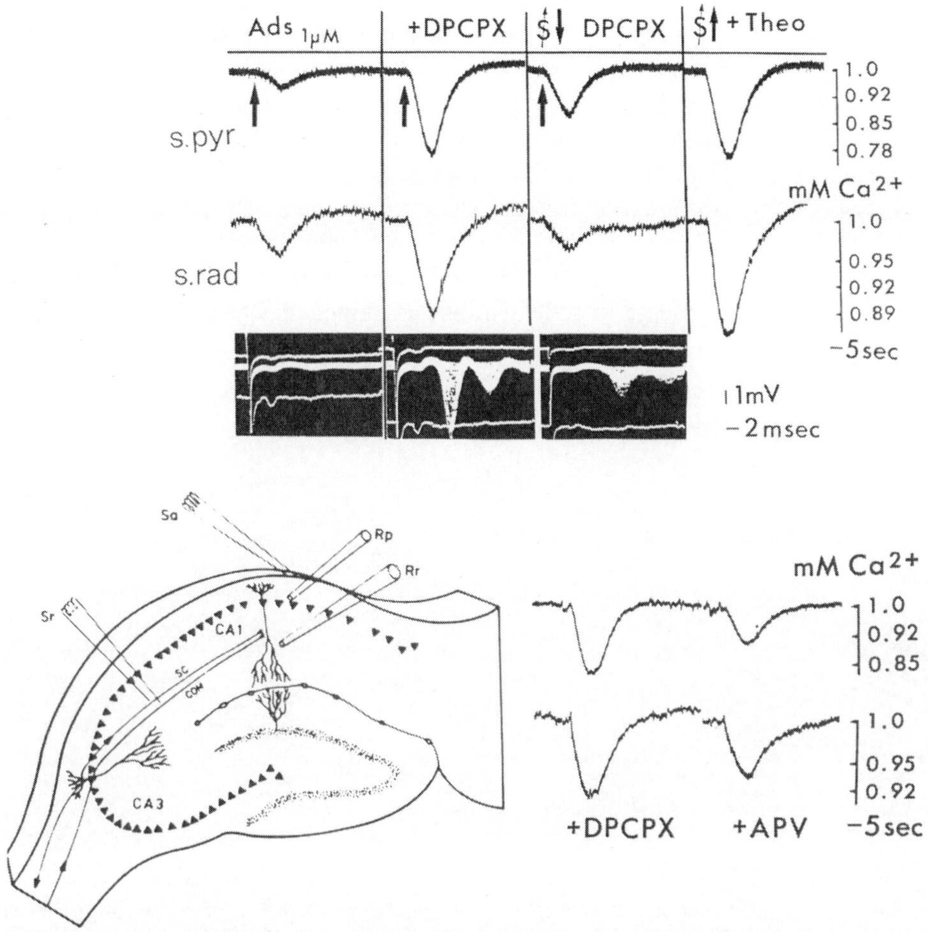

Fig. 3. Effect of a blockade of endogenous adenosine action on synaptically-evoked neuronal Ca^{2+} influx as determined in a rat hippocampal slice preparation superfused with a 1 mM Ca^{2+} containing medium. CAl neurons were repetitively activated by stimulating the Schaffer collaterals *(SC)* and commissural fibers *(COM)* via Sr with a 20 Hz/10 sec orthodromic stimulus train. The evoked decreases of the extracellular Ca^{2+} concentration were measured with combined recording/ion -sensitive electrodes located in the synaptic area *(Rr)* and in the CA1 pyramidal cell layer *(Rp)*. For preparation of electrodes see Heinemann *et al.*, 1977. Addition of the A1 receptor-selective antagonist DPCPX (50 nM) led to the early generation of action potentials during the stimulus train (+DPCPX) which was not seen in the presence of 1 μM adenosine action ($Ads_{1\mu M}$). The enhanced frequency potentiation in DPCPX was accompanied by a marked increase of the stimulus train-evoked Ca^{2+} fluxes in the synaptic area *(s.rad)* and in the soma layer of the postsynaptic pyramidal neurons *(s.pyr)*. When the intensity of afferent stimulation was reduced, an increased somatic Ca^{2+} influx and generation of action potentials was still observed (S DPCPX). Addition of the less selective adenosine receptor antagonist theophylline on top of DPCPX was ineffective (S +Theo) indicating that the enhancement of evoked neuronal Ca^{2+} influx results from a blockade of the A1 and not of the A2 receptors (from Schubert 1988b). Recordings below right: The enhanced neuronal Ca^{2+} influx seen in the presence of DPCPX contains a significant proportion which is blocked by further addition of the NMDA receptor antagonist 2-APV. This was not the case if endogenous adenosine was allowed to act

The increase of the neuronal Ca^{2+} influx in the presence of DPCPX was accompanied by a steeply developing DC shift measured in the synaptic region. This means that a given input leads to a more pronounced membrane depolarization. In the presence of adenosine action, the depolarization developed much more slowly during repetitive activation and the DC shift did not reach the same amplitude. We conclude that endogeneous adenosine exerts a powerful control on the efficiency of a repetitive input; it limits postsynaptic depolarization and synaptically evoked neuronal Ca^{2+} influx.

Since the degree of membrane depolarization determines also the number of NMDA receptors which are activated by a given input, one should expect that NMDA receptor activation and the related additional Ca^{2+} influx are controlled by endogeneous adenosine. This is in fact the case as shown in the next series of experiments (Schubert, submitted). We monitored an involvement of NMDA receptors by testing whether the stimulus train-evoked Ca^{2+} fluxes contained a proportion which can be blocked by an NMDA receptor antagonist. If this experiment was performed under normal conditions, this means in the presence of endogenous adenosine action, no changes were observed. Addition of the NMDA receptor blocker 2-amino-5-phosphono-valeric acid (APV) did not change the amount of Ca^{2+} influx in the synaptic region elicited by a 20 Hz/10 sec synaptic activation. This was different when the action of adenosine was blocked. The increased synaptic Ca^{2+} influx, elicited upon synaptic activation in the presence of DPCPX, contained a large proportion (about 40%) which was blocked by APV (Fig. 3). Thus, the same synaptic input strength which was previously ineffective generates a marked NMDA receptor-mediated Ca^{2+} influx as soon as the adenosine action is prevented.

There are also other membrane potential- dependent mechanisms which are relevant for determining the evoked firing pattern of a neuron. For example, specific K^+ currents are turned on when one action potential

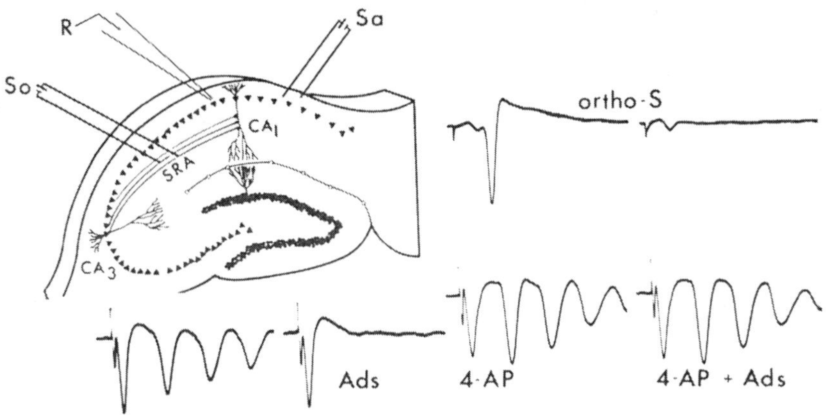

Fig. 4. Non-synaptic antiepileptic effect of adenosine. If slices were superfused with a medium in which the Ca^{2+} concentration was reduced to 0.2 mM, synaptic transmission was blocked. Orthodromic activation of CA1 neurons via *So* by a single stimulus pulse only generated a presynaptic fiber potential whereas a population spike was no longer elicited (traces above, *ortho-S*). Direct antidromic stimulation of the CA1 pyramidal neurons via *Sa* led to the generation of burst discharges (below, anti-S). The afterdischarges were blocked upon addition of 40 μM adenosine to the superfusion medium *(Ads)*. This adenosine effect could not be elicited in the presence of 50 μM 4-aminopyridine *(4-AP + Ads)*. *R* = recording electrode, *SRA* = substratum radiale

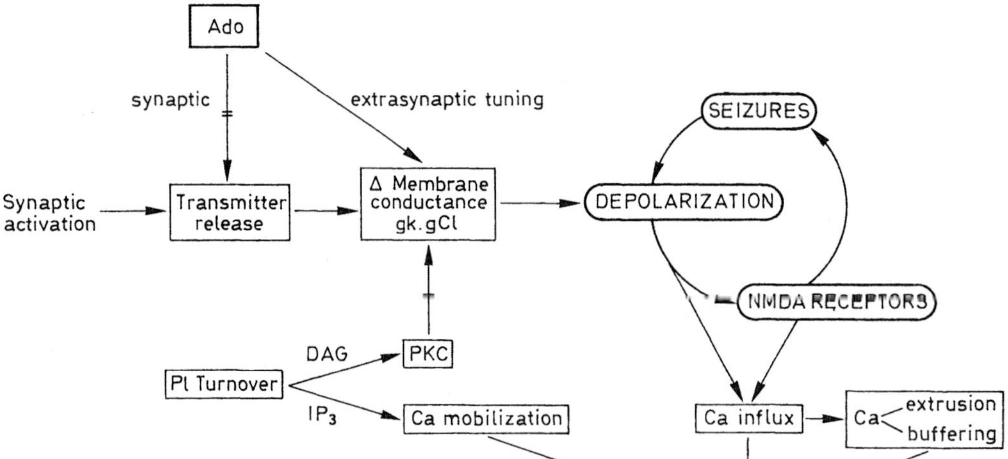

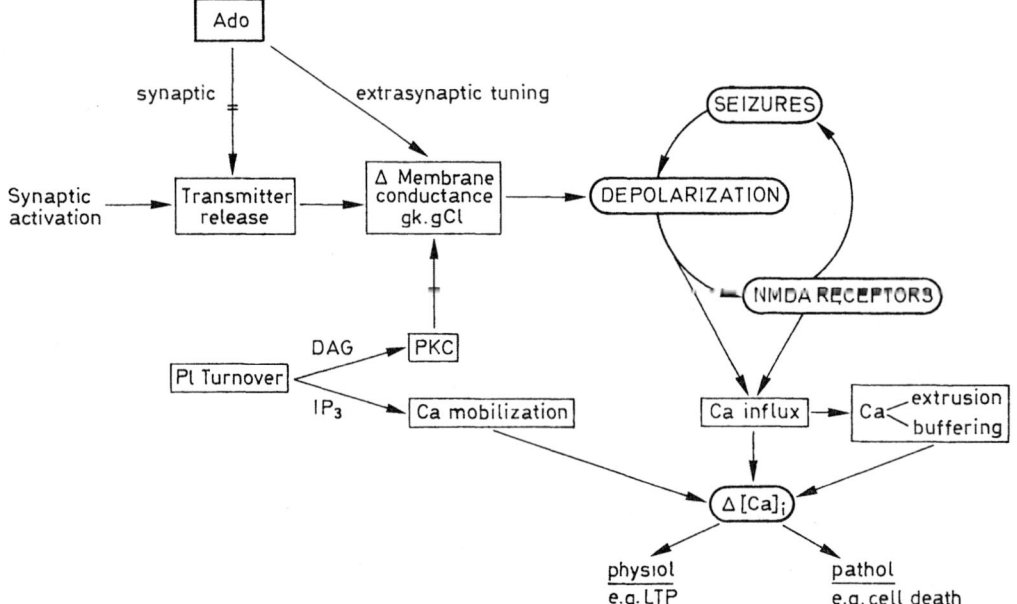

Fig. 5. Schematic illustration of proposed neuroprotection by adenosine *(Ado)*; for further explanation see Summary

has been generated. This leads to a membrane hyperpolarization and prevents consecutive discharges. One of these K^+ currents is the A current which is highly voltage-sensitive and which is inactivated when the membrane potential is depolarized beyond a critical level (Segal *et al.*, 1984). Adenosine tends to keep the membrane potential in a range that a depolarization-induced inactivation of these K^+ currents does not occur. We think, this is another indirectly achieved protective effect of adenosine; the nucleoside hinders neurons from becoming epileptic.

A frequently used model for studying epilepsy is the generation of burst discharges in a hippocampal slice by superfusing a medium with a reduced Ca^{2+} concentration (Fig. 4). When the pyramidal neurons were activated by direct antidromic stimulation, they responded with burst discharges instead of single action potentials. Addition of adenosine did not affect the first action potential but depressed the afterdischarges (Lee *et al.*, 1984). Adenosine was ineffective in the presence of the K^+ channel-blocker 4-aminopyridine (4-AP) indicating that the possibility to turn on the A current is a prerequisite for the observed depression of burst discharges by adenosine (Fig. 4; Schubert and Lee 1984). Since synaptic transmission is almost blocked in low Ca^{2+} medium, this antiepileptic effect of adenosine has to be attributed to a non-synaptic action which is presumably mediated by the extrasynaptic receptors.

In this context it is interesting that the depression of burst discharges was only achieved if adenosine was applied locally to the apical dendrites. Application of adenosine to the soma layer had almost no effect. The findings suggest that adenosine acts by tuning the dendritic membrane potential. We have evidence from ongoing 'patch clamp' experiments that this effect is achieved by enhancing the steady state membrane conductance for chloride ions (Mager *et al.*, 1989). As a result, the dendritic cable properties should be changed influencing the signal propagation from the synapse to the soma. This would explain the reduced efficiency of a repetitive synaptic input in the presence of adenosine action. Such a dendritic tuning effect would also add to the stabilization of the soma membrane preventing the generation of burst discharges.

Summary

Neuroprotection by adenosine results apparently from a concerted cooperation of different actions (see Fig. 5). A presynaptic depression of transmitter release will antagonize an excessive release of the excitatory transmitter glutamate. In addition, postsynaptic and non-synaptic mechanisms seem to be operative. Endogeneous adenosine concentrations were found to control effectively the power of a repetitive input limiting the amount of synaptically evoked neuronal Ca^{2+} influx. This is ascribed to a reduction of the postsynaptic membrane depolarization which will allow less voltage-operated Ca^{2+} channels to be opened in response to afferent activation. By tuning the membrane potential, adenosine controls the activation of the voltage-dependent NMDA receptors which are known to play a significant role in the generation of seizures and postischemic nerve cell death. Tuning of the dendritic membrane prevents also a depolarization-induced inactivation of voltage-sensitive K^+ currents which adds to the antiepileptic effect of adenosine. The latter is still seen in the absence of synaptic transmission underlining that this effect is due to a non-synaptic action and presumably mediated by extrasynaptic A1 receptors. When the adenosine-mediated tuning of the postsynaptic dendritic membrane is blocked, seizure generation and postischemic nerve cell death will be favored.

In this context it is interesting that a blockade of adenosine actions may not only result from external pharmacological interference but also from endogeneous metabolic regulation. If hippocampal slices were pretreated for 1–2 hours with phorbol esters known to activate protein kinase C (Castagna *et al.*, 1982), the above described effects of adenosine could no longer be elicited (Schubert 1988a). This is in accordance with the findings reported by the Snyder group (Worley *et al.*, 1987). Specifically, the marked depression of the stimulus train evoked Ca^{2+} signals as observed after the addition of 20–40 μmol adenosine was not seen in phorbol ester-treated slices. Also the increase of the Ca^{2+} signals upon removal of the endogeneous adenosine action with ADA or theophylline was not found after phorbol ester-treatment. This suggests that the protective neuromodulation by adenosine is itself subject of modulation and controlled by the protein kinase C. Whether an abnormal activation of the protein kinase C is involved in the generation of pathological hyperactivity and/or postischemic nerve cell death needs further elucidation.

Acknowledgement

This study has been supported by the Deutsche Forschungsgemeinschaft (SFB 220).

The technical help of Maria Köber is gratefully acknowledged. We thank Dr. Carola Haas for critical reading of the manuscript.

References

1. Castagna M, Takai Y, Kaibuchi K, Sano K, Kikkawa U, Nishizuka Y (1982) Direct activation of calcium-activated, phospholipid-dependent protein kinase by tumor-promoting phorbol esters. J Biol Chem 257: 7847–7851
2. DeLeo J, Toth L, Schubert P, Rudolphi K, Kreutzberg GW (1987) Ischemia-induced neuronal cell death, calcium accumulation, and glial response in the hippocampus of the Mongolian gerbil and protection by propentofylline (HWA). J Cereb Blood Flow Metabol 7: 745–751
3. DeLeo J, Schubert P, Kreutzberg GW (1988a) Propentofylline (HWA285) protects hippocampal neurons of Mongolian gerbils against ischemic damage in the presence of an adenosine antagonist. Neurosci Lett 84: 307–311
4. DeLeo J, Schubert P, Kreutzberg GW (1988b) Protection against ischemic brain damage using propentofylline in gerbils. Stroke 19: 1535–1539
5. Dux E, Mies G, Hossmann KA, Laszlo S (1987) Calcium in the mitochondria following brief ischemia of gerbil brain. Neurosci Lett 78: 295–300
6. Evans MC, Swan JH, Meldrum BS (1987) An adenosine analogue, 2-chloroadenosine, protects against long term development of ischaemic cell loss in the rat hippocampus. Neurosci Lett 83: 287–292
7. Fredholm BB, Hedqvist P (1980) Modulation of neurotransmission by purine nucleotides and nucleosides. Biochem Pharmacol 29: 1635–1643
8. Heinemann U, Lux HD, Gutnick NJ (1977) Free extracellular calcium and potassium during paroxysmal activity in cerebral cortex of the cat. Exp Brain Res 27: 237–243
9. Kashiwa HK, Atkinson WG (1963) The applicability of a new Schiff base glyoxal bis-(2-hydroxy-anil), for the cytochemical localization of ionic calcium. J Histochem Cytochem 11: 258–264
10. Kirino T (1982) Delayed neuronal death in the gerbil hippocampus following ischemia. Brain Res 239: 257–269
11. Kreutzberg GW, Barron K, Schubert P (1978) Cytochemical localization of 5′ nucleotidase in glial plasma membranes. Brain Res 158: 247–257
12. Lee KS, Schubert P, Heinemann U (1984) The anticonvulsive action of adenosine: a postsynaptic dendritic action by a possible endogeneous anticonvulsant. Brain Res 321: 160–164
13. Lee KS, Schubert P, Reddington M, Kreutzberg GW (1985) The distribution of 5′ nucleotidase and adenosine A1 receptors: evidence for diversification and conservation in the hippocampi of several commonly employed experimental animals. Acta Histochem [Suppl] 31: 47–51
14. Lewin E, Bleck V (1979) Uptake and release of adenosine by cultured astrocytoma cells. J Neurochem 33: 365–367
15. Lohse MJ, Klotz KN, Lindenborn-Fotinos J, Reddington M, Schwabe U, Olsson RA (1987) 8-cyclopentyl-1,3-dipropylxanthine (DPCPX) a selective high affinity antagonist radioligand for A1 adenosine receptors. Naunyn Schmiedebergs Arch Pharmacol 336: 204–210
16. Mager R, Ferroni 5, Schubert P (1990) Adenosine modulates a voltage dependent membrane chloride conductance in hippocampal neurons. Brain Res 532: 58–62
17. Meldrum B, Griffiths T, Evans M (1982) Hypoxia and neuronal hyperexcitability. A clue to mechanisms of brain protection. In: Wauquier A, Borgers M, Amery W (eds) Protection of tissues against hypoxia. Raven, New York, pp 276–286
18. Mody I, Lambert JD, Heinemann U (1987) Low extracellular magnesium induces epileptiform activity and spreading depression in rat hippocampal slices. J Neurophysiol 57: 869–888
19. Nicoll RA, Kauer JA, Malenka RC (1988) The current excitement in long term potentiation. Neuron 1: 97–103
20. Nowak L, Bregestovski P, Ascher P, Herbet A, Prochiantz A (1984) Magnesium gates glutamate-activated channels in mouse central neurones. Nature 307: 462–465
21. Rudolphi KA, Keil M, Hinze HJ (1987) Effect of theophylline on ischemically induced hippocampal damage in Mongolian gerbils: a behavioral and histopathological study. J Cereb Blood Flow Metabol 7: 74–81
22. Segal M, Rogawski MA, Barker IL (1984) A transient potassium conductance regulates the excitability of cultured hippocampal and spinal neurons. Neurosci 4: 604–609
23. Schubert P (1988a) Potentiation of synaptic transmission by phorbol esters is accompanied by a reduction of the physiological adenosine action. In: Haas HL, Buzsaki G (eds) Synaptic plasticity in the hippocampus. Springer, Heidelberg New York Berlin Tokyo, pp 61–64
24. Schubert P (1988b) Physiological modulation by adenosine: selective blockade of A1 receptors with DPCPX enhances stimulus train-evoked neuronal Ca influx in rat hippocampal slices. Brain Res 458: 162–165
25. Schubert P, Mayer R (1991) The critical input frequency for NMDA receptor-mediated neuronal Ca^{2+} influx depends on endogenous adenosine. Int J Purine Res 2: 11–16
26. Schubert P, Lee KS (1984) Non-synaptic modulation of repetitive firing by adenosine is antagonized by 4-aminopyridine in a rat hippocampal slice. Neurosci Lett 67: 334–338
27. Schubert P, Lee K, West M, Deadwhyler S, Lynch G (1976) Stimulation-dependent release of ^{3}H-adenosine derivatives from central axon terminals to target neurones. Nature 260: 541–542
28. Schubert P, Lee KS, Tetzlaff W, Kreutzberg GW (1985) Postsynaptic modulation of neuronal firing pattern by adenosine. In: Changeux P, Hucho E (eds) Molecular basis of nerve activity. de Gruyter, Berlin, pp 283–293
29. Schubert P, DeLeo J, Rudolphi K, Kreutzberg GW Selective protection against ischemia-induced hippocampal nerve cell death by the NMDA receptor antagonist MK-801: in preparation
30. Tetzlaff W, Schubert P, Kreutzberg GW (1987) Synaptic and extrasynaptic localization of adenosine binding sites in the rat hippocampus. Neuroscience 21: 869–875
31. Worley PF, Baraban JM, McCarren M, Snyder SH (1987) Cholinergic phospatidylinositol modulation of inhibitory, G protein-linked, neurotransmitter actions: electrophysiological studies in rat hippocampus. Proc Natl Acad Sci USA 84: 3467–3471
32. Zetterstrom T, Vernet L, Ungerstedt U, Tossmann U, Jonzon B, Fredholm B (1984) Purine levels in the intact rat brain: studies with an implanted perfused hollow fibre. Neurosci Lett 29: 111–115
33. Zimmermann H, Whittaker VP (1974) Effect of electrical stimulation on the yield and composition of vesicles from the cholinergic synapses of the electric organ of Torpedo. J Neurochem 22: 435–450

Correspondence and Reprints: Peter Schubert, Max Planck Institute for Psychiatry, Department of Neuromorphology, Am Klopferspitz 18a, D-8031 Martinsried, Federal Republic of Germany.

Acta Neurochir (1993) [Suppl] 57: 89–93

Effects of Acute Isotonic Saline Administration on Serum Osmolality, Serum Electrolytes, Brain Water Content and Intracranial Pressure*

H. E. James[1,2] **and S. Schneider**[1]

[1] Division of Neurosurgery and [2] Department of Pediatrics, School of Medicine, University of California, San Diego, U.S.A.

Abstract

Albino rabbits who had undergone a cryogenic insult over the left parieto-occipital cortex 24 hours previously were analyzed for serum osmolality, serum electrolytes, brain water content, intracranial pressure (ICP), following a 3 hour baseline intravenous infusion of above maintenance isotonic saline, and compared to sham operated controls.

In the acute setting there was no difference in the serum osmolality and electrolytes between the subgroups. There was a significant increase in the water content of the white matter of the left hemisphere in the cold lesion group when compared to sham operated controls. Despite the intravenous fluid challenge, the ICP did not rise during the 3 hour experimental trial when compared to pre-trial values.

It is concluded that in the acute setting an isotonic fluid load is compensated without significant disturbances of the above measured parameters, and should not alter therefore intracranial dynamics per se, in acute resuscitation measures following brain insults.

Keywords: Vasogenic brain edema; intracranial hypertension; isotonic saline infusion; cerebral water content.

Introduction

In the situations where intracranial dynamics may be altered such as following trauma, cerebral ischemia or hypoxia, concern exists for further disturbances of intracranial volume by intravenous fluid therapy. This is of importance in the situation of multiple trauma or dehydration and hypovolemia, where urgent and immediate fluid replacement is necessary[1,2].

Studies have been made to look at the types of intravenous fluid regimes to be employed, the fluid solutions, and the rapidity of fluid replacement[3–8]. More recently hyperonic fluid solutions have been employed with the objective of correcting systemic hypovolemia and correcting intracranial hypertension[9,10]. No uniform consensus exists in the literature as to the effectiveness of these various regimes.

In an attempt to address these concerns and to gain information that would assist in management decisions in the acute setting, the following studies were undertaken. In albino rabbits with controlled mechanical ventilation, with and without a cryogenic brain insult, intravenous isotonic was administered acutely. The serum osmolality and the electrolytes were measured during the period of infusion. At the end of the 3 hour infusion the brain water content was measured. Comparison of the ICP between the groups was performed during the experimental trials.

Materials and Methods

Two groups of albino rabbits were studied, one with a cryogenic insult to the left parieto-occipital cortex, and the other with no lesion (sham-operated)[11]. Above maintanence intravenous isotonic saline was administered over 3 hours.

Albino rabbits weighing 2 to 3 kg were anesthetized with Halothane (2%), placed in a stereotactic head holding device, and the scalp was incised under local anesthesia (marcaine 1%). The calvarium was exposed and with a 12.5 mm diameter circular trephine stainless steel probe equilibrated in liquid nitrogen, a 90-second cold injury was performed over the intact dura. The bone remnant was sutured to the skull, the scalp incision closed, and topical antibiotic ointment (bacitracin) applied to the suture line. All animals received 1 ml of a 2% Evans blue solution intravenously for subsequent documentation of the extent of blood brain barrier breakdown. They also were given 50 mg/kg of ampicillin (intraperitoneal). The second group of animals received a similar procedure, but no cryogenic insult. They then constituted the (sham-operated) control group.

Twenty-four hours following the operation at the time of maximal brain edema[12], the surviving animals underwent the experimental trials. They were reanesthetized with halothane (2%), intubated with

* This work was supported in part by Research Funds from the Division of Neurosurgery and by the Foundation for Pediatric and Laser Neurosurgery, Inc., San Diego, California.

an endotracheal tube, and continuously ventilated with a mixture of oxygen (50%), nitrous oxide anesthesia (50%) and halothane (0.5%). Arterial and central venous (femoral) lines were placed with local anesthesia (marcaine 1%), and the blood gases continuously monitored to maintain a constant $PaCO_2$ (37–43 torr), by adjusting the settings on the small animal ventilator (Harvard Apparatus, Harvard Instruments, Framingham, Mass.). The animals were then placed in a stereotactic head holding device, a 19-gauge needle was inserted into the cisterna magna. The scalp incision was opened and the calvarium exposed. Two platinum electrodes were placed in the skull on each side of the midline for continuous EEG recording. The cisterna magna needle, the arterial and the venous lines were connected to appropriate strain-gauge transducers for a continuous display of ICP, systemic arterial pressure and central venous pressure on a multichannel polygraph (Hewlett-Packard, Model 7758B System, Palo Alto, Ca.). Following stabilization of the $PaCO_2$ the intravenous infusions were initiated. The animals received an intravenous infusion of 0.45 normal saline at 15 ml/hour, as baseline, with additional fluids being administered to maintain patency of arterial and venous lines. The controls received 79 ml ± 4 and the cold lesion group 75 ml ± 3,5. Samples of serum osmolality, sodium and potassium were taken prior to infusion and at 60, 120 and 180 minutes, from the beginning of the infusion.

Upon completion of the 3 hour infusate, the animals were killed by intravenous air embolization, rapid craniectomy allowed for prompt removal of the brain, and it was placed in cold kerosene for gravimetry studies[13]. The samples were taken in front and behind the areas of hemorrhage and necrosis from the cold lesion, where the Evans blue extravasation indicated a disturbance of the blood brain barrier, on the left side. An equivalent area was sampled from the right hemisphere. In the sham operated subgroups homologous samples were taken from both parieto-occipital regions[11]. There were 8 animals per group.

The data obtained was submitted to statistical analysis and is presented in the next section.

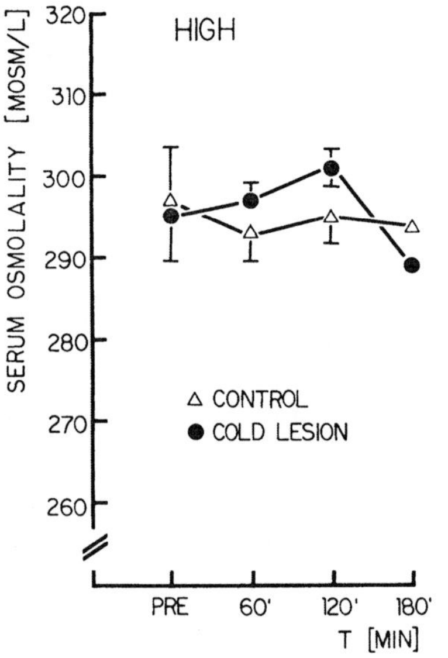

Fig. 1. Serum osmolality during the experimental trials in the both groups

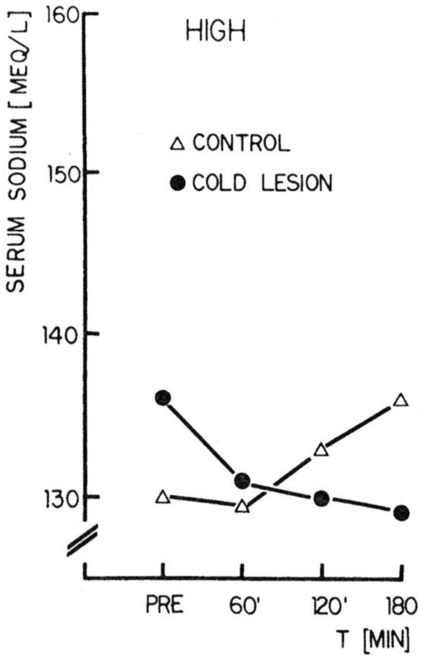

Fig. 2. Serum sodium during the experimental trials (see Materials and Methods)

Results

All animals that had a systolic blood pressure of less than 80 torr were discarded from the study. The systolic blood pressure ranged from 85 to 120 torr in all groups, and the central venous pressure from 0 to 4 torr. The EEG demonstrated the previously described slow wave activity over the left hemisphere in the cyrogenic lesion subgroups[11]. The sham-operated group had symmetrical fast wave activity over both hemispheres[11].

Serum Osmolality

The pre-trial serum osmolality in the control (sham-operated) group was 294 ± 8 (SD) mOsm/L, and in the cold lesion group it was 295 ± 6 (SD) mOsm/L. The serum osmolality values at 60, 120 and 180 minutes are illustrated in Fig. 1. The pre-trial and subsequent values in the cold injury group are also seen in Fig. 1. There is no statistical difference between the groups.

Serum Sodium

The pre-trial sodium in the cold lesion and control groups was 130 ± 9.6 (SD) mEq/L and 136 ± 11.2 (SD) mEq/L, respectively. The values at 60, 120 and 180 minutes are presented in Fig. 2. There is no statistical difference between these.

Serum Potassium

The serum potassium was 3.6 ± 0.5 (SD) mEq/L for the cold lesion group, and 3.0 ± 0.4 (SD) mEq/L for

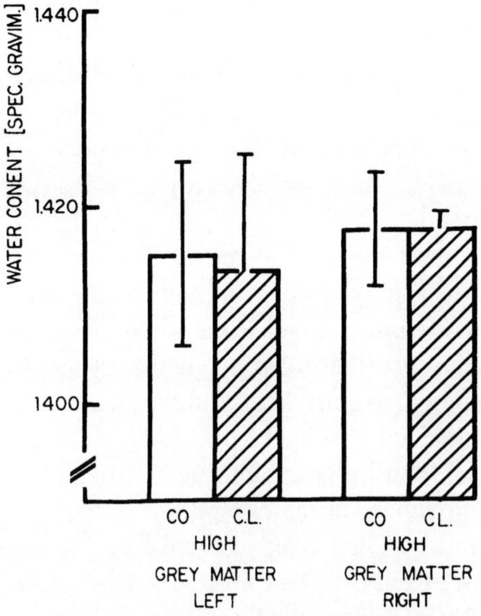

Fig. 3. Specific gravity of the gray matter of the experimental groups upon completion of the 3 hour trials

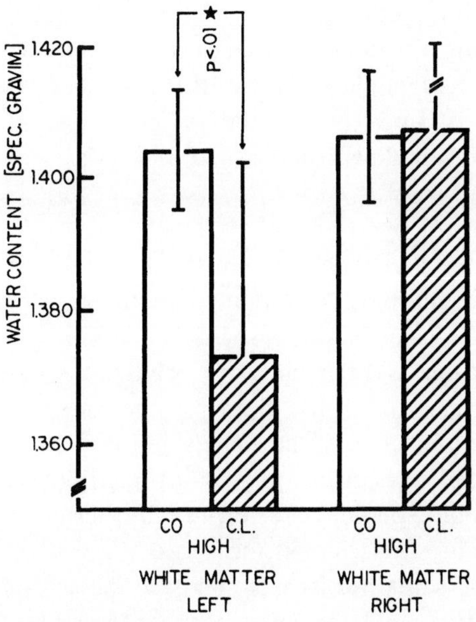

Fig. 4. Specific gravity of the white matter of the experimental groups upon completion of the 3 hour trials. A decrease in the value indicates an increase in water content

control. There was no statistical difference from these to the values obtained at 60, 120 and 180 minutes into the experimental run.

Brain Water Content

The gravimetry studies reflect the brain water content, and a reduction of the gravimetry values signifies an increase in brain water. The specific gravity for the gray matter in the left hemisphere in the control (sham-operated) was 1.0417 ± 0.0009 (SD) and in the cyrogenic group it was 1.0416 ± 0.0009 (SD). This difference is not significant (Fig. 3).

The gravimetry value for the white matter of the left hemisphere of the sham-operated group was 1.0404 ± 0.0009 (SD) and for the equivalent hemisphere in the cold lesion group it was 1.0372 ± 0.0036 (SD). This difference is significant ($p < 0.01$) and the data presented in Fig. 4. The difference is expected due to the edema generated in the left hemisphere by the cold injury[11]. The right hemisphere gravimetry white matter values are presented in Fig. 4.

Intracranial Pressure

The mean pre-trial ICP for the controls was 1.9 ± 1.2 (SD) torr. The values at 60, 120 and 180 minutes after the start of infusion are illustrated in Fig. 5. The mean ICP for the cold lesion group was 3.5 ± 2.9 (SD) torr.

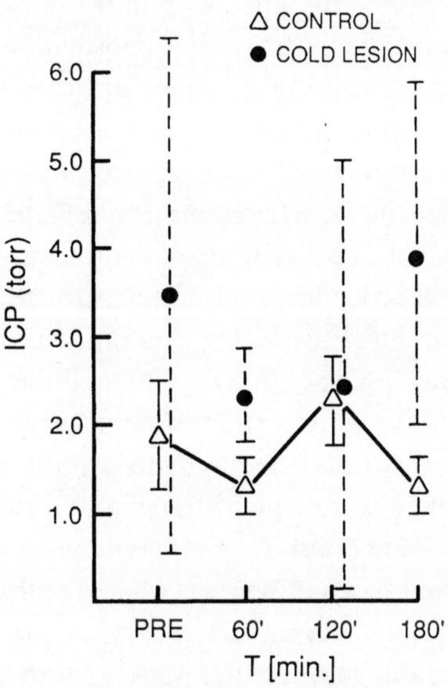

Fig. 5. Intracranial pressure (ICP) in the experimental groups during the experimental trials. The animals were mechanically ventilated at a $PaCO_2$ at a range of 37–43 torr

There is no statistical difference within the group between the pre-trial and the 60, 120 and 180 minute values (Fig. 5).

Discussion

Management of fluid and electrolyte replacement in patients with acute brain insults is a subject of concern and debate for fear of worsening intracranial dynamics. This is particularly important in the first few hours following a brain insult, due to the fact that the brain is more susceptible to changes in the intracranial volume that may interfere with cerebral perfusion, by reducing the cerebral perfusion pressure, secondary to elevated ICP[14–16].

In situations of hypovolemia such as is seen with multiple trauma and shock, or pre-shock states, the fluid replacement most commonly employed is a crystalloid solution of 5% dextrose in lactated Ringer's by rapid infusion[3,7]. If this does not stabilize the patients systemic status, it has been recommended to acutely administer more of the same solution[3,7]. As a general rule amongst trauma and critical care physicians, crystalloid solutions are felt to be superior to colloid for acute resuscitation measures[5,6,17]. We selected isotonic saline for its isotonicity to the serum and to simplify the volume replacement of the rabbit, without creating changes secondary to lactate.

In an opposite approach others have emphasized that there may be a need for fluid restriction, or at least some degree of fluid restriction, when a brain insult is present[8,18,19]. This is felt to be particularly important in the first 3 days following head injury, when an excessive secretion of antidiuretic hormone may occur[20]. There is general consensus that salt should be administered in the fluid replacement, to prevent or alleviate the hyponatremia that follows brain insults and postoperative neurosurgical procedures[20–22]. The presence of a low sodium serum level following a brain insult may already indicate an expanded extracellular space[23], or it may reflect a salt depletion. Both of these extremes can create as an end result hypo-osmolar states, that may permit easy passage of free water into the brain and deteriorate intracranial dynamics[8]. Shenkin *et al.*[24], reported series of patients treated with different fluid and electrolyte regimes. In the last of the series they maintained the adult postoperative neurosurgical patient with an average of 1500 ml of 0.45% sodium chloride solution, with 2.5% dextrose, every 24 hours. With this regime they noted that contrary to their previous experience the serum osmolality remained within 1.5% of their average preoperative level of 290 mOsm/L. With this level of fluids they observed that the patients tended to increase their osmolality rather than reduce it, during the postoperative period, with less pendular swings of osmolality and serum sodium. Though they emphasized this technique, they also made it just as clear that patients with brain insults need to be followed closely because of rapid fluctuations that can occur from hormonal derangement. This has also been demonstrated in head injured patients with alcohol intoxication[25].

The water requirements for the resting rabbit range from 4 to 7 ml/hours[26]. In the current experiments the animals were given an amount above the maintenance requirements, and the mean was even higher at the end of the experimental run. This was due to the irrigants needed to maintain the patency of the arterial and venous lines. The average amount administered was of 24–27 ml/hour. Despite fluid loads the control animals (sham-operated), as well as those that had a cryogenic brain insult, did not have any change in serum osmolality, at least in the acute period of this study (Fig. 1).

The serum sodium in the rabbit ranges from 110–147 mEq/L, with the mean values reported in the 125–131 mEq/L range[26]. There was no change during the fluid infusions in either group (Fig. 2). The serum potassium in the rabbit is around 3.0 mEq/L[26]. This is similar to that which was obtained in the subgroups of the current series, and there was no change during the trials.

The finding that the fluid load administered during the acute period of these studies with normal saline solution, even though at larger volumes than were necessary, did not create a disturbance of serum osmolality or electrolytes, suggests that in the initial management of the acute brain insult this form of cristalloid prevents a hypo-osmolar and hyponatremic disturbance.

The development of brain edema with or without ICP increase following a brain insult is a source of concern in clinical practice. This may follow hypoxia, ischemia or traumatic contusions[11,15,18]. It may be aggravated by acute fluid administration[4], and by increasing intracranial volume, further elevating ICP or raising ICP if not already above normal values. The current study shows that the gray matter water content was not significantly changed in the control or cryogenic insult animals with either method of fluid administration, in the acute period of the experiment (see Fig. 4). In the white matter where the brunt of the edematous process takes place following a cryogenic insult[11,27], there was an increase in the water content in the left hemisphere. This increase in water content accounts for the increase in ICP in the cold injury group when compared to the sham operated group[11].

In clinical practice ICP is monitored as one of the parameters for assessment of the patient's clinical condition and for management decisions. In the current series the administration of large amounts of isotonic saline intravenously to the control and the cryogenic groups, did not significantly elevate ICP (Fig. 5), during the 3 hour period of the study. This indicates that in the anesthetized and mechanically ventilated animal, an isotonic saline fluid challenge does not alter intracranial pressure or brain water significantly, or jeopardizes a normal cerebral perfusion pressure. Therapy decisions in reference to the use of diuretics, hypertonic fluids, steroids and other agents, can be addressed understanding that fluid challenges with isotonic crystalloids will not in themselves interfere or counteract the objectives of these agents.

References

1. Downes JJ, Raphaely RC (1979) Anesthesia and intensive Care. In: Ravitch MM, Welch KJ, Benson CD, *et al* (eds) Pediatric surgery, Vol 2. Year Book Medical Publishers, Chicago, 1979, pp 12–38
2. Perkin RM (1985) Pathophysiology of childhood shock syndromes. In: James HE, Anas NG, Perkin RM (eds) Brain insults in infants and children. Grune and Stratton, Orlando, pp 191–198
3. Carrico CJ, Canizaro PC, Shires GT (1976) Fluid resuscitation following injury: rationale for the use of balanced salt solutions. Crit Care Med 4: 46–54
4. Fishman RA (1953) Effects of isotonic intravenous solutions on normal and increased intracranial pressure. Arch Neurol Psychiatry 70: 350–360
5. Lucas CE, Weaver D, Higgins RF, Ledgerwood AM, Johnson SD (1978) Effects of albumin versus non-albumin resuscitation on plasma volume and renal excretory function. J Trauma 18: 564–570
6. Lucas CE, Ledgerwood AM, Higgins RF, Weaver DW (1980) Impaired pulmonary function after albumin resuscitation from shock. J Trauma 20: 446–451
7. Rowe MI, Arango A (1975) The choice of intravenous fluid in shock resuscitation. Pediatr Clin North Am 22: 269–274
8. Shenkin HA, Bezier HS, Bouzarth WF (1976) Restricted fluid intake. Rational management of the neurosurgical patient. J Neurosurg 45: 432–436
9. Gunnar W, Jonasson O, Merlotti G, Stone J, Barrett J (1988) Head injury and hemorrhagic shock: studies of the blood brain barrier and intracranial pressure after resuscitation with normal saline solution, 3% saline solution, and dextran-40. Surgery 103: 398–408
10. Tommasino C, Moore S, Todd MM (1988) Cerebral effects of isovolemic hemodilution with crystalloid or colloid solutions. Crit Care Med 16: 862–868
11. James HE, Laurin RA (1981) Intracranial hypertension and brain oedema in albino rabbits. Part 1: experimental models. Acta Neurochir (Wien) 55: 213–226
12. Herrmann HD, Neuenfeldt D (1972) Development and regression of a disturbance of the blood brain barrier and of edema in tissue surrounding a circumscribed cold lesion. Exp Neurol 34: 115–120
13. Ferszt R, Hahm H, Cervos-Navarro J (1980) Measurement of the specific gravity of the brain as a tool in brain edema research. In: Cervos-Navarro J, Ferszt R (eds) Advances in neurology, Vol 28. Brain edema. Raven, New York, pp 15–26
14. James HE, Langfitt TW, Kumar VS, Ghostine SY (1977) Treatment of intracranial hypertension: analysis of 105 consecutive, continuous recordings of intracranial pressure. Acta Neurochir (Wien) 36: 189–200
15. Langfitt TW (1972) Pathophysiology of increased ICP. In: Brock M, Dietz H (eds) Intracranial pressure. Springer Berlin Heidelberg New York, pp 361–364
16. Rowan JO, Johnston IH, Harper AM, Jennett WB (1972) Perfusion pressure in intracranial hypertension. In: Brock M, Dietz H (eds) Intracranial pressure. Springer, Berlin Heidelberg New York, pp 165–170
17. Virgilio RW, Smith DE, Zarins CK (1979) Balanced electrolyte solutions: experimental and clinical studies. Crit Care Med 7: 98–106
18. Fishman RA, Chan PH (1980) Metabolic basis of brain edema. In: Cervos-Navarro J, Ferszt R (eds) Advances in neurology, Vol 28. Brain edema. Raven, New York, pp 207–215
19. Bakay L, Crawford JD, White JC (1954) Effects of intravenous fluids on cerebrospinal fluid pressure. Surg Gynecol Obstet 99: 48–52
20. Fox JL, Falik JL, Shalhoub RJ (1971) Neurosurgical hyponatremia: the role of inappropriate antidiuresis. J Neurosurg 34: 506–514
21. McLaurin RL, King LR, Elam EB, Budde RB (1960) Metabolic response to craniocerebral trauma. Surg Gynecol Obstet 110: 282–288
22. Wise BL (1965) Fluid and electrolytes in neurological surgery. Thomas, Springfield, Ill
23. Shenkin HA, Bouzarth WF, Tatsumi T (1968) The analysis of body water compartments in postoperative craniotomy patients. 1. The effects of major brain surgery alone. 2. The effects of mannitol administered preoperatively. J Neurosurg 28: 417–428
24. Shenkin HA, Gutterman P (1969) The analysis of body water compartments in postoperative craniotomy patients. The effects of dexamethasone. J Neurosurg 31: 400–407
25. Steinbok P, Thompson GB (1978) Metabolic disturbances after head injury: abnormalities of sodium and water balance with special reference to the effects of alcohol intoxication. Neurosurgery 3: 9–15
26. Aikawa JK (1950) Fluid volumes and electrolyte concentrations in normal rabbits. Am J Physiol 162: 695–702
27. Klatzo I, Wisniewski H, Steinwall O, Streicher E (1976) Dynamics of cold injury edema. In: Klatzo I, Seitelberger F (eds) Brain edema. Springer, Berlin Heidelberg New York, pp 554–563

Correspondence and Reprints: Hector E. James, 7930 Frost Street, Suite 304, San Diego, California 92123, U.S.A.

Acta Neurochir (1993) [Suppl] 57: 94–101

Ischemia as an Excitotoxic Lesion: Protection Against Hippocampal Nerve Cell Loss by Denervation

N. H. Diemer, F. F. Johansen, H. Benveniste, T. Bruhn, M. Berg, E. Valente, and **M. B. Jørgensen**

Cerebral Ischemia Research Group and Pharmabiotec Research Center, Institute of Neuropathology, University of Copenhagen, Denmark

Abstract

There are several indications for an involvement of neuroexcitatory mechanisms in ischemic neuron damage. Since we forwarded the hypothesis in 1982 that the transmitter glutamate is playing a key role, several lines of evidence have substantiated this: there is a pronounced transmitter release induced by ischemia and there is uptake of Ca^{++} via NMDA-operated calcium channels. Under certain circumstances postischemic neuron death can be impaired by administration of either NMDA-antagonists or calcium blockers.

Further proof for the induction of harmful excitatory mechanisms by ischemia has been obtained by preischemic denervation of the vulnerable nerve cells. After transient cerebral ischemia in rats or gerbils, there are signs of irreversible damage (eosinophilia) of neurons in the dentate hilus (somatostatin-positive cells) after 2–3 hours and of hippocampal pyramidal neurons after 2–3 days (delayed neuron death). In the first case, removal of the (main) input to hilus cells by degranulation (colchicine selectively eliminates granule cells) protects these. In the case of pyramidal neurons removal of Schaffer collaterals/commisurals or input from the entorhinal cortex have a protective effect.

Recently, we have measured glutamate and calcium in CA1 of denervated rats during 10 min of ischemia, and it turns out that there is almost no extracellular glutamate release or lowering of calcium in contrast to ischemic animals with intact innervation.

Also in the postischemic period there are indications of a continuation of the damaging processes induced by ischemia. Besides the well known postischemic hypoperfusion, a prolonged release of glutamate has been reported, as well as burst firing in some models. If an immediately postischemic denervation of CA1 neurons is performed, there is a partly protection of these cells.

The GABA-ergic interneurons, which are lying among the pyramidal neurons in CA1 are always resistant to ischemia; receptor autoradiography indicates that they have glutamate receptors of the kainate/quisqualate type but no (or few) of the NMDA-type.

Keywords: Ischemic neuronal damage; excitotoxicity; denervation; NMDA-receptor.

Introduction

The neuron loss in cerebral ischemia is characterized by selectivity: within even small regions one type of nerve cell can be resistant and another one vulnerable. This difference in vulnerability between even neighbouring cells (Johansen *et al.*, 1984) eliminates local intraischemic and postischemic flow differences as being responsible for the selective neuron loss, since flow regulation influences groups of cells. Cerebral ischemia has been studied in a number of animal models, mainly in rats; although the most popular models differ somewhat with respect to neuropathological changes, often a hierarchy of vulnerable cells is indicated. The development of morphological changes (acidophilia, eosinophilia) takes from a few hours (dentate hilus) to 4 days or longer (hippocampal CA1 region, delayed neuronal death). More severe ischemia (i.e. longer occlusion) results in earlier occurring eosinophilia and vice versa, the so-called maturation phenomenon (Ito, 1975; Schmidt-Kastner and Hossmann, 1988). Even in the different rat models of complete forebrain ischemia the neuropathological damage is variable, i.e. in the Siemkowicz-Hansen model (carotid artery occlusion plus hypovolemic hypotension) 14 minutes of ischemia results in the same damage to hippocampus (Diemer and Siemkowicz, 1981) as 20 min in the Pulsinelli-Brierley (1979) model (4-VO-model). Since the neuronal damage induced by ischemia has a multifactorial pathogenesis, where the involved factors, if given enough impact, in itself is capable to destroy vulnerable neurons, it is advisable to study as short periods of ischemia as possible when the effect of a single factor has to be evaluated. In the case of the influence of the excitatory input to the ischemia vulnerable CA1 neurons of hippocampus, which is the main subject of the present review, the use of short ischemia periods revealed the significance of this factor (Diemer *et al.*, 1987).

The Neuropathological Lesion

A neuropathological study of the brain several days after ischemia reveals that a number of cell types are involved. The hematoxylin-eosin stain is suitable for the detection of eosinophilic neurons, but the Fink-Heimer impregnation shows both dead nerve cell bodies as well as their processes, and is superior if the distribution of ischemic damage has to be mapped. Recently it has been shown that also ^{45}Calcium-autoradiograms can be used for mapping of the damage (Dienel, 1986; Benveniste and Diemer, 1988; Joo, personal communication).

Some of the sensitive nerve cell types together with their proposed transmitters are listed below:

- modified pyramidal cells, multipolar neurons of dentate hilus (glutamate, somatostatin)
- pyramidal neurons of subiculum (glutamate)
- pyramidal cells of hippocampal CA1 (glutamate)
- neurons of lateral reticular nucleus of thalamus (GABA)
- septal neurons (Ach)
- purkinje cells (GABA)
- medium sized neurons of striatum (GABA)
- cortical layers 3 and 5 (glutamate)
- substantia nigra, pars reticulata (dopamine, GABA)

One of the most studied brain regions is the hippocampal CA1 (Fig. 1). In the rat, more than 8 min of ischemia destroys the pyramidal cells whereas the interneurons (about 10–15% of the total neuron population in this region) are resistant (Johansen *et al.*, 1984; Johansen, unpublished; Crain *et al.*, 1988). Concomitant with the nerve cell loss there is an increase in glial cell number, mainly due to invasion by microglial cells (rod-shaped nucleus). For a neuropathological study, perfusion fixation is a prerequisite in order to avoid confusion between the dark neuron artifact of Cammermeyer and ischemically damaged (eosinophilic) neurons.

A study of the different stages of ischemic cell injury in the Gerbil (Ito *et al.*, 1975) revealed a maturation phenomenon, i.e. the denser the ischemia, the faster the development of these stages. The size of the neuron is not determining the maturation speed; thus small sized striatal neurons and large hilar neurons are among the first to disintegrate after ischemia. As shown by Kirino (1985) in the gerbil, three different types of cell changes could be observed: in the hilus the cell changes were similar to the ischemic cell change, in CA2 there were changes of the reactive type (Ito *et al.*, 1975) and in the CA1 they described changes typical for delayed neuronal death after shorter periods of ischemia, whereas longer periods shifted the pattern into the ischemic cell change. Electron microscopically, stratum radiatum of CA1 4 days after ischemia shows damage to dendrites whereas the axons (and astrocytes) are spared (Johansen *et al.*, 1985). A similar pyramidal cell lesion can be seen after i.v.

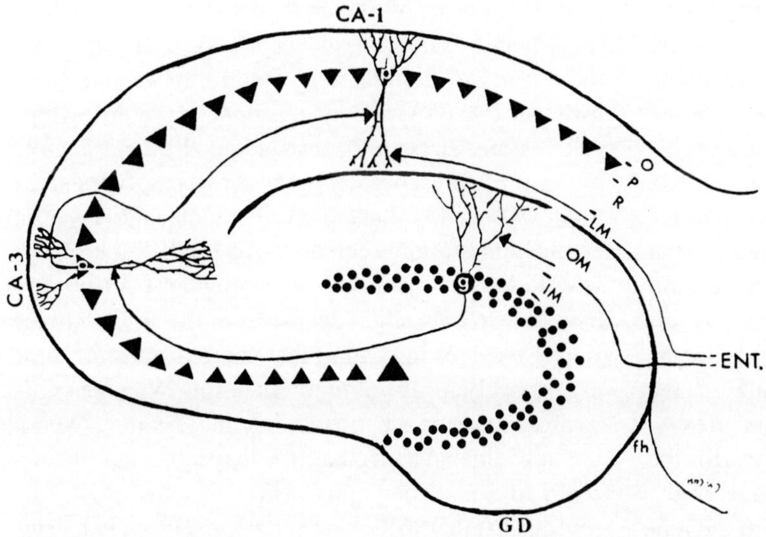

Fig. 1. Schematic drawing of a frontal section through dorsal hippocampus. The pyramidal cell layer is indicated by triangles and the granule cell bodies by filled circles *(g)*. A dendritic tree is shown for each of the neuron types in CA1, CA3 and dentate fascia, respectively. The axons of components of the trisynaptic pathway, referred to in the text, are shown: perforant path from the entorhinal cortex to dentate gyrus, mossy fibers from granule cells to CA3 and Schaffer collaterals from CA3 to CA1; *O* = stratum oriens; *p* = pyramidal layer; *R* = stratum radiatum; *LM* = stratum lacunosum moleculare; *OM* = outer molecular layer; *IM* = inner molecular layer; *ENT* = eutorhinal cortex; *GD* = dentate gyrus

injection of the glutamate analogue, Kainic acid, which also results in axon sparing and selective neuron death. From this resemblance between the ischemic lesion and the kainic acid lesion came the idea that excitotoxic mechanisms could be operating in ischemia (Jørgensen and Diemer, 1982).

Hippocampal "Wiring" and Glutamate Receptor Distribution

A major contribution to the description of hippocampal anatomy was given already by Cajal in 1893. Lorente de No (1934) classified the hippocampus into four subfields, CA1, CA2, CA3 and CA4 based on the morphology of neuronal processes. The internal architecture is somewhat similar in the frequently used rodents (mice, gerbils and rats) and in the human and the ischemia vulnerability likewise strikingly similar. The macroscopic structure of hippocampus can be compared to a bent tube with the hippocampus and the dentate fascia inserted into each other. Thereby a ventral and a dorsal component is seen on most coronal sections. In most ischemia models only the dorsal part is damaged. Compared to neocortex, the light microscopical structure of hippocampus is much simpler; it has only three layers with a distinct organization of somata in stratum pyramidale and dendrites in stratum oriens and stratum radiatum/lacunosum moleculare.

The main wiring of the hippocampal formation and the entorhinal cortex is described as a "trisynaptic pathway" (Fig. 1). Neurons in the entorhinal cortex innervate via the perforant path the dentate granule cells. These in turn innervate via the mossy fibers the CA3 pyramidal cells. The CA3 neurons innervate the CA1 ipsilateral pyramidal cells via the Schaffer collaterals and the contralateral CA1 pyramidal cells via the commissural fibers. Fink-Heimer impregnations for degenerated axons after a Schafferotomy show an equal contribution to stratum radiatum and to stratum oriens.

The most apical dendrites of the pyramidal cells, situated in the stratum lacunosum-moleculare have also a direct innervation from the entorhinal cortex.

The output from CA1 is mainly directed towards the subiculum and septum. The morphological axis of the trisynaptic circuit is oriented perpendicular to the longitudinal axis of the hippocampal formation (lamellar structure). The border between the CA1 and CA3 regions is not always easy to distinguish in the Nissl (cresyl violet) stain, whereas the Timm-stain for free zinc content shows that only CA3 has a mossy fiber input. The question whether the CA2 cells (Spielmeyer's sector) are resistent to ischemia has not been fully settled.

About 10% of the hippocampal nerve cells are interneurons. Thereby the region is resembling neocortex, but is different from e.g. striatum which is also ischemia sensitive, but where most neurons are inhibitory interneurons. Most of the GABAergic hippocampal interneurons are also peptidergic (Somatostatin, Cholecystokinin, Vasointestinal Polypeptide, Neuropeptide Y, etc). The inhibitory terminals are arranged as basket terminals as wells as rows of dendritic inhibitory terminals (symmetric synapses with elongated vesicles). Parvalbumin staining of interneurons in CA1 display their dendrites distinctly and reveal that they are running parallel to the dendrites of the pyramidal cells and reaching the stratum lacunosum-pyramidale.

Generally, the use of quantitative receptor autoradiography shows that the density of glutamate receptors is high in hippocampus (Halpain *et al.*, 1983; Greenamyre *et al.*, 1983; Monaghan *et al.*, 1983). However the subtypes are located somewhat differently. The following applies to the rat brain. The kainic acid (KAIN) preferring receptor has its highest density in stratum lucidum of CA3 and in the supragranular layer of the dentate fascia whereas the most ischemia sensitive subregions, CA1, and dentate hilus have only moderate densities of KAIN-receptors. On the other hand there is a high density of kainic acid receptors in the lateral reticular nucleus of thalamus, which is very sensitive to ischemia (GABA-ergic neurons). The two other main types of glutamate receptors, the N-methyl-D-aspartate (NMDA) and the quisqualate preferring type (QUIS) are especially localized to the CA1 region, where the highest density of these receptor types in the brain is found, followed by the molecular layer of cerebellum (Purkinje cell dendritic field). Experiments with lesions of pyramidal cells in either CA1 or CA3 have shown only slight decreases of the binding to the KAIN-receptor indicating that some or most of these are probably localized to interneurons (Valente *et al.*, in press). On the other hand, similar lesioning experiments evaluated with quantitative receptor autoradiography of a non competitive NMDA-ligand (MK 801) reveals that the NMDA receptor is exclusively located postsynaptically, on the pyramidal cells. In case of the QUIS preferring receptor, a CA3 lesion reveals that it is not located presynaptically in CA1; on the other hand a CA1 pyramidal cell lesion decreases only the density to about 50%, indicating that the QUIS receptor is occurring both postsynaptically at the CA1 pyramidal cell and at the interneurons. The division of the glutamate receptors into these 3 subtypes seems not to be sufficient any longer; thus the QUIS receptor can be divided into two types, one which by stimulation mediates an inward calcium flux through voltage sensitive calcium channels (receptor antagonists: CNQX (Ferrosan), AMNH (PharmaBiotec, P. Krogsgaard-Larsen) and another type, the so-called metabotropic QUIS receptor (Sugyama *et al.*, 1986) which via inositoltrisphosphate (IP3) induces calcium release from the endoplasmic reticulum (no known antagonists at the receptor site).

Selective Pathway Cuttings

Due to the relatively simple trilaminar structure of hippocampus (archicortex) this region has been one of the most used for denervation studies. Thus the ischemia vulnerable CA1 cells can be (subtotally) denervated by a parasagittal cut of Schaffer collaterals and commissurals or after destruction of CA3 (with kainic acid, ibotenic acid etc). Bilateral lesion of CA3, as it is seen after intraventricular injection of kainic acid, removes about 80% of the presynaptic terminals on the CA1 dendrites. As earlier mentioned, the apical tips of the CA1 pyramidal cells are innervated by the perforant path (see also Fig. 2), and this

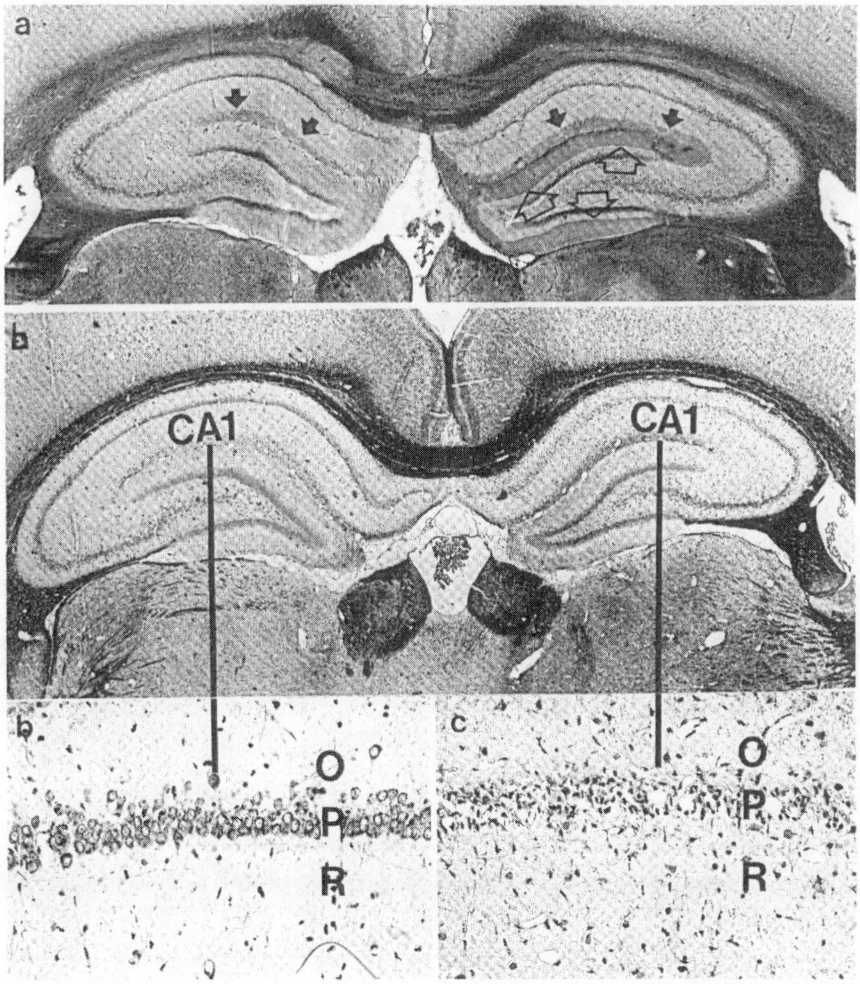

Fig. 2. (a) Fink-Heimer impregnation of frontal section from a rat with unilateral entorhinotomy. Degenerated axon terminals are seen bilaterally in the stratum lacunosum moleculare (filled, small arrows) and ipsilaterally in the outer 2/3 of stratum moleculare of the dentate gyrus (open, large arrows). (b) Kluwer-Barrera stained frontal sections from a rat with pre-ischemic unilateral (left) entorhinotomy; postischemic survival time was 4 days. The higher power magnifications from each CA1 region show only a slight loss of pyramidal cells at the left side, whereas the right side (c) shows total pyramidal cell loss, a slight astrogliosis and numerous rod-shaped microglial cells. *P* pyramidal layer; *O* stratum oriens; *R* stratum radiatum

innervation can be eliminated by a cut through the angular bundle or removal of the proper parts of entorhinal cortex (Fig. 2). Franck *et al.* (1988) reported that the CA1 pyramidal neurons became hyperexcitable after a CA3 lesion, however up to several months after a CA3 lesion, no damage to CA1 cells was observed (Jørgensen and Diemer, unpublished). The granule cells of the dentate fascia can be selectively removed by local injection of colchicine or radiation early in life.

Effect of Denervation on Loss of CA1 Pyramidal Cells After Ischemia

The first study on the effect of denervation on excitatory transmitter mediated nerve cell damage was done by Kohler *et al.* (1978). In rats an injection of kainic acid into the dentate fascia produced loss of granule cells. If the animals had a perforant path transsection, the injection caused no loss. Since the lesion produced by ischemia has similarities to that produced by i.v. kainic acid, a further proof of the excitatory hypothesis would be a protective effect of a denervation of CA1. The first lesion we studied in the 4-vessel occlusion rat model with 20 min ischemia was unilateral injection of kainic acid in the CA3 region, destroying it at the part where the Schaffer collaterals to the vulnerable CA1 cells arises. It turned out that some of the animals had a protective effect of this lesion (Jørgensen and Diemer, 1984, unpublished). However, further studies were needed because the lesion did not have a protective effect in all of the animals. It was later found out that if ischemia was reduced to 10 min in the same model, all rats displayed a protective effect (Benveniste *et al.*, 1989).

In 1985 Wieloch and coworkers described the protective effect of removal of some of the input from the entorhinal cortex, the angular bundle (Wieloch *et al.*, 1985). We also studied other lesions of the trisynaptic pathway, removal of the granule cells (Johansen *et al.*, 1985) or of the entorhinal cortex (Fig. 2; Jørgensen *et al.*, 1986) where we found 80 and 85% protection, respectively, after 20 min of 4-VO ischemia. Both the preischemic Schafferotomy and the degranulation reduces the CA1 pyramidal cell loss from about 90% (the remaining 5–10% neurons in the pyramidal cell layer are interneurons) to about 20%. (Diemer *et al.*, 1987). Of further interest for the influence of events taking place postischemia was the finding that even an (immediate) post-ischemic Schafferotomy reduced the CA1 damage to about 50% (Johansen *et al.*, 987).

Effect of Denervation on Transmitter Release in Hippocampus

One of the pathological mechanisms which is involved in hippocampal ischemic damage is the release of the excitatory amino acids glutamate and aspartate combined with the arrest of their energy dependent uptake mechanisms (Benveniste *et al.*,1984; Hagberg *et al.*, 1985). They are released at 3.5 to 8 fold their normal level (together with GABA and taurine) in both hippocampus (Fig. 3) and striatum whereas it is not known whether they are released in non-vulnerable regions. However, if there is a relationship between density of glutamate receptors and release of transmitter glutamate, the highest release should take place in CA1 and in the molecular layer of cerebellum, whereas a region like thalamus should have a low release. There is both a calcium dependent and a calcium independent component of glutamate release. During the first 10 min of ischemia only the calcium dependent pool is released (transmitter glutamate), but later also glutamate from the metabolic pool is released (see also Kauppinen *et al.*, 1988). Thus it was demonstrated that a longer period of ischemia resulted in higher concentrations of glutamate and aspartate. It takes 5–10 min before extracellular glutamate is cleared postischemia (Benveniste *et al.*, 1989).

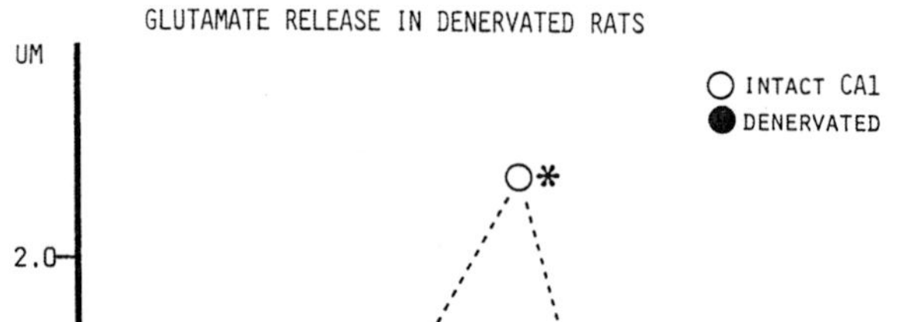

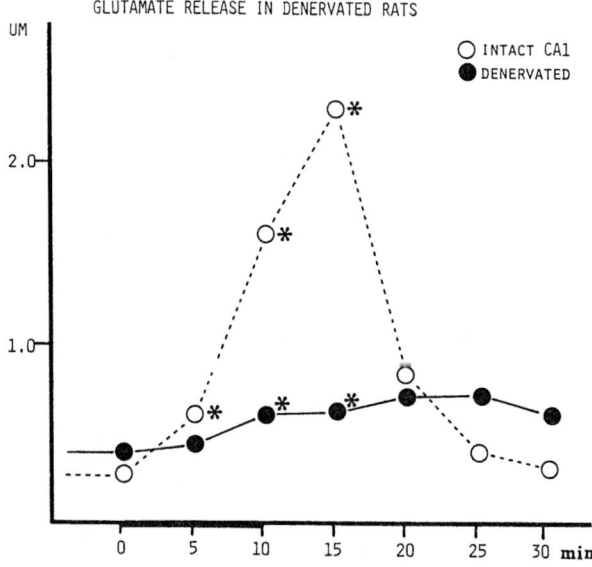

Fig. 3. Microdialysis measurements of extracellular glutamate release during ischemia in control rats (open circles, n = 9) and in rats with a bilateral CA3 lesion (filled circles, n = 7). The graph shows extracellular conc in umoles vs. time after induction of ischemia (which was of 10 min duration). The asterisks indicate significant differences between the two groups

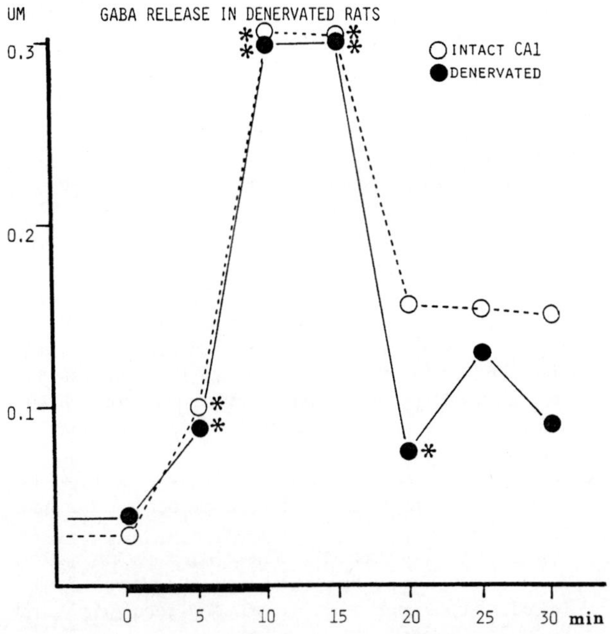

Fig. 4. Microdialysis measurements of extracellular GABA release during ischemia in control rats (open circles, n = 9) and in rats with a bilateral CA3 lesion (filled circles, n = 7). The graph shows extracellular conc. in umoles vs. time after induction of ischemia. Asterisks indicate statistically significant differences from the baseline levels of the two groups

In the denervated rat (kainic acid induced destruction of CA3 and its axons to CA1) 10 min of ischemia did not produce the dramatic increase of extracellular glutamate or aspartate seen in non-lesioned animals (Benveniste *et al.*, 1984) and extracellular calcium did not decrease significantly (Benveniste *et al.*, 1988). However, GABA showed the same pronounced increase as in non-lesioned ischemic rats (Fig. 4), illustrating that GABA (which is stored in interneurons) can be released by potassium which is increasing slowly during ischemia in the denervated rats (Benveniste *et al.*, unpublished results).

Effect of Denervation on Postischemic Glucose Metabolism

The pattern of regional cell loss is not correlated to the preischemic (normal) blood flow or glucose metabolism (CMRglu) as shown by autoradiography (Sakurada *et al.*, 1969, Sokoloff *et al.*, 1977) indicating that the preischemic level of regional energy metabolism does not determine the outcome of a region. Immediately after postischemic recirculation is initiated, however, ^{14}C-2-deoxyglucose (2-DG) autoradiograms show a striking pattern (Fieschi *et al.*, 1978, Choki *et al.*, 1983, Diemer and Siemkowicz 1980, Pulsinelli *et al.*, 1982). Using the neck cuff/hypotension 10 min ischemia model (normoglycemic rats) and an isotope circulation period from 5 to 15 min, increased accumulation of 2-DG was found in hippocampal CA1, cerebellar molecular layer, substantia nigra and globus pallidus (Diemer and Siemkowicz, 1980a,b). All other brain regions showed about the same accumulation. However, in hyperglycemic rats, the 2-DG accumulation pattern was totally uniform, 15 min after ischemia. In the 4-VO model, using a circulation period of 45 min starting at 60 min after ischemia a somewhat similar pattern was found. The calculated values of the glucose metabolism were generally decreased by about 30% except in hippocampus and dentate fascia (Jørgensen *et al.*, 1989). If the animals had a CA3 lesion with denervation of the dorsal CA1 sector, the postischemic glucose metabolism was uniformly low throughout the brain except for the gliotic CA3 area which showed unchanged deoxyglucose accumulation, most possible due to anaerobic glycolysis of the macrophages in the region.

In most of the studies on the postischemic glucose metabolism, the usual 45 min tracer circulation period has been used, but since cerebral blood flow is changing after ischemia it can be debated whether the calculated CMRglu values obtained immediately after ischemia are correct, but still it is striking that only the vulnerable regions have increased and later normal glucose metabolism while all other regions are depressed up to several days after ischemia (Jørgensen *et al.*, 1989).

The high postischemic deoxyglucose accumulation in CA1 (Diemer and Siemkowicz, 1980) could be due to increased Na^+-K^+ pumping. After ischemia there is shortly a further decrease in pH (Hossmann, 1985) probably due to a formation of lactate, which can be the result of glycolytic energy production used to normalize the disturbed ion gradients (Andersen and Marmarou 1989). Pappius (1988) found that the general decrease in glucose metabolism after brain injury (freezing lesion) was due to an increase of serotonin metabolism and that p-chlorophenylalanine, a 5-HT synthesis blocker, normalized the cortical LCGU in the traumatized hemisphere. It was also found that the noradrenergic system was involved by using the noradrenergic blocker, prazosine. These findings indicate an inhibitory role of serotonin and norepinephrin in cerebral cortex after certain injuries, maybe including ischemia; a possibility which is now tested out in our laboratory.

Comments

In hippocampus, cerebral ischemia is accompanied by a pronounced increase of excitatory amino acids in the extracellular space. This is due to a combination of release from terminals and inhibition of reuptake processes, which normally controls the level of glutamate. Removal of the excitatory afferents to the CA1 pyramidal cells has a protective effect against ischemia induced damage most probably due to the lack of glutamate and aspartate release. A further proof is that substitution of glutamate by injection into CA1 during ischemia in such denervated animals abolish the protection otherwise provided by the denervation (Benveniste *et al.*, 1989).

The elevated concentration of excitatory amino acids stimulates the QUIS and NMDA receptors in CA1. This in turn leads to depolarization and suspension of the Mg^{++} block of the NMDA operated calcium channel with calcium influx into the cell (Benveniste *et al*, 1988). In some ischemia models the non-competitive NMDA antagonist, MK 801, has a protective effect on CA1, and although excessive stimulation of the NMDA receptor leading to calcium influx is a tempting explanation, it has not been proven that this source of Ca^{++} is the main mechanism in the process of cell damage. Calcium influx starts already at about 2 min after occlusion but if ischemia is terminated 8 min after occlusion there is no cell loss in CA1 (Diemer, unpublished). Furthermore, if the cortex is superfused with 10 mM Ca^{++} for 2 hours before 10 min of forebrain ischemia, there is still no cell loss in the vulnerable cortical layers. Another mechanism of increasing intracellular free calcium is by activation of the metabotropic QUIS receptor which leads to an inositoltrisphosphate mediated calcium release from endoplasmic reticulum. There are several indications that this mechanism is operating also in the post-ischemic period. The main excitatory synapse (Schaffer collateral terminations) in CA1, seems to be of the QUIS (or non-NMDA) type (Herreras *et al.*, 1989). Immediate postischemic cutting of the Schaffer collaterals has a protective effect on ischemia induced CA1 damage (Johansen *et al.*, 1987), and administration of QUIS-antagonists in different models of neurotoxicity and ischemia (Garthwaite and Garthwaite, 1989; Sheardown *et al.*, in press; Diemer *et al.*, unpublished results) has a pronounced protective effect in hippocampus. Until now there are no consistent reports on increased postischemic release of excitatory amino acids or on upregulation of QUIS (or other glutamatergic) receptors in hippocampus, but the mechanism could also be an ischemia-induced malfunction at the second messenger level.

References

1. Andersen, BJ, Marmarou A (1989) Energy compartmentalization in neural tissue. J Cereb Blood Flow Metabol 9 [Suppl 1]: S386
2. Auer RN, Siesjö BK (1988) Biological differences between ischemia, hypoglycemia, and epilepsy. Ann Neurol 24: 699–707
3. Benveniste H, Drejer J, Schousboe A, Diemer NH (1984) Elevation of the extracellular concentrations of glutamate and aspartate in rat hippocampus during transient cerebral ischemia monitored by intracerebral microdialysis. J Neurochem 43: 1369–1374
4. Benveniste H, Diemer NH (1988) Early Postischemic 45 Ca accumulation in rat dentate hilus. J Cereb Blood Flow Metabol 8: 713–719
5. Benveniste H, Jørgensen MB, Diemer NH, Hansen AJ (1988) Calcium accumulation by glutamate receptor activation is involved in hippocampal cell damage after ischemia. Acta Neurol Scand 78: 529–536
6. Benveniste H, Jørgensen MB, Sandberg M, Hagberg H, Diemer NH (1989) Ischemic damage in hippocampal CA1 is dependent on glutamate-release and intact innervation from CA3. J Cereb Blood Flow Metabol 9: 629–639
7. Buchan AM, Pulsinelli WA (1989) Fimbria-fornix lesions: The temporal profile for protection of CA1 hippocampus against ischemic injury. J Cereb Blood Flow Metabol 9 [Suppl 1]: S749
8. Cajal SR y (1893) Über die feinere Struktur des Ammonshornes. Z Wiss Zool 56: 619–663
9. Choki J, Greenberg J, Reivich M (1983) Regional cerebral glucose metabolism during and after bilateral cerebral ischemia in the gerbil. Stroke 14: 568–574
10. Crain BJ, Westerkam WD, Harrison AH, Nadler JV (1988) Selective neuronal death after transient forebrain ischemia in the Mongolian gerbil: a silver impregnation study. Neuroscience 27: 387–402
11. Diemer NH, Siemkowicz E (1980a) Increased 2-deoxyglucose uptake in hippocampus, globus pallidus and substantia nigra after cerebral ischemia. Acta Neurol Scand 61: 56–63
12. Diemer NH, Siemkowicz E(1980b) Regional glucose metabolism and nerve cell damage after cerebral ischemia in normo- and hypoglycemic rats. In: Spatz M, Mrsjulja BB, Rakic LJ, Lust WD (eds) Circulatory and developmental aspects of brain metabolism. Plenum, New York, pp 23–32
13. Diemer NH, Siemkowicz (1981) Regional neuron damage after cerebral ischemia in the normo- and hypoglycemic rat. Neuropath Appl Neurobiol 7: 217–227
14. Diemer NH, Jørgensen MB, Johansen FF (1987) Significance of intra- and postischemic pathophysiological processes for development of ischemic nerve cell loss. In: Raichle ME, Powers WJ (eds) Cerebrovascular diseases. Raven, New York
15. Diemer NH, Sandberg M, Jørgensen MB, Benveniste H (1989) Ischemia-induced release of glutamate in the hippocampal CA1 region is decreased after removal of the excitatory input from CA3. J Cereb Blood Flow Metabol 9 [Suppl 1]: S747
16. Dienel GA, Pulsinelli WA (1986) Uptake of radiolabelled ions in normal and ischemia-damaged brain. Ann Neurol 19: 465–472
17. Fieschi C, Sakurada O, Sokoloff L (1978) Local cerebral glucose utilization during resolution of embolic experimental ischemia. In: Cervos-Navarro J, *et al* (eds) Advances of neurology, 20. Raven, New York, pp 223–229

18. Franck JE, Kunkel DD, Baskin DG, Schwartzkroin PA (1988) Inhibition in kainate-lesioned hyperexcitable hippocampi: Physiologic, autoradiographic, and immunocytochemical observations. J Neurosci 8: 1991–2002
19. Garthwaite G, Garthwaite J (1989) Quisqualate neurotoxicity – a delayed, CNQX sensitive process triggered by a CNQX-insensitive mechanism in young rat hippocampal slices. Neurosci Lett 99: 113–118
20. Greenamyre JT, Young AB, Penney JB (1983) Quantitative autoradiography of L-3H-glutamate binding to rat brain. Neurosci Lett 37: 155–160
21. Hagberg H, Lehmann A, Sandberg M, Nystrom B, Jacobson I, Hamberger A (1985) Ischemia-induced shift of inhibitory and excitatory amino acids from intra- to extracellular compartments. J Cereb Blood Flow Metabol 5: 413–419
22. Halpain S, Parsons B, Rainbow TC (1983) Tritium-film autoradiographic distribution of L-3H-glutamate binding sites in rat central nervous system. J Neurosci 4: 2133–2144
23. Herreras O, Menendez N, Herranz AS, Solis JM, Martin del Rio R (1989) Synaptic transmission at the Schaffer-CA1 synapse is blocked by 6,7-dinitro-quinoxaline-2,3,-dione. An *in vivo* brain dialysis study in the rat. Neurosci Lett 99: 119–124
24. Hossmann K-A (1985) post-ischemic resuscitation of the brain: selective vulnerability versus global resistance. In: Kogure K, Hossmann K-A, Siesjö BK, Welsh FA (eds) Progress in brain research, Vol 63. pp 3–17
25. Ito U, Spatz, M, Walker JT Jr, Klatzo I (1975) Experimental cerebral ischemia in Mongolian Gerbil. I. Light microscopic observations. Acta Neuropath 32: 209–233
26. Johansen FF, Jørgensen MB, Diemer NH (1983) Resistance of hippocampal CA1 interneurons to 20 min of transient cerebral ischemia in the rat. Acta Neuropath 61: 135–140
27. Johansen FF, Jørgensen MB, von Lubitz DKJE, Diemer NH (1984) Selective dendrite damage in hippocampal CA1 stratum radiatum with unchanged axon ultrastructure and glutamate uptake after transient cerebral ischemia in the rat. Brain Res 291: 373–377
28. Johansen FF, Jørgensen MB, Diemer NH (1987) Ischemia-induced neuronal death in the CA1 hippocampus is dependent on intact glutamatergic innervation. In: Hicks TP, Lodge D, McLennan H (eds) Excitatory amino acid transmission. Liss, New York, pp 245–248
29. Johansen FF, Zimmer J, Diemer NH (1987) Early loss of somatostatin neurons in dentate hilus after cerebral ischemia in the rat precedes CA1 pyramidal cell loss. Acta Neuropath 73: 110–114
30. Jørgensen MB, Diemer NH (1982) Selective neuron loss after cerebral ischemia in the rat possible role of transmitter glutamate. Acta Neurol Scand 66: 536–546
31. Jørgensen MB, Johansen FF, Diemer NH (1987) Removal of the entorhinal cortex protects hippocampal CA1 neurons from ischemic damage. Acta Neuropath 73: 189–194
32. Jørgensen MB, Wright DC (1988) The effect of unilateral and bilateral removal of the entorhinal cortex on the glucose utilization in various hippocampal regions in the rat. Neurosci Lett 87: 227–232
33. Jørgensen MB, Wright DC, Diemer NH (1989) The effect of CA1 lesion and CA3 lesion on the postischemic glucose metabolism in the rat brain. J Cereb Blood Flow Metabol 9 [Suppl 1]: S552
34. Kameyama M, Wasterlain CG, Ackermann RF, Finch D, Lear J, Kuhl DE (1983) Neuronal response of the hippocampal formation to injury: Blood flow, glucose metabolism and protein synthesis. Exp Neurol 79: 329–346
35. Kauppinen RA, McMahon HT, Nicholls DG (1988) $Ca2^+$ dependent and $Ca2^+$ independent glutamate release, energy status and cytosolic free $Ca2^+$ concentration in isolated nerve terminals following metabolic inhibition – possible relevance to hypoglycemia and anoxia. Neuroscience 27: 175–182
36. Kirino T, Tamura A, Sano K (1985) Selective vulnerability of the hippocampus to ischemia – reversible and irreversible types of ischemic cell damage. In: Kogure K, Hossmann K-A, Siesjö BK, Welsh FA (eds) Progress in brain research, Vol 63. pp 39–58
37. Kohler C, Schwarz R, Fuxe K (1978) Perforant path transections protect hippocampal granule cells from kainate lesion. Neurosci Lett 10: 241–246
38. Korf J, Klein HC, Venema K, Postema F (1988) Increases in striatal and hippocampal impedance and extracellular levels of amino acids by cardiac arrest in freely moving rats. J Neurochem 50: 1087–1096
39. Lee KS, Kreutzberg GW (1987) The role of adenosine neuromodulation in postanoxic hyperexcitability. In: Gerlach E, Becker BF (eds) Topics and perspectives in adenosine research. Springer, Berlin Heidelberg New York Tokyo, pp 574–585
40. Linden T, Kalimo H, Wieloch T (1987) Protective effect of lesion to the glutamatergic cortico-striatal projections on the hypoglycemic nerve cell injury in rat striatum. Acta Neuropath (Berl) 74: 335–344
41. Martins E, Inamura K, Themner K, Malmquist KG, Siesjö BK (1988) Accumulation of calcium and loss of potassium in the hippocampus following transient cerebral ischemia: a proton microprobe study. J Cereb Blood Flow Metabol 8: 531–538
42. Monaghan DT, Holets RV, Toy DW, Cotman CW (1983) Anatomical distributions of four pharmacologically distinct 3H-glutamate binding sites. Nature 306: 176–179
43. Murphy SN, Miller RJ (1988) A glutamate receptor regulates $Ca2^+$ mobilization in hippocampal neurons. Proc Natl Acad Sci USA 85: 8737–8741
44. Novelli A, Reilly JA, Lysko PG, Henneberry RC (1988) Glutamate becomes neurotoxic via the N-methyl-D-aspartate receptor when intracellular energy levels are reduced. Brain research 45: 205–212
45. Pappius HM (1988) Significance of biogenic amines in functional disturbances resulting from brain injury. Metab Brain Dis 3: 303–310
46. Pulsinelli, WA, Brierley, JB (1979) A new model of bilateral hemispheric ischemia in the unanesthetized rat. Stroke 10: 267–272
47. Pulsinelli WA, Levy DE, Duffy TE (1982) Regional cerebral blood flow and glucose metabolism following transient forebrain ischemia. Ann Neurol 11: 499–509
48. Schmidt-Kastner R, Hossmann K-A (1988) Distribution of ischemic neuronal damage in the dorsal hippocampus of rat. Acta Neuropath (Berl) 76: 411–421
49. Sokoloff L, Reivich M, Kennedy C, Des Rosiers MH, Patlak CS, Pettigrew KD, Sakurada O, Shinohara M (1977) The (^{14}C) deoxyglucose method for the measurement of local cerebral glucose utilization: theory, procedure, and normal values in the conscious and anesthetized albino rat. J Neurochem 28: 897–916
50. Sugiyama H, Ito I, Hirono C (1987) A new type of glutamate receptor linked to inositol phospholipid metabolism. Nature 325: 531–533
51. Wieloch T, Lindvall O, Blomquist P, Gage FH (1985) Evidence for amelioration of ischemic neuronal damage in the hippocampal formation by lesions of the perforant path. Neurol Res 7: 24–26

Correspondence and Reprints: N. H. Diemer, Institute of Neuropathology, University of Copenhagen, 11 Frederik V's vej, DK-2100 Copenhagen, Denmark.

Acta Neurochir (1993) [Suppl] 57: 102–109

Recovery of Brain Function Following Ischemia

L. Symon

Gough-Cooper Department of Neurological Surgery, Institute of Neurology, London, U.K.

Abstract

Experimental evidence conveys clear suggestions that early reperfusion following at least focal cerebral ischemia in the primate is accompanied by a return of function demonstrably suspended during the ischemic period.

Complete and permanent arrest of the cerebral circulation has been known within seconds to lead to depression of brain electrical activity, and within minutes to gross disruption of the normal energy metabolism with failure of ionic homeostatic mechanisms. There is irreversible cell change and death within 5 to 10 minutes.

Very much more protracted periods of ischemia have been shown more recently to be associated with potential viability of neuronal function, and in clinical neurosurgery we have known for years that patients with established cerebral vascular occlusion and a dense neurological deficit may show quite evident improvement over months or years. In these protracted recoveries, the potential for re-learning in nervous circuits may play a part, but in more acute circumstances, for example in the progressive recovery from vasospasm, re-learning is clearly not a factor, and this demonstrates quite evidently that neurons at one moment apparently non-functioning, can again within a few minutes recover function even after hours of apparent suppression.

The experimental evidence is fairly well known. In this symposium and elsewhere we have presented a model of experimental occlusion of the middle cerebral artery in primates demonstrating irreversible recovery of electrical function after some 20 minutes of middle cerebral artery occlusion, and reversible recovery of ionic homeostasis after periods of up to an hour. Measurements of regional blood flow has established thresholds for electrical function of around 16 ml/100 g × min and for ionic homeostasis of around 10 ml/100 g × min.

Interestingly, pH changes appear to occur in the region of the higher flow threshold for electrical failure, and the same applies to the early movements of water as assessed by accumulation in tissue.

Many of the biochemical experiments from Hossmann's group demonstrate disintegration of biochemical aspects, such as protein synthesis during ischemia and the question of free radical generation remains uncertain.

This paper addresses a number of cases of careful monitoring during aneurysm surgery or during the recovery from sub-arachnoid hemorrhage associated with vasospasm in which protracted periods of dysfunction with definable occlusion as assessed by either direct operative observation, angiography, or bedside blood flow measurements with transcranial Doppler. The studies have shown unequivocal recovery in a time scale, which attests to the validity of the experimentally generated hypotheses in man.

Keywords: Focal cerebral ischemia; ischemic flow thresholds; recovery from ischemia.

Introduction

Within seconds of complete arrest of cerebral circulation there is depression of the brain electrical acticityl and within minutes gross disruption of the normal energy metabolism with failure of ionic homestatic mechanisms. Such disruption produces irreversible cell change and death follows within 5–10 minutes[7,8,23,26]. Considerable interest in recent years, however, has developed in the viability of neuronal function following much more protracted periods of ischaemia[13,20]. Clinically, neurosurgeons have long known that patients with established cerebral vascular occlusion and dense neurological deficit may show quite evident improvement over months or years. While the potential for re-learning in nervous circuits may play a part in such prolonged recovery, in more acute circumstances, as for example, progressive recovery from anaesthesia or resolution of neurological deficit from aneurysm surgery or vasospasm, it is clearly possible for neurones to be at one time apparently non-functioning and yet under conditions of improved perfusion return to normal.

This paper presents some views about the recovery of various brain functions after transient ischaemia and illustrates the scientific concept with a number of clinical examples.

Disturbances of Brain Function During the Development of Brain Ischemia

Research has been directed towards the analysis of a whole variety of aspects of nervous function during ischaemia. The most helpful observations in relation to human physiology have arisen from experimental models designed to produce vascular occlusion in defined territories severe enough to produce disturbance of function yet recoverable on reperfusion and which bears some resemblance to the characteristics of human cerebral vascular disease. On the other hand, a good deal of research has also employed methods of complete circulatory arrest which have a parallel only in similar circumstances in man and in which the disruption of neural function is more absolute than graded. Nevertheless in certain specific terms, such may be analysed in a helpful fashion.

From the point of view of electrical function, studies have been made of the differential effects of ischaemia on the electrophysiological activity of presynaptic terminals and synaptic transmission[5,6]. The experimental primate model of occlusion of the middle cerebral artery has been extensively studied as producing a graded reduction in blood flow in which the relationship between electrical functional failure and recovery on reperfusion may be studied in detail. Most of our personal observations of progression and irreversibility in brain ischaemia have been made in relation to this experimental model. Occlusion of the middle cerebral artery simulates the production of an acute clinical stroke[11,24] producing an ischaemic lesion whose extent is restricted to known regions of the cerebral hemisphere and which selectively interferes with neural pathways. Such ischaemia is more dense in the centre of the middle cerebral field and progressively diminishes towards the collateral bounds of anterior and posterior cerebral arteries[24]. The presence of such a gradient along the sensory motor strip enables assessment of the effects of ischaemia on a peripherally stimulated somatosensory evoked potential whose central transmission is unaffected by middle cerebral occlusion, the thalamic nuclei being outwith the area of ischaemia[16,18]. This technique enables measurement of regional blood flow as ischaemia is graded following the occlusion by progressive reduction in blood pressure. Flow is measured by the hydrogen clearance method[19] in the immediate area of the evoked potential electrode, during progressive ischaemia and reperfusion. An apparent threshold of around 16 ml/100 g min emerged from this study[4].

Using ion sensitive microelectrodes in the extracellular space to measure calcium, potassium and pH[2,3,9,10], it has also been possible to demonstrate similar threshold relationships to flow for each of these variables. While pH clearly changes early in ischaemia in relation to progressive exhaustion of buffering capacity and changes in brain lactate (at or around the level of failure of electrical response), breakdown in ionic homeostasis attends very much lower levels of flow. Increase in extracellular potassium concentration and decrease in extracellular calcium concentration did not occur unless local blood flow was reduced to the area of 10 ml /100 g min. At this point, sudden massive changes in potassium concentration indicated the flux of potassium from the intracellular space, and the rapid disappearance of calcium at only slightly lower levels from the extracellular space has also been shown to be associated with a considerable increase in intracellular calcium, a highly dangerous situation from the point of view of cell metabolism[12].

The resolution of ionic abnormalities has been studied in a number of experiments; our own group have shown normalisation of extracellular potassium activity after one hour of regional ischaemia in the monkey, and extracellular pH has been shown to recover to normal within 30 min in normo-glycaemic rats. Once again, effects of hyperglycaemia have been not only to produce more severe ischaemic and post ischaemic acidosis but also more prolonged acidosis which may never recover.

Such evidence that we have would suggest that calcium recovers more slowly than potassium but nevertheless will normalise within an hour of recirculation in the primate.

Ischaemic and Post Ischaemic Oedema

It is of particular interest to study the effects of ischaemia on water movement.

Work from our own laboratory[22,14] has indicated a definite relationship between the development of cerebral swelling and the intensity of brain ischaemia. In the primate model, for example, assessing brain water content by brain impedance measurements during life together with intracranial pressure changes or by the use of graded density kerosene-bromobenzene columns following the sacrifice of the animal, we found that significant ischaemia was associated at 1.5 hours with an increase in water content in the most densely ischaemic zones but also in the area of the penumbra where potassium movement was absent although evoked responses had been abolished. Movement of water was detected in our preparation when flow fell below 20 ml/100 g min.

When ischaemic areas were reperfused after 1.5 hours of the regional ischaemia of the primate modell it was associated with an increase rather than a decrease in ischaemic oedema. Our observations and those of Hossmann indicate that the degree of post-ischaemic oedema is determined by the initial flow reduction after vascular occlusion in the early phase of brain oedema which we have regarded as cytotoxic rather than vasogenic. Hossmann has also shown that following one hour of complete ischaemia there is a significant increase in tissue osmolality[15] from 308–353 mOsm, creating a gradient of about 50 mOsm between brain and blood. We have suggested that this increase in osmolality may be associated with failure of synthesis of large neural transmitter molecules, smaller precursors remain free significantly increasing osmolality as a result. The correspondence of the apparent threshold for the initiation of brain oedema and for the failure of the evoked response somewhat supported this hypothesis.

The Relationship of Blood Flow to Tissue Recovery

In almost all preparations in which ischaemia has been induced, reperfusion is accompanied by an initial brief period of hyperaemia followed by a prolonged phase of cerebral hypoperfusion. The complex problem of no-reflow remains to some extent unsettled. Initial experiments of Ames[1] have been regarded as flawed but there seems no doubt that vascular swelling and micro-aggregation within vessels during the period of ischaemia do result in irregular patchy perfusion in certain circumstances. The question of vascular occlusion by physical factors such as endothelial swelling remains controversial, but both increased vascular smooth muscle tone as a result of vasoactive substances released during ischaemia, or permeability changes resulting in pericapillary oedema could produce the increased cerebral vascular resistance demonstrated in certain circumstances in post-ischaemic reperfusion.

Functional Recovery of Tissue Following Ischaemia

In the experimental middle cerebral model, the degree of ischaemia in a given region determined whether or not the evoked response was significantly depressed. When the local blood flow fell below the critical level of about 16 ml/100 g min[6] there was a very severe electrophysiological depression. If this degree of ischaemia was maintained for longer than 15 minutes then it was unlikely that the EP would recover at least over the following first hour. During this period local blood flow could be observed to be restored but tissue PO_2 remained reduced on average to levels well below control. These experiments suggested distinct post-ischaemic hypometabolism, despite hyperaemic blood flow.

On the other hand if the critical level of ischaemia was not obtained, EP was much less depressed and in a subsequent post clip phase there was rapid recovery of the EP together with normal flow and a strong suggestion from PO_2 measurements of tissue hyperoxia suggesting the phenomenon of luxury perfusion as described by Lassen[17]. During EP recovery tissue PO_2 tended to be greater than normal probably associated with vasodilatation or local metabolic acidosis[17,27]. Such phenomenon persisted through the post-ischaemic recovery when the tissue flow was not on the average above normal and could perhaps be below it, suggesting that the presence of hyperoxia might well be a necessary condition for altered or complete recovery of function.

Experimental Evidence of Infarct Recovery

Occlusion of the middle cerebral artery in the baboon produces a typical clinical course and a variable hemiparesis. In all instances, however, distinct neurological deficit could be detected and in its most dense portions usually affecting the face, this deficit remained detectable for periods of up to 3 years[24]. However, even in the most severely affected animals, leg weakness would persist in a severe form for only a day or two following the experimental infarct and would thereafter rapidly recover, and when animals were observed free-walking which was possible in the majority after three or four months, it was almost impossible to detect any weakness of the leg and the animals would leap to a height of 3 or 4 feet without difficulty.

False circling has often been noted in experimental middle cerebral occlusion both in dogs and primates[24,26]. Later on, however, the circling became a much more casual affair and appeared to rise only from visual inattention. Persistent hemianopia was a characteristic of this preparation. The hemiparesis produced by middle cerebral artery occlusion with involvement of the perforating bearing segment was usually fairly dense in the arm. Complete abolition of arm movements even in the first few days following surgery was rare and from 10 days onwards most of our animals showed reasonable movement in the proximal muscles and evident recovery in the muscles of the elbow and wrist.

The differentiation of stroke density was best performed by considering the time scale of recovery of finger movements and of reaching and placing reactions, but none of the animals ever recovered these completely. Only 1 of 10 animals recovered to such an extent that it was possible to watch it for a few minutes and be uncertain from its upper limb movements which was the hemiparetic side.

Following such single vessel occlusion in the primate the significant neurological examination did not change from 4 months following the episode, over an observation period extending to three years.

Human Clinical Correlates

In human terms, we are now in a position to consider how scientific generalisations fit into the evident pattern of recovery from ischaemia displayed by patients.

Case 1

The first instance cited is recovery from short-term ischaemia which may very well indicate that profound functional loss does indeed follow extremely dense ischaemia after a relatively short time in man.

The case is that of a 36 year old school teacher who noted over a period of 2 years that his handwriting on the blackboard was becoming smaller and that his right hand was losing to some extent its manipulative dexterity. He was a highly intelligent man, a teacher of Mathematics, but put up with the disability for many months before consulting his doctor. His doctor and one of my colleagues, a close friend and an experienced neurologist, thought that the evident mixture of pyramidal and extra-pyramidal signs which he showed, a little stiffness in the arm with positive Hoffman's sign, and slight increase in reflexes in the leg, suggested the early development of Idiopathic Parkinson's disease, and indeed he was demonstrated as such to a post-graduate audience to the evident satisfaction of staff and students. CT scan led to a rapid reassessment of the situation. He had a giant middle cerebral aneurysm arising from the trunk of the artery lateral to the deep perforating vessels of the lenticulo-striate group, and proximal to the major trifurcation of the middle cerebral.

Angiography showed that the filling portion of the aneurysm was relatively small in relation to the total mass, the remainder of the sac being occupied by clot.

In the course of operation, which was of course conducted with continuous monitoring of electrical function, the middle cerebral artery was occluded both proximal and distal to the aneurysm for a period of 6 minutes and 30 seconds. Within 2 minutes of application of the clips which excluded the lenticulo-striate vessels entirely from the circulation, the evoked response from the middle cerebral cortex disappeared. It remained substantially absent over the next hour although transient middle cerebral occlusion only was employed, but had to some extent recovered by the return of the patient to the ward. At that time he was aphasic extending the arm and leg and we were deeply concerned about the production of irreversible neurological damage after a period of dense ischaemia in the basal nuclei of no more than 6 minutes and 30 seconds. However, later that day flexion returned to the arm and leg and by that evening he was speaking. Within 48 hours his neurological signs had cleared completely and at the time of his discharge from hospital his I.Q. was determined as 140, which we believed to be not substantially different from his premorbid I.Q. although that was not formally assessed. The giant aneurysm was obliterated by the application of a single clip to its neck, and the sac evacuated by the use of the ultrasonic dissection apparatus.

The lesson of this case is that under circumstances of very severe ischaemia, in which one could predict that the residual blood flow would be less than 10 ml/100 g

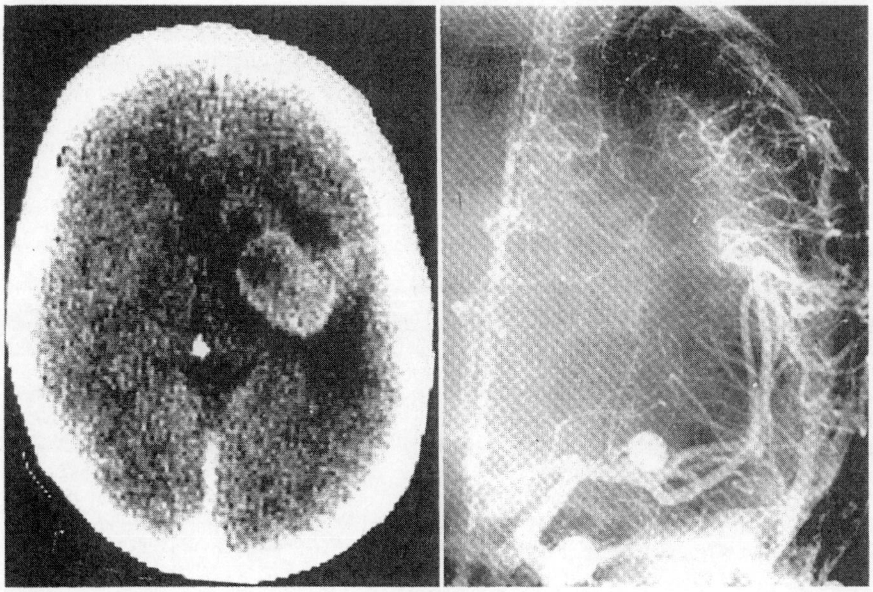

a b

Fig. 1. (a) Enhanced CT scan of a giant middle cerebral aneurysm, the clinical history as detailed in case 1. (b) A P angiogram of a giant middle cerebral aneurysm. Clinical history as detailed in case 1

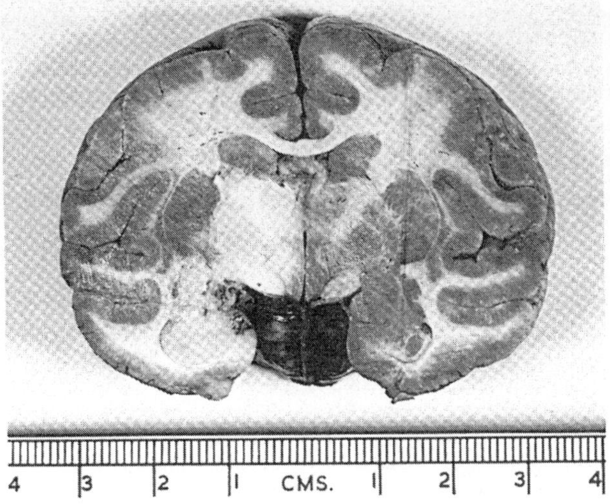

Fig. 2. Carbon black perfusion of a primate brain following occlusion of the perforating branches of the middle cerebral artery. The lack of collateral circulation to the amygdala in the central areas of capsular lentiform nucleus are well seen

min in view of the paucity of collateral circulation to the basal nuclei (Fig. 2), we have the possibility of apparently complete neurological recovery after a period of 6 minutes and 30 seconds of ischaemia. It is possible from observation in other patients, that a period of about 10 minutes of very dense if not total ischaemia may be sustainable with complete recovery of function as assessed over periods of up to 4 years, because it is 4 years since this gentleman had his operation and he remains well. The possible remote effects of such ischaemia of course must be considered and will be addressed subsequently.

Case 2

The second clinical example addresses the problem of the more gradual recovery from very dense neurological deficit which one may see in circumstances of brain ischaemia.

The patient, a lady of 53, had an anterior communicating aneurysm and had bled 3 times, the last bleed precipitating her into coma. She had no intracerebral haematoma, but she was stable at a level of flexion of both arms with gross weakness of both legs and was operated upon some 8 days after haemorrhage by myself, I was much younger then. The aneurysm was successfully occluded but the patient remained comatose with gross leg weakness and poor arm function. The young surgeon discharged her from the hospital to long-term care and took the lesson into his consciousness grade 4 patients do not do well, the acquisition one might regard, of surgical maturity. A CT scan was not at that time available. Two years later a patient appeared in Outpatients with a folio of cuttings from her local newspaper. She carried a stick because of some residual weakness of one leg, she was voluble but articulate and not dysphasic. The cuttings which she carried documented her miraculous recovery from a period of six months unresponsive in her local hospital, her final state of capacity showed her to be managing her housework, to be apparently, to her family, slightly more cheerful than heretofore but infinitely more acceptable socially, to be able to remember her shopping lists and to function as an independent housewife without problem.

Once again although it is now nearly 20 years since the operation was performed, and the lady remains clinically well, we have no absolute assurance of the degree of recovery of all damaged brain, although we have clear evidence that major portions of brain which could not have been functioning at the time of her discharge from the neurosurgical unit, recovered function after a period of 6 months. It would indeed have been fascinating to have blood flow examinations at that time and although such a striking case has not presented since, a somewhat similar case could be detailed in more recent years.

Case 3

This again was a middle-aged patient, a farmer of 55 years with a severe coma producing anterior communicating artery aneurysm, operated on in grade 3 with a stable slight weakness of one leg. Serious post-operative vasospasm was associated with fluctuating changes in the electrical conduction in both hemispheres indicating the fluctuation in state of perfusion, and CT scanning showed extensive low density bi-frontally maximal in one instance associated with non filling post-operatively of a distal anterior cerebral artery. The aneurysm was successfully occluded from the circulation, and at surgery it was not thought that extensive impairment of any peripheral vessel had occurred, but post-operative angiography proved otherwise. His troubled post-operative course went on for several weeks and then as so many of us have seen, he improved. The CT scan showed undoubted bimedial frontal tissue loss but his hemiparetic signs cleared completely, his intellectual function recovered almost to normal so that he was able to conduct his activities as farm manager without problem and he remains stable some years later.

Once again, the clear implication of such a case study is that brain non-functioning immediately after an ischaemic episode in this instance due to vasospasm compounded by surgical insult, may recover.

Case 4

A fourth case drawn again from our records of subarachnoid haemorrhage is a rather younger woman of 35, again with an anterior communicating aneurysm this time occluded in grade 2 without neurological deficit. Some days post-operatively, however, hypo-perfusion developed and a fluctuating hemiparesis associated with clear evidence of delay in central conduction indicative of hemispheral ischaemia. Cerebral blood flow showed decline in flow in both hemispheres of apparently similar extent to just under 30 ml/100 g min, but clearly the resolving capacity of the flow determinations were insufficient to pick up the most dense area of ischaemia just below a functional threshold (Figs. 3a,b). Again, the induction of hypervolemic hypertension resulted in an improvement in CBF and a resolution of the neurological deficit and over

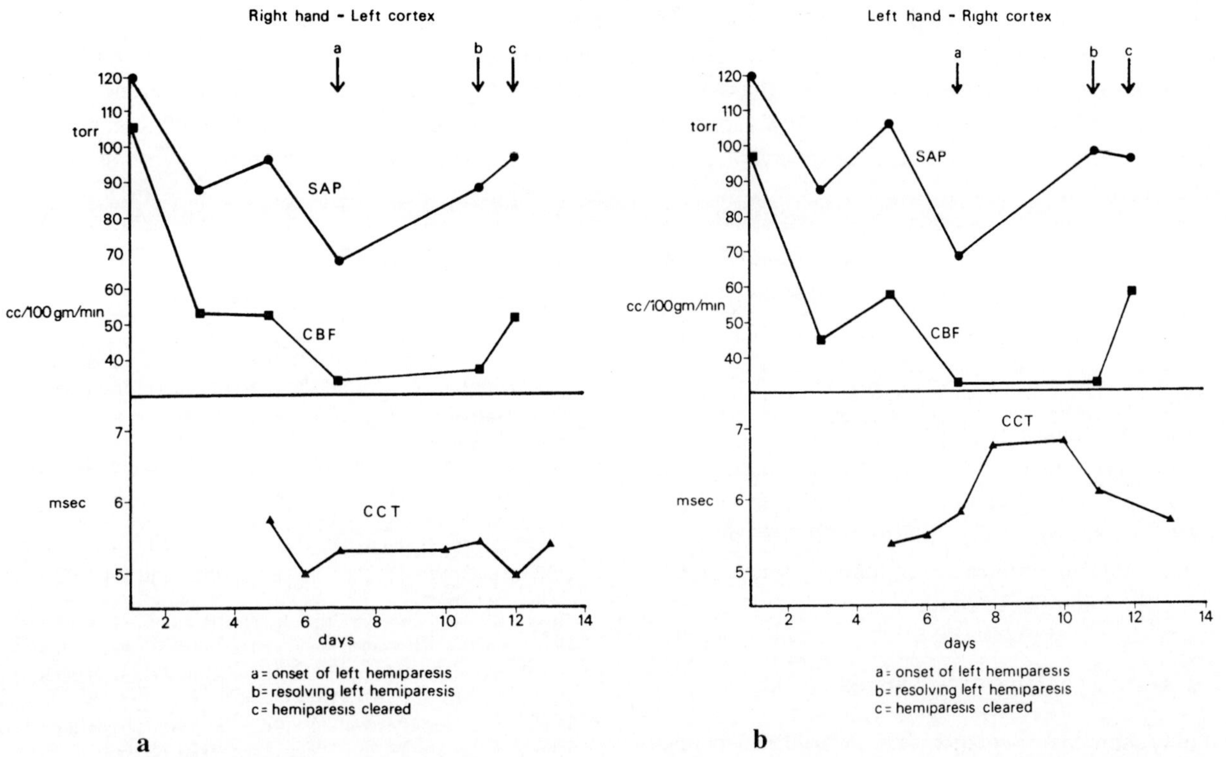

Fig. 3. (a, b) Case 4. Systemic arterial pressure, cerebral blood flow and central conduction time in the two hemispheres of this case during recovery from anterior communicating artery aneurysm surgery. CBF formed similar levels in the two hemispheres but conduction time is elevated only in the hemisphere responsible for the hemiparesis, and recovers with hypervolemic hypertension

nearly 2 weeks this lady's blood pressure had to be sustained by metaraminol infusion. Whenever the blood pressure was allowed to fall, hemiparesis recurred, as the blood pressure was elevated once more, hemiparesis resolved. The capacity, therefore, of brain to retain the ability to recover and yet to be a functionally silent for many hours, is adduced by such case.

Case 5

An exactly similar case pertains to a fifth case, a man in his forties who had suffered a severe subarachnoid haemorrhage from a large posterior communicating artery aneurysm around whose neck the anterior choroidal artery was draped. Operated on in grade 2, he developed over the post-operative period a hemiparesis related to spasm of the anterior choroidal artery associated with low density in the deep nuclear distribution of this vessel. Blood flow determined by a 6 channel xenon clearance system showed levels in several counters of below 20 ml/100 g min but once again, hypervolemic hypertension using a fluid load with dextran and metaraminol infusion resulted in resolution of the deficit. It also resulted in the disappearance of the low density and the patient was discharged without neurological deficit. The metaraminol infusion had to be maintained for five days.

Case 6

A sixth and revealing case is that of a middle-aged woman operated on by a colleague many years before, for an inner third sphenoidal ridge meningioma in the dominant hemisphere. Intense involvement of the middle cerebral artery resulted in a severe post-operative stroke. She was aphasic and hemiplegic. Over the next 3 or 4 years the power in her leg recovered, not unusual one might think in view of the anterior cerebral collateral circulation, but over the next few years, appreciable recovery in her arm also appeared. With the recovery in arm movement she began to recover increasing speech. At first, her speech recovery was in her native language. She was of Central European origin, an immigrant to Britain in the thirties, and it was only many months later that her dysphasia, so evident in English, improved. When I saw her 10 years after the operation she was still detectably dysphasic but could manage normal conversation and coped with her moderate residual hemiparesis well.

This type of very prolonged recovery, raised speculation about possible retraining mechanisms within the central nervous system but here we are right to doubt such a hypothesis. Consider the evolution of recovery in this particular instance, not an unusual evolution except in the protracted period of time over which it occurred. We have first the recovery of function in areas which could be regarded as closest to increasing collateral blood supply, that is the leg. Then we have recovery of a highly differentiated motor skill, the use of the arm, and finally improvement and function in the most highly developed facet of neurological function which it is possible for us to test, speech. It is of particular interest that the improvement of speech function occurred in that neuronal engram, if we may borrow Penfield's phrase, which was most deeply ingrained, the speech of childhood and adult life.

The more lately acquired speech facility, the second language recovered only later. As Gillett (personal communication) points out in a discussion of brain bi-section and personal identity, the loss of extensive portions of brain in hemispherectomy procedures results in severe physical and mental disability. There is no blithe continuation of personal life with largely intact pyschological function and no convincing evidence that retraining can occur other than in circumstances of extreme youth. Plasticity of the nervous system, while undoubted in childhood, scarcely extends to the brain of the adult.

Long-Term Problems of Ischaemia

It may seem that in relation to recovery from ischaemia the developments of recent years have indicated an optimistic prospect. A cautionary note, however, must be included in relation to the remote effects of brain damage.

Evidence is emerging from the long-term studies of patients with brain trauma, from head injury or from pugilism, of accelerated dementia in later life. One of the major causes of dementia has been classified as multi-infarct dementia.

The work of Strong and Tomlinson *et al.*[21] would suggest that while gross neuropathological abnormalities in the area of an ischaemic penumbra are unlikely, some degree of cellular loss appears to be the norm at any rate in the animal species studied by them (the cat). CT scanning of patients with multi-infarct dementia leaves no doubt of cumulative functional loss as a result of repetitive episodes of ischaemia, and mixed pictures of dementia of Alzheimer type and multi-infarct dementia do occur.

It would not be surprising, therefore, if every episode of ischaemia to which the brain was subject, resulted in some loss of neuronal reserve and possibly in a contribution to a later dementing process.

The conclusion must be, therefore, that while we have good evidence that transient cerebral ischaemia can recover apparently fully, we must maintain caution in attributing complete safety to techniques which require diminution of the blood supply or the metabolism of the brain. Our task in the coming years will be more fully to analyse the impact of every technique, surgical, medical or anaesthetic, upon the function of the nervous system with particular reference to its possible cumulative and long-term effects and, of course, to avoid unnecessary manoeuvres as far as possible.

References

1. Ames A III, Gurian BS (1963) Effects of glucose and oxygen deprivation on function of isolated mammalian retina. J Neurophysiol 26: 617–634
2. Astrup J, Symon L, Branston NM, Lassen NA (1977) Cortical evoked potential and extracellular K^+ and H^+ at critical levels of brain ischaemia. Stroke 8: 51–57
3. Branston NM, Strong AJ, Symon L (1977) Extracel lular potassium activity, evoked potential and tissue blood flow, relationship during progessive ischaemia in baboon cerebral cortex. J Neurol Sci 32: 305–321
4. Branston NM, Symon L, Crockard HA, Pasztor E (1974) Relationship between the cortical evoked potential and local cortical blood flow following acute middle cerebral artery occlusion in the baboon. Exp Neurol 45: 195–208
5. Collewijn H, van Harreveld A (1966) Intracellular recording of spinal motoneurones during acute asphyxia. J Physiol 185: 1–14
6. Collewijn H, van Harrevald A (1966) Membrane potential of cerebral cortical cells during spreading depression and asphyxia. Exp Neurol 15: 425–436
7. Dennis C, Kabat H (1939) Behaviour of dogs after complete temporary arrest of the cephalic circulation. Proc Soc Exper Biol Med 40: 559–561
8. Grenell RG (1946) Central nervous system resistance. I. The effects of temporary arrest of cerebral circulation for a period of two to ten minutes. J Neuropath Exper Neurol 5: 131–154
9. Harris RJ, Symon L, Branston NM, Bayhan M (1981) Changes in extracellular calcium activity in cerebral ischaemia. J Cereb Blood Flow Metabol 1: 203–209
10. Harris RJ, Richards PG, Symon L, Habib A-HA, Rosenstein J (1987) pH, K^+ and PO_2 of the extracellular space during ischaemia of primate cerebral cortex. J Cereb Blood Flow Metabol 7: 599–604
11. Harvey J, Rasmussen T (1951) Occlusion of the middle cerebral artery. Arch Neurol 66: 20–29
12. Hass WK (1981) Beyond cerebral blood flow, metabolism and ischaemic thresholds: an examination of the role of calcium in the initiation of cerebral infarction. In: Meyer JS (eds) Cerebral vascular disease 3. Excerpta Medica, Amsterdam, pp 3–17
13. Hossmann KA, Sato K (1970) The effect of ischemia on sensorimotor cortex of cat. Z Neurol 198: 33–45
14. Hossmann KA, Schuier FJ (1979) Metabolic (cytotoxic) type of brain oedema following middle cerebral artery occlusion in cats. In: Price T, Nelson E (eds) Cerebrovascular diseases. Raven, New York, pp 141–165
15. Hossmann KA, Takagi S (1976) Osmolality of brain in cerebral ischaemia. Exp Neurol 51: 124–131
16. Kaplan HA, Ford DH (1966) The Brain Vascular System. Elsevier, London
17. Lassen NA (1966) The luxury-perfusion syndrome and its possible relation to acute metabolic acidosis localized within the brain. Lancet 2: 113–115
18. Lazorthes G, Campan L (1964) La cirulation cerebrale. Editions Sandoz, Paris

19. Pasztor E, Symon L, Dorsch NWC, Branston NM (1973) The hydrogen clearance method in assessment of blood flow in cortex, white matter and deep nuclei of baboons. Stroke 4: 556–567
20. Przybylski A (1971) Activity pattern of visceral cortex neurons during asphyxia. Exp Neurol 32: 12–21
21. Strong L, Tomlinson B, Venables G, Gibson G, Hardy J (1983) The cortical ischaemic penumbra associated with occlusion of the middle cerebral artery in the cat. 2. Studies of histopathology, water contents, *in vitro* neurotransmitter uptake. J Cereb Blood Flow Metabol 3: 97–108
22. Symon L, Branston NM, Chikovani O (1979) Ischaemic brain oedema following middle cerebral artery occlusion in baboons. Relationship between regional cerebral water content and blood flow at 12 hours. Stroke 10: 184–191
23. Symon L, Branston NM, Strong AJ, Hope TD (1977) The concepts of thresholds of ischaemia in relation to brain structure and function. J Clin Path 30 [Suppl 11]: 149–154
24. Symon L, Dorsch NWC, Crockard HA (1975) The production and clinical features of a chronic stroke model in experimental primates. Stroke 6: 476–481
25. Symon L, Pasztor E, Branston NM (1974) The distribution and density of reduced cerebral blood flow following acute middle cerebral artery occlusion: an experimental study by the technique of hydrogen clearance in baboons. Stroke 5: 355–364
26. Waltz AG (1969) Red venous blood: occurrence and significance in ischaemic and non-ischaemic cerebral cortex. J Neurosurg 31: 141–148
27. Zwetnow NN (1970) Effects of increased cerebrospinal fluid pressure on the blood flow and on the energy metabolism of the brain. Acta Physiol Scand Suppl 339: 1–31

Correspondence and Reprints: Lindsay Symon, Gough-Cooper Department of Neurological Surgery, Institute of Neurology, Queen Square, London WC1N 3BG, U.K.

Acta Neurochir (1993) [Suppl] 57: 110–121

Systematic Development of Cerebral Resuscitation After Cardiac Arrest. Three Promising Treatments: Cardiopulmonary Bypass, Hypertensive Hemodilution, and Mild Hypothermia*

P. Safar, F. Sterz, Y. Leonov, A. Radovsky, S. Tisherman, and **K. Oku**

International Resuscitation Research Center, University of Pittsburgh, Pittsburgh, PA, U.S.A.

Abstract

Since 1970 we have investigated postischemic anoxic encephalopathy and potential treatments for cerebral resuscitation after cardiac arrest by cardiopulmonary-cerebral resuscitation (CPCR). The post-resuscitation syndrome has been studied at the levels of cell, organ, organism and community. Short-term and long-term models in rats, dogs, and monkeys have been developed, and an international multicenter randomized clinical trial mechanism was established. Clinical studies disproved the 5-min limit of reversible cardiac arrest and yielded other valuable data on treatments and prognostication. Thiopental loading or calcium entry blocker therapy (lidoflazine) gave no significant improvement in patients. Free radical scavengers are under investigation in the laboratory. We hypothesize that post-arrest perfusion failure and necrotizing cascades require etiology-specific combination treatments. Standard (control) therapy in a current dog model of cardiac arrest (no flow) of 12.5–20 min, reperfusion with cardiopulmonary bypass, and intensive care for 72–96 h has consistently resulted in survival with brain damage. After ventricular-fibrillation (VF) arrest of 17 min, moderate hypothermia (28–32°C) inconsistently improved cerebral outcome. After VF arrest of 12.5 min, hypertension plus hemodilution normalized the local (multifocal) cerebral hypoperfusion post-arrest and, again, inconsistently improved cerebral outcome. Additional mild hypothermia (34–36°C), however, consistently improved cerebral outcome, whether induced before or during and after arrest.

Keywords: Cardiac arrest; cerebral resuscitation; multimodality treatment.

Background

About 25% of all deaths occur before old age and without incurable, lethal disease (Safar and Bircher 1988). About 50% of the victims (250,000 per year in the U.S.A.) are candidates for cardiopulmonary-cerebral resuscitation (CPCR) attempts with basic, advanced, and prolonged life support (Safar and Bircher, 1988; Natl Acad Sci Res Council (NAS-NRC) 1966, 1974, 1980, 1986). Present community-wide results of cardiopulmonary resuscitation (CPR) attempts are suboptimal. Hospital discharge rates after prehospital CPR have been < 10% overall, but 25–40% when CPR was initiated within 4 min (Eisenberg *et al.*, 1986; Safar 1988a). Between 10% and 40% of survivors suffer permanent brain damage (Abramson *et al.*, 1986, 1989; Levy *et al.*, 1985; Longstreth *et al.*, 1983; Snyder and Tabbaa, 1987).

In the 1950s, modern CPCR techniques and guidelines were developed (Safar and Bircher, 1988; NAS-NRC, 1966,1974,1980, 1986). Since 1960 the delivery of CPCR was improved (Amer Soc Anesthesiol 1968; Safar *et al.*, 1974). Since 1970, attempts have been made to extend the reversible period of normothermic cardiac arrest beyond the assumed 5-min limit (Safar, 1978, 1986, 1988a; Safar and Elam, 1977; Safar *et al.*, 1987). Ames and Nesbett (1983) reported that retinal neurons can tolerate up to 20 min anoxia. Hossmann and Kleihues (1973; Hossmann, 1988) showed that most cerebral neurons can often recover electric and chemical activity after up to 60 min of normothermic complete global ischemia. Studies by Siesjö (1981, 1988) indicated that treatable pathogenetic factors are involved. Our research in cardiac arrest has examined the following hypotheses:

1) Outcome is determined not only by the duration of the no flow insult and the temperature (T) during cardiac arrest, but also the pattern of dying. For example, arrest caused by asphyxiation is more injurious (Vaagenes *et al.*, 1988b), while that caused by exsanguination (Kirimli *et al.*, 1968; Negovsky *et al.*, 1983)

* Submitted 1989.

may be less injurious (Safar, 1988a) to the brain than the same period of normovolemic ventricular fibrillation (VF) arrest.

2) The secondary brain damage during and after reperfusion, the postresuscitation syndrome (Negovsky *et al.*, 1983), is the result of transient or prolonged multiple yet treatable organ derangements (Safar, 1985, 1988a; Safar and Vaagenes, 1986).

3) The secondary brain damage has a multifactorial pathogenesis, and therefore requires tailored, etiology-specific combination treatments (Safar, 1978, 1986, 1988a; Safar and Elam, 1977; Safar and Vaagenes, 1986; Safar *et al.*, 1987).

4) The secondary derangements have at least four components that interact (Safar, 1986):

- Cerebral and extracerebral (multifocal, inhomogeneous) post-arrest perfusion failure (perfusion inadequate to meet metabolic demands). Perfusion failure follows a pattern of initial no reflow (Ames *et al.*, 1968; Nemoto and Frinak, 1978) followed by brief global hyperemia and then protracted global (Lind *et al.*, 1975; Snyder *et al.*, 1975) and multifocal (Kagström *et al.*, 1983; Wolfson *et al.*, 1988; Latchaw *et al.*, 1989; Safar *et al.*, 1989a) hypoperfusion (caused by complex, still unclear factors) either leading to resolution and improvement, permanent disability, or brain death with "false" hyperemia (Beckstead *et al.*, 1978).
- Reoxygenation injury generating chemical cascades that lead to cell necroses (Safar 1986, 1988a; Ernster 1988).
- Self-intoxication from post-anoxic viscera (Negovsky *et al.*, 1983).
- Blood dyscrasias after stasis (Hallenbeck *et al.*, 1982; Kochanek *et al.*, 1987).

The protracted post-arrest reduction in CBF is accompanied by an equal reduction in cardiac output, even with MAP controlled at normotensive levels by catecholamines and fluids (Kampschulte *et al.*, 1969a; Cerchiari *et al.*, 1992).

After exploration of short-term neurologic recovery of dogs after cardiac arrest (Kampschulte *et al.*, 1969b), Lind *et al.* (1975) and Snyder *et al.* (1975) in 1971 conducted experiments on dogs with an aortic occlusion model. They documented that after cardiac arrest there is a protracted reduction of global cerebral blood flow (gCBF), with no increase in intracranial pressure (ICP). In 1974, Safar demonstrated that cerebral outcome can be improved by post-arrest therapy using a combination treatment of intracarotid hemodilution, hypertension and heparinization to improve perfusion (Safar *et al.*, 1976). In 1976, a global brain ischemia monkey model was developed (Nemoto *et al.*, 1977). Its first use revealed a beneficial effect on cerebral outcome of immediate post-arrest thiopental loading (Bleyaert *et al.*, 1978). Barbiturate therapy was based on a multifactorial rationale that remains valid (Safar, 1980). Repeated evaluation of thiopental loading in the same laboratory, with the same model, using a different subspecies of monkeys and different reperfusion pressure patterns and intensive care protocol, demonstrated no measurable improvement in outcome (Gisvold *et al.*, 1984a). Several years of expensive outcome experiments with long-term (3–7 d) post-ischemic intensive care were needed to establish which extracerebral variables must be controlled to minimize the many factors that can influence outcome, beyond insult and the specific treatment tested (Safar *et al.*, 1982).

In the early 1980s, White *et al.* (1983) and Siesjö (1981) suggested the use of calcium entry blockers after arrest. Vaagenes *et al.* (1984) conducted the first outcome study in animals on a calcium entry blocker given immediately after cardiac arrest (lidoflazine), using a reproducible dog model of VF no flow of 10 min, and found significant improvement in outcome. Gisvold introduced the monkey neck tourniquet model developed in Pittsburgh (Nemoto *et al.*, 1977; Bleyaert *et al.*, 1978; Gisvold *et al.*, 1984a) at the Mayo Clinic to study a different calcium entry blocker (nimodipine) in a different species and without heart ischemia. He observed exactly the same improvement in cerebral outcome (Steen *et al.*, 1985). After asphyxiation cardiac arrest, lidoflazine therapy gave no beneficial cerebral effect, and histologic damage was worse than after VF arrest (Vaagenes *et al.*, 1986, 1988a). In contrast, anti-reoxygenation injury (free radical scavenger) therapies have not yet shown any improvement in outcome after VF arrest (Vaagenes *et al.*, 1986; Reich *et al.*, 1988a), but one such protocol improved CBF (Cerchiari *et al.*, 1987) and outcome (Vaagenes *et al.*, 1986) after asphyxial arrest. The excitatory neurotransmitter blocker MK-801, given after cardiac arrest, did not improve cerebral outcome after VF of 15 or 17 min in dogs (Sterz *et al.*, 1989), although it gave promising results in focal cerebral ischemia experiments by others.

In 1979, we initiated an international clinical study mechanism of cardiac arrest and CPCR, the "Brain Resuscitation, Clinical Trial" (BRCT) (Abramson *et al.*, 1986, 1989b). Its main objective has been to conduct randomized trials of novel drug therapies. BRCT phase I (1979–84) evaluated thiopental loading, which gave a negative treatment result (Abramson *et al.*, 1986). BRCT phase II (1984–89) evaluated lidoflazine therapy, which also gave a negative treatment result (Abramson *et al.*, 1989b). Analyses of subgroups and study of the literature suggest that the great variability between patients in underlying disease, response to CPR, severity of insult, details of post-arrest therapy (in spite of adherence to protocol), and other unknown factors may have prevented these clinical trials from revealing cerebral resuscitation effects of single drugs. That is, the "noise" in clinical conditions may obscure treatment results unless there is a breakthrough effect. Such a breakthrough effect may be achievable only with a multifaceted treatment regimen. Nevertheless, the clinical study mechanism has been very valuable for pursuing other hypotheses on the limits of reversibility of clinical death, standard therapy potentials, and prediction of outcome (Safar 1988a,b; Safar *et al.*, 1989; Abramson *et al.*, 1991; Jastremski *et al.*, 1989). The BRCT also provides a mechanism for clinical feasibility studies of novel treatment protocols, and for a case registry to study standard CPCR results.

After prolonged no flow, standard external CPR basic life support produces low cerebral perfusion pressures and blood flows. Epinephrine by causing systemic vasoconstriction improves flow during CPR. With or without epinephrine, external CPR is unpredictable in its ability to restore stable spontaneous circulation promptly. Modifications of external CPR were studied by Safar *et al.* and others around 1960 and again around 1980, and were not found to be physiologically superior to standard external CPR – at least not for basic life support (cf. Safar *et al.*, 1976; Safar and Bircher, 1988; NAS-NRC, 1966, 1974, 1980, 1986; Bircher and Safar, 1981, 1984). The CPR modifications of ventilation simultaneously with chest compressions ("new CPR") raise jugular vein pressure and thereby can reduce cerebral perfusion pressure and oxygenation (Bircher and Safar, 1981). Krischer *et al.* (1989) found that this method did not improve outcome in a randomized clinical trial. Around 1980, Bircher and Safar reconfirmed the physiologic superiority of open-chest CPR (Bircher and Safar, 1984). Unfortunately, this physician-dependent invasive method is unlikely to become popular again. Therefore, from 1982 to 1988, we conducted 10 studies in dogs to evaluate an even more promising, physician-dependent, but semi-invasive advanced life support method: emergency cardiopulmonary bypass (CPB) without thoracotomy (using portable equipment) for CPR-resistant cardiac arrest (Safar *et al.*, 1990). This method provides full control over pressure, flow, composition, and temperature of blood. CPB enhanced cardiac resuscitability and provided prolonged assisted circulation early after arrest. It resulted in better survival and neurologic recovery than CPR advanced life support attempts only. With CPB and standard intensive care in dogs, it was possible to reverse normothermic VF cardiac arrest (no flow) of up to 10 min to survival without neurologic deficit. After VF of 20 min there was consistent survival, but with brain damage (Safar, 1986, 1988a; Safar *et al.*, 1990; Reich *et al.*, 1988b). After no flow of 30 min (VF) spontaneous circulation could be restored transiently, but there was always secondary cardiac death from what seemed to be myocardial reperfusion injury (Reich *et al.*, 1988b).

Current Methods and Models

We are currently engaged in simultaneous coordinated studies at the molecular, cellular, organ, organism, and community levels. Studies at the *molecular level* (Ernster, 1988) include brain, blood and CSF chemistry. In an asphyxial cardiac arrest model in rats (Hendrickx *et al.*, 1984a,b) we established reproducible post-arrest brain cytosolic enzyme loss (Katz *et al.*, 1989), which correlates with early post-arrest morphological neuronal changes (Hosain, Radovsky, unpublished). In dogs and patients, Vaagenes *et al.*, (1988a) have found post-arrest CSF enzyme activity peaks at 48–72 h that correlated with insult and outcome, and thus can be used as adjuncts for outcome prediction. Dog experiments with free radical therapy (Reich *et al.*, 1988a) gave no conclusive results with measurements of free iron, conjugated dienes and hypoxanthine in CSF, blood and brain in comparison to controls (Basford, Diven, unpublished). Upon the suggestion of Ernster (1988), we are in the process of applying chemiluminescence (Cadenas, 1988) in our rat model for the detection of free radical reactions *in vivo* after cardiac arrest. We are also searching for viscera-generated neurotoxins.

At the *cellular level,* we are determining the distribution, density and characteristics of microscopic lesions in brain and viscera after temporary global brain ischemia in a variety of models (Nemoto *et al.*, 1977; Bleyaert *et al.*, 1978; Safar *et al.*, 1982; Vaagenes *et al.*, 1984, 1986, 1988a; Steen *et al.*, 1985; Reich *et al.*, 1988a; Cerchiari *et al.*, 1987; Sterz *et al.*, 1992).

At the *organ level,* Pretto *et al.* (1985) are studying the effects of temporary ischemia and reperfusion on isolated rat heart and liver, which are perfused with either crystalloid solutions or blood, using modifications of the Langendorff preparation. We are using invasive short-term experiments (< 24 h), with careful control of extracerebral life support variables (Safar *et al.*, 1982), for the study of mechanisms of brain and heart damage. Some of these experiments include:

1) monitoring of ICP, sagittal sinus pressure, and cerebral and systemic arterio-venous gradients for oxygen, glucose and lactate (Oku, unpublished);
2) direct and indirect brain temperature measurements;
3) global CBF (Cerchiari *et al.*, 1987);

4) regional CBF by stable Xenon-computer tomography (Xe-CT) (Wolfson *et al.*, 1988; Latchaw *et al.*, 1989; Safar *et al.*, 1989);

5) EEG spectra and evoked potentials (Cerchiari *et al.*, 1987); and

6) brain tissue chemistry (Vaagenes *et al.*, 1988a; Hendrickx *et al.*, 1984b; Katz *et al.*, 1989).

In collaboration with Paschen and Hossmann of the Max Planck Institute in Cologne, we demonstrated the feasibility of *in vivo* brain freezing in dogs. It resulted in reliable pH and ATP values even in the brain stem (Hoel, unpublished). The function of individual organs (e.g., cardiovascular-pulmonary system) is also monitored continuously or intermittently in our long-term dog outcome models (Safar *et al.*, 1982).

In rats, asphyxial cardiac arrest of > 5 min yielded a low survival rate, in part because of the inability to monitor invasively and to provide life support. Also, neurologic deficit scoring in rats is less sensitive than brain histologic damage scoring (Siesjö, 1988; Hendrickx *et al.*, 1984a,b). Therefore, we are currently using rat asphyxiation and exsanguination cardiac arrest models with 6 h life support, for treatment testing. EEG activity recovery time and either brain cytosolic enzyme loss or ischemic neuronal changes (primarily in the hippocampus) at 6 h post-arrest are used as end points.

At the *organism level*, outcome post-arrest has been evaluated in monkeys (Nemoto *et al.*, 1977; Bleyaert *et al.*, 1978; Safar, 1980; Safar *et al.*, 1982; Gisvold *et al.*, 1984a) and dogs (Safar *et al.*, 1982, 1988; Vaagenes *et al.*, 1984, 1986, 1988a; Reich *et al.*, 1988a; Sterz *et al.*, 1989, 1992; Leonov *et al.*, 1990). We evaluate outcome at 3–7 d after insult in terms of neurologic deficit (ND) scores (Table 1) and overall performance categories (OPC) (Table 2). ND and OPC usually reach their best values by 72 h post-arrest. Occasional worsening between 72 and 96 h reflects secondary neurologic deterioration, which usually can be traced to extracerebral organ malfunction. At 96 h, final OPC and ND are determined. Then the re-anesthetized animals are killed by perfusion fixation for the determination of histopathologic damage (HD) scores for the brain (Table 3) and extracerebral organs. HD scores correlated well with the duration of no-flow and the final OPC and ND (Cerchiari *et al.*, 1992; Nemoto *et al.*, 1977; Bleyaert *et al.*, 1978; Safar, 1980; Safar *et al.*, 1982; Gisvold *et al.*, 1984a; Vaagenes *et al.*, 1984, 1986, 1988a; Reich *et al.*, 1988; Cerchiari *et al.*, 1987; Leonov *et al.*, 1990).

There are crucial *requirements for outcome models:* studies using mortality as the end point cannot be used for valid evaluation of potential treatments in cerebral resuscitation. Only mortality before 72–96 h due to primary brain death may be included in the analysis. In previous studies we found such primary brain death extremely rare, occurring only after global brain ischemia of ≥ 20 min (Nemoto *et al.*, 1977; Safar *et al.*, 1982). After cardiac arrest with no flow of ≥ 15 min in normothermic dogs, brain death can be caused secondarily by cardiovascular-pulmonary failure, usually as a result of errors in life support. Thus, outcome studies of cerebral resuscitation must provide in the standard (control) experiments a quality of extracerebral life support that achieves survival for at least 72 h after arrest, and outcome with neurologic deficit. The severity of the insult must allow treatments known to be effective (e.g., mild pre-cooling, see later) or slight reduction in arrest time to achieve complete neurologic recovery (OPC No. 1; Table 2). Only survivors without life support errors must be included in outcome evaluation. The 72–96 h of intensive care is necessary to give the postischemic-anoxic encephalopathy time to "mature" and stabilize (Safar *et al.*, 1982).

Our current dog model is the most reproducible outcome model to date: VF cardiac arrest with no flow for 12.5–20 min, reperfusion by CPB (for restoration of spontaneous circulation within 5 min, and assisted circulation as needed to 4 h), intermittent positive pressure ventilation to 20 h, and intensive care to 96 h. With this model, all 28 dogs with standard (control) therapy with life support according to the protocol, survived to 96 h, and had neurologic deficits (Safar, 1988a; Safar *et al.*, 1990). There were no brain deaths. The same CPB protocol after VF 10 min in pilot experiments led to survival with no neurologic deficit. After VF 12.5 or 15 min arrest, mild pre-intra-cooling achieved complete recovery (OPC No. 1) (Safar 1988a). Thus, our current VF 12.5 min CPB model fulfills the above requirements. This paper is a preliminary report of studies using this reproducible model concerning cardiopulmonary bypass (Safar *et al.*, 1990), hypertensive hemodilution (Sterz *et al.*, 1992) and hypothermia (Leonov *et al.*, 1990). In short-term (< 24 h) experiments we applied the new non-invasive method of assessing multifocal lCBF by stable Xe-CT. Measurement of sagittal sinus and arterial blood contents of O_2, lactate and glucose simultaneously with lCBF permitted monitoring of changes in gCBF, $gCMRO_2$, gCMR glucose, and gCMR lactate.

Table 1. *Neurologic Deficit (ND) Scoring for Long-Term Outcome Experiments in Dogs Maximal (Worst) Points = 5 × 100 = 500 Points = ND Score 100% (R = Right; L = Left)*

Level of consciousness			*Respiration*		
Normal		0	Normal		0
Cloudy		30	Hyperventilation		25
Delirium		45	Abnormal		50
Stupor		60	Apnae		100
Coma		100			
Subtotal (worst = 100)			Subtotal (worst = 100)		
Cranial nerve function					
Pupil size	R	0 2 5	Ciliospinal reflex	R	0 2 5
	L	0 2 5		L	0 2 5
Light reflex	R	0 2 5	Oculocephalic reflex		0 2 5
	L	0 2 5	Menace reflex		0 2 5
Eye position	R	0 2 5	Auditory reflex		0 5 10
	L	0 2 5	Gag reflex		0 5 10
Lid reflex	R	0 2 5	Carinal reflex		0 5 10
	L	0 2 5			
Corneal reflex	R	0 2 5			
	L	0 2 5			
			Subtotal (worst = 100)		
Motor and sensory function			*Behavior*		
Motor stretch reflex		0 10 25	Drinking		0 15
			Chewing		0 15
Motor response to pain		0 5 10 25	Sitting		0 15
			Standing		0 15
Positioning		0 10 25	Walking		0 15 30
Muscle tonus		0 10 25	Cleaning		0 15
Subtotal (worst = 100)			Subtotal (worst = 100)		
Total points					
% score (total points divided by 5)					

Results and Discussion

Hypertensive Hemodilution

After VF no-flow of 12.5 min in dogs, multifocal CBF (lCBF) by Xe-CT (n = 22) showed transient hy-

Table 2. *Overall Performance Categorization (OPC) – Dog*

		1 Normal	2 Moderate disability	3 Severe disability	4 Coma, veg. state	5 Brain death
		conscious			unconscious	
Awareness	fully aware	+	(+)			
	aware		+			
	unaware			+	+	+
Motion	walks	+				
	stands	+				
	sits	+	+			
	rights itself	+	+			
Tonus and clonus	normal	+	+	+		
	running movements			(+)	(+)	
	hyperventilation				(+)	
	seizures			(+)	(+)	
	opisthotonus				(+)	(+)
Noise	reaction	+	+			
	reflex			+	(+)	
Pain	reaction	+	+	+		
	reflex			(+)	(+)	(+)
Food	feeds self	+	+			
	reflex			+		
Viscal	reaction (looks)	+	(+)			
	reflex (menace)	+	(+)	(+)		
Pupil light reflex		+	+	+	+	

+ Must be present, (+) may be present.

Table 3. *Brain Histopathologic Damage (HD) Scoring for Long-Term Outcome Experiments in Dogs or Monkeys*

Brain areas (R, L)	
1. Frontal cortex and white matter	11. Globus pallidus
2. Parietal cortex and white matter	12. Thalamus
3. Occipital cortex and white matter	13–14. Midbrain, including substantia nigra
4. Temporal cortex and white matter	15. Pons
5. Insular cortex and white matter	16. Medulla
6. Hippocampus	17–19. Cerebellum, including Purkinje cells, granule cells, and dentate nucleus
7. Dentate gyrus	
8. Central white matter	
9. Caudate nucleus	
10. Putamen	

Severity	Lesion type × assigned number
0 No change	Infarction × 4
1 Minimal	Ischemic neuronal changes × 2
2 Moderate	Edema × 1
3 Marked	
4 Maximal	

Total score = (severity number) × (number assigned each type of lesion).
Total max. score possible for each area = (4 × 4) + (4 × 2) + (4 × 1) = 28.
Total max. score possible for entire brain = 28 × 19 (areas) × 2 (right and left) = 1,064.

peremia without evidence of sustained no-reflow foci, followed by global, very inhomogeneous hypoperfusion for >4 h. Post-arrest mean arterial pressure (MAP) and hematocrit (Hct) were normal. There were multiple foci of trickle flow (lCBF 5–10 ml/100 g/min) and low flow (10–20 ml/100 g/min), which were not present pre-arrest. Hypertension induced immediately after arrest by norepinephrine (MAP > 150 reduced to 110 mm Hg over 4 h) (n = 3) resulted in initial homogeneous hyperemia, followed by global CBF higher than in normotensive controls, and without trickle flow areas (Sterz *et al.*, 1992). With subsequent normotension, however, trickle flow areas appeared. Additional normovolemic hemodilution to Hct 20%, starting with reperfusion (n = 2), however, resulted in normalization of lCBF patterns and normal gCBF values after arrest (Sterz *et al.*, 1992) (Table 4). At 5 min post-arrest, all brain areas had flows greater than 40 ml/100 g/min, and from 30 min to 4 h there were almost no trickle or low flow areas in grey matter.

To confirm the beneficial effects of this flow-promoting therapy of hypertensive hemodilution on outcome, we used the same VF 12.5 min model with CPB for reperfusion, and life support to 96 h. In the control group with normotension and normal Hct, all 6 dogs remained severely disabled (OPC No. 3) (Table 5). In the group treated with hypertension alone (MAP >150 to >110 mm Hg), 3 of 6 achieved good outcome (OPC No. 2). In the group treated with the same degree and duration of hypertension plus normovolemic hemodilution (Hct 20%) with dextran 40/Ringers solution 50:50, again 3 of 6 animals achieved good outcome (OPC No. 2) (Leonov *et al.*, 1992). Thus, in the outcome model, we could see no difference between hypertension with vs. without hemodilution.

Mild Hypothermia Before and During Arrest: Protection and Preservation

We retrospectively analyzed (Safar, 1988) the dichotomous outcomes of several previous series of cardiac arrest experiments in normothermic dogs with reperfusion by CPR advanced life support (Cerchiari *et al.*, 1992; Safar *et al.*, 1990) or CPB (Sterz *et al.*, 1992; Safar *et al.*, 1990; Reich *et al.*, 1988) and standard life support to 72 or 96 h (control protocols). There was no difference between dogs with poor vs. good neurologic outcome, in pre-arrest glucose, MAP, blood gas values or anesthesia; but there was a difference in temperature. After VF cardiac arrest of 12, 12.5, or 15 min (no flow), 10 of 13 dogs that achieved good outcome (OPC No. 1 or 2) had unintentional mild hypothermia (core temperature of 35.0–36.9°C) throughout VF. With the same duration of VF no-flow,

Table 4. *Effect of Hypertensive Hemodilution (HT & HD) on Global and Multifocal (Local) Cerebral Blood Flow (lCBF) Post-Ventricular Fibrillation (VF) Arrest of 12.5 min, as Determined by the Xe-CT Method in One Control Therapy Dog (#16) and One with Hypertensive Hemodilution (#15)*

lCBF	Dog	Normotension controls	Post-arrrest approx. 1 h normotension normal hct	Post-arrest approx. 1 h hypertension (MAP > 140–120 mmHg) plus hemodilution (hct 20%)
gCBF* brain slice ml/100 g per min	control HT & HD	43 [100%] 42 [100%]	25 [58%] –	– 42 [100%]
No flow areas % of voxels of slice area with lCBF 0–5 ml/100g per min	control HT & HD	0% 0%	9% –	– 0%
Trickle flow areas %of voxels of slice area with lCBF 0–10 ml/100 g per min	control HT & HD	1% 0%	15% –	– 2%
Low flow areas % of voxels of slice area with lCBF 0–20 ml/100 g per min	control HT & HD	9% 7%	43% –	– 10%

*gCBF: One coronal slice parieto-temporal, 0.5 mm thick, lCBF foci calculated from 1 mm voxels; spatial resolution 6 mm diameter. From Wolfson *et al.*, 1988; Latchaw *et al.*, 1989; Sterz *et al.*, 1992.

Table 5. *Effect of Hypertension with or Without Hemodilution on Outcome as Overall Performance Categories (OPC, Table 2).* Each circle (o) represents one dog. VF cardiac arrest (no flow) 12.5 min, CPB ≤ 5 min, IPPV 20 h, intensive care 96 h

Best OPC 24–96 h	Normotension normal hct controls	Hypertension MAP > 150–110 normal hct	Hypertension and hemodilution hct 20%
OPC 5 Brain death	–	–	–
OPC 4 Coma Vegetative state	–	o	–
OPC 3 Severe disability Conscious	oooooo	oo	ooo
OPC 2 Moderate disability Conscious	–	ooo	ooo
OPC 1 Normal	–	–	–
Compared with controls:	$p < 0.05$		

From Leonov *et al.*, 1992.

3 of the 13 mildly hypothermic dogs and all 27 dogs with normothermia at start of VF (37–38°C) achieved poor cerebral outcome (OPC No. 3–5). In the VF 15 min group only (Reich *et al.*, 1988b) the 3/10 that achieved OPC No. 1 and the 2/10 that achieved OPC No. 2 had temperatures of 35.0–36.9°C at start of VF whereas the 5/10 that achieved OPC No. 3 had temperatures of 37.2–38.2°C ($p < 0.002$). After VF cardiac arrest of 20 min (no flow), all 9 dogs had poor outcome (OPC No. 3 or 4) although 7/9 were mildly hypothermic. Thus, mild precooling protected the brain against 12–15 min of no flow, but not against a longer ischemia time.

The cerebral protective (pretreatment) effect of *moderate* hypothermia (30°C) is well known (Bigelow *et al.*, 1950; Rupp and Severinghaus, 1986). After we discovered a protective effect of *mild* pre-cooling (Safar 1988; Safar *et al.*, 1990) we obtained supporting evidence by reports of other investigators who found more rapid EEG recovery (Hossmann, 1988) or reduced histopathologic brain damage (Hirsch and Müller, 1962; Berntman *et al.*, 1981; Busto *et al.*, 1987) after global ischemia with mild precooling. In 1989, Siesjö and Minimisawa reported the protective effect of mild pre-cooling on brain histologic outcome in their rat model of incomplete forebrain ischemia. Variability in outcomes of past experimental series found by us and others might well have been caused by slight variations in brain temperature during the no flow or low flow states. In recent studies, when we controlled tympanic membrane and pulmonary artery temperature at exactly 37.5°C, neurologic outcome was narrowly reproducible (Leonov *et al.*, 1990; Sterz *et al.*, 1992; Safar *et al.*, 1990).

Mild Hypothermia During and After Arrest: Preservation and Resuscitation

In 1968, Kampschulte and Safar (1969) tried to resuscitate two dogs after VF cardiac arrest (no flow) of 15 min, by use of moderate total body hypothermia (30°C) immediately after arrest. Because it led to early post-CPR intractable VF, this method was not pursued. In 1982, Gisvold and Safar (1984) used a monkey model of global brain ischemia (Nemoto *et al.*, 1977) to evaluate a combination of post-insult hypothermia, hemodilution, barbiturate and steroid; it seemed to mitigate the neurologic deficit somewhat, but inconsistently. In 1987, we conducted the first controlled animal outcome study of *moderate* cooling (28–32°C) after cardiac arrest (VF no flow of 17 min in dogs) (Leonov *et al.*, 1990). Hypothermia was started with reperfusion by CPB. Some of the outcome data were statistically better, but clinically unimpressive; no survivor achieved OPC No. 1.

Thus, we hypothesized that the potentially beneficial

effects of moderate hypothermia at the molecular and cellular levels might be offset by hypothermia-induced microcirculatory blood sludging. A beneficial post-arrest effect might be achieved with mild cooling (34–36°C) combined with hypertensive hemodilution. We further speculated that in order to exert the beneficial effect within the first few seconds of reoxygenation, the brain must be at least slightly cooled during the cardiac arrest (no flow) and CPR (low flow) states. We had previously shown that it is feasible to cool the brain during no flow to an epidural temperature of approximately 34°C in 5–7 min and 30–32°C in 10 min of arrest, either by total body immersion in ice water (Alfonsi *et al.*, 1982; Tisherman *et al.*, 1985) or by external cooling of the head only (Brader, 1988; Brader *et al.*, 1985). When a head cooling method was applied to dogs during CPR in a short-term study, there was a suggestion of improved outcome (Brader *et al.*, 1985).

In 1988/89, we conducted the first controlled animal outcome study of *mild intra-plus-post-arrest cooling* (Table 7). We used randomized concurrent controls, the same team, and blinded outcome evaluation. We found a significant mitigating effect of mild brain cooling (34°C) on postischemic ND, OPC and HD, with or without hypertensive hemodilution, as compared to randomized (concurrent) controls with the same Hct and MAP (Table 7). The head and neck were immersed in ice water from VF (no flow) of 3 min to the end of VF 12.5 min, which lowered epidural temperature by 4°C and tympanic membrane temperature by 2°C. During reperfusion, until 1 h post-arrest tympanic membrane and core temperatures were controlled at 34°C by CPB with low flow through heat exchanger. In the normothermic control groups, tympanic membrane and pulmonary artery temperature were controlled throughout at 37.5°C. With normal MAP and Hct, 10 of 12 dogs in the normothermic control group achieved OPC No. 3 or 4 (poor outcome) (Table 7). In the mild hypothermia group with normotension and normal Hct 5/10 achieved OPC No. 1 ($p < 0.01$) and 4/10 OPC No. 2. One animal with OPC 3 had cardiac damage. In the normothermic control group with hypertension (with or without hemodilution) (Leonov *et al.* 1992), 6 of 12 animals achieved OPC 2 and the others OPC 3 or 4; none achieved OPC 1. In contrast, all 10 dogs treated with mild intra-post-cooling in addition to hypertensive hemodilution had good outcome – 6/10 achieved OPC 1 ($p < 0.01$) and 4/10 OPC 2 ($p < 0.05$) (Table 7). Best and final ND scores and total brain histopathologic damage scores showed similar trends. However, none of the brains examined

Table 6. *Effect of Mild Hypothermia (H) (35.0–36.9°C) Before and During Arrest (Protection and Preservation) vs. Normothermia (N) (37–38°C) on Outcome as Overall Performance Categories (OPC, Table 2).* Normotension, normal hct. Each (N) or (H) represents one dog. VF cardiac arrest in dogs, CPR or CPB, IPPV 20 h, intensive care 72 or 96 h

	Ventricular fibrillation (VF)	Cardiac arrest	No flow period
Outcome Best OPC 24–96 h	12 or 12.5 min	15 min	20 min
OPC 5 Brain death	N		
OPC 4 Coma Vegetative state	NNNN NNNNN		N HH
OPC 3 Severe disability Conscious	NNNN NNNN	NNNN NNNNN	N HHHHH
OPC 2 Moderate disability Conscious	NNN H	HH	
OPC 1 Normal	HHHH	HHH	

histologically were entirely free of injury, even in the hypothermic group. There was no evidence of selective protection of the hippocampus or caudoputamen. The beneficial effect of mild hypothermia seemed distributed uniformly to all brain regions. The lesions were predominantly ischemic neuronal changes. Microinfarcts were minimal or absent in all groups.

The effect on outcome of mild cooling during and after cardiac arrest had not been studied before. The mechanisms of beneficial effects are unclear, but probably multifactorial. Reduced cerebral oxygen consumption by hypothermia alone (5–8% per °C reduction in temperature) (Bigelow *et al.*, 1950; Rupp and Severinghaus, 1986) cannot explain the above beneficial pre-cooling or intra- and post-cooling effects of mild hypothermia. Additional mechanisms might include:

1) During ischemia (preservation), hypothermia slows degradation of ATP (Thorn *et al.*, 1958; Kramer *et al.*, 1968; Michenfelder *et al.*, 1970); mitigates abnormal ion fluxes including influx of calcium (Astrup *et al.*, 1981); reduces excitatory neurotransmitter release (Busto *et al.*, 1989); slows destructive enzymatic reactions (Ernster, 1988); and protects the fluidity of lipoprotein membranes by stiffening them, as suggested by Ernster (Keough and Davis, 1984; Cossins and Shinitzky, 1984).

Table 7. *Effect of Mild Hypothermia (34°C) During and After Arrest (Preservation and Resuscitation) vs. Normothermia (37–38°C) on Outcome as Overall Performance Categories (OPC, Table 2) with and without Hypertension (MAP > 150–120 mmHg) and Hemodilution (hct 20%).* Each circle (o) represents one dog. VF cardiac arrest (no flow) 12.5 min, CPB ≤ 5 min, IPPV 20 h, intensive care 96 h.

Best OPC 24–96 h	Normothermia normotension normal hct controls	Hypothermia normotension normal hct	Normothermia hypertension hemodilution controls	Hypothermia hypertension hemodilution
OPC 5 Brain death	–	–	–	–
OPC 4 Coma Vegetative state	ooo	–	o	–
OPC 3 Severe disability Conscious	ooo oooo	o +	ooo oo	–
OPC 2 Moderate disability Conscious	oo	oo oo	ooo ooo	oo oo
OPC 1 Normal		ooo $p < 0.01$ oo		ooo $p < 0.01$ ooo
Compared with controls		$p < 0.05$		$p < 0.05$

+ Massive myocardial necroses.
From Sterz *et al.*, in preparation.

2) During and after reperfusion and reoxygenation (resuscitation), hypothermia may slow free radical reactions (Ernster, 1988) and the propagation of lipid peroxidation cascades (Safar, 1986, 1988a; Ernster, 1988) and may reduce metabolic demands (Bigelow *et al.*, 1950; Rupp and Severinghaus, 1986), thereby protecting brain regions with low flow and trickle flow (Kagström *et al.*, 1983; Wolfson *et al.*, 1988; Latchaw *et al.*, 1989; Safar *et al.*, 1989) during the early postischemic phase of global cerebral hypoperfusion (Lind *et al.*, 1975; Snyder *et al.*, 1975; Kagström *et al.*, 1983; Wolfson *et al.*, 1988; Latchaw *et al.*, 1989; Safar *et al.*, 1989; Beckstead *et al.*, 1978).

Conclusions

Extending the maximal duration of clinical death that is reversible to survival without neurologic deficit is both important and feasible. Both focused, physiologic mechanisms-oriented and global, multi-level outcome-oriented studies are needed. Animal models with reproducible outcome after cardiac arrest and intensive care, although expensive and difficult to manage, are available. Use of such models in an established CPCR laboratory with animal intensive care permits the timely demonstration of significant, therapeutic effects on outcome of a given therapy. Positive results from such animal trials would lead to clinical feasibility trials, which are more expensive and less likely to give clear answers on outcome, which, however, are clinically more valid. The multifactorial pathogenesis of postischemic-anoxic encephalopathy needs tailored, etiology-specific, multifaceted treatments. Post-arrest moderate hypertension and moderate hemodilution should be incorporated into standard therapy protocols for animal outcome experiments and clinical trials. These treatments seem to overcome low flow after arrest, which would prevent physical or pharmacologic treatments from reaching the neurons. Mild brain cooling during and early after cardiac arrest mitigates brain damage, without the risks and management problems associated with moderate hypothermia with spontaneous circulation. Methods of optimal brain and heart cooling and degrees and durations of hypothermia should be developed.

Acknowledgements

Drs. Norman Abramson, Syed Hosain, Ernesto Pretto, and Harvey Reich helped with the experiments reported and made valuable suggestions. William Stezoski and Henry Alexander coordinated animal intensive care. Alan Abraham, Kathy Bickerstaff-Swales, Scott Nagel, Annette Pastula, and Mark Ritchey helped with experiments. Eric Brader, M.D. gave ideas on head cooling. Sheryl Kelsey, Ph.D. and John Wilson, Ph.D. helped with statistical analyses. Drs. Richard Latchaw, David Johnson, Steven Hecht, and Robert Tarr helped with the multifocal cerebral blood flow measurements. Lisa Cohn helped with editing. Gale Foster helped with preparation of the manuscript.

Supported by NIH grants NS24446 and NS15295 (USA), the A.S. Laerdal Foundation (Norway), and the E. Schroedinger Foundation (Austria).

References

1. Abramson NS, Safar P, Detre K (1986) Brain Resuscitation Clinical Trial (BRCT) I Study Croup Randomized clinical study of thiopental loading in comatose survivors of cardiac arrest. N Engl J Med 314: 397–403
2. Abramson NS, Safar P, Detre K, Brain Resuscitation Clinical Trial (BRCT) II Study Group (1991) A randomized clinical study of a calcium-entry blocker (lidoflazine) in the treatment of comatose survivors of cardiac arrest. N Engl J Med 324: 1225–1231
3. Abramson N, Safar P, Detre K (1989b) Brain Resuscitation Clinical Trial (BRCT) II Study Group Lidoflazine administration to survivors of cardiac arrest. Ann Emerg Med 18: 478 (Abstract)
4. Alfonsi G, Gilbertson L, Safar P, Stezoski W, Bircher N (1982) Cold water drowning and resuscitation in dogs. Anesthesiology 57: A80 (Abstract)
5. American Society of Anesthesiologists, Committee on Acute Medicine (Safar P, Chairman) (1968) Community-wide emergency medical services. JAMA 204: 595–602
6. Ames III A, Wright RL, Kowada M, Thurston JM, Majno G (1968) Cerebral ischemia. II. The no-reflow phenomenon. Am J Pathol 52: 437–453
7. Ames III A, Nesbett FB (1983) Pathophysiology of ischemic cell death. I. Time of onset of irreversible damage; importance of the different components of the ischemic insult. Stroke 14: 219–226
8. Astrup J, Moller-Sorenson P, Rahbeck-Sorenson H (1981) Inhibition of cerebral oxygen and glucose consumption in the dog by hypothermia, pentobarbital and lidocaine. Anesthesiology 55: 263–268
9. Beckstead JE, Tweed WA, Lee J, MacKeen WL (1978) Cerebral blood flow and metabolism un man following cardiac arrest. Stroke 9: 569–573
10. Berntman L, Welsh FA, Harp JR (1981) Cerebral protective effect of low-grade hypothermia. Anesthesiology 55: 495–498
11. Bigelow WG, Lindsay WK, Harrison RC, Gordon RA, Greenwood WF (1950) Oxygen transport and utilization in dogs at low body temperature. Am J Physiol 160: 125
12. Bircher N, Safar P (1981) Comparison of standard and "new" closed-chest CPR and open-chest CPR in dogs. Crit Care Med 9: 384–385
13. Bircher N, Safar P (1984) Manual open-chest cardiopulmonary resuscitation. Ann Emerg Med 13: 770–773
14. Bleyaert AL, Nemoto EM, Safar P, Stezoski SW, Mickell J, Moossy J, Rao G (1978) Thiopental amelioration of brain damage after global ischemia in monkeys. Anesthesiology 49: 390–398
15. Brader EW (1988) Cerebral preservation during cardiac arrest. Am J Emerg Med 6: 151–156
16. Brader EW, Jehle D, Safar P (1985) Protective head cooling during cardiac arrest in dogs. Ann Emerg Med 14: 510 (Abstract)
17. Busto R, Dietrich WD, Globus MY-T, Valdes I, Scheinberg P, Ginsberg MD (1987) Small differences in intraischemic brain temperature critically determine the extent of ischemic neuronal injury. J Cereb Blood Flow Metabol 7: 729–738
18. Busto R, Dietrich WD, Globus MY-T, Ginsberg MD (1989) The importance of brain temperature in cerebral ischemic injury. Stroke 20: 1113–1114
19. Cadenas E (1988) Biological chemiluminescence. In: Quintanilha A (ed) Reactive oxygen species in chemistry, biology, and medicine. Plenum, New York, p 117
20. Cerchiari EL, Holl TM, Safar P, Scalabassi R (1987) Protective effects of combined superoxide dismutase and deferoxamine on recovery of cerebral blood flow and function after cardiac arrest in dogs. Stroke 18: 869–878
21. Cerchiari E, Safar P, Klein E, Cantadore R, Pinsky M (1992) Cardiovascular function and neurologic outcome after cardiac arrest in dogs. The cardiovascular post-resuscitation syndrome. Resuscitation: in press
22. Cossins AR, Shinitzky M (1984) Adaptation of membranes to temperature, pressure, and exogenous lipids. In: Shinitzky M (ed) Physiology of membrane fluidity, Vol 1. CRC Press, Boca Raton/FL
23. Eisenberg MS, Bergner L, Hallstrom AP, Cummins RO (1986) Sudden cardiac death. Sci Am 254: 37–47
24. Ernster L (1988) Biochemistry of reoxygenation injury. Crit Care Med 16: 947–953
25. Gisvold SE, Safar P, Hendrickx H, Rao G, Moossy J, Alexander H (1984a) Thiopental treatment after global brain ischemia in pigtailed monkeys. Anesthesiology 60: 88–96
26. Gisvold SE, Safar P, Rao G, Moossy J, Kelsey S, Alexander H (1984b) Multifaceted therapy after global brain ischemia in monkeys. Stroke 15: 803–812
27. Hallenbeck JM, Leitch DR, Dutka AJ, Greenbaum LJ, McKee AE (1982) Prostaglandin I_2, indomethacin, and heparin promote postischemic neuronal recovery in dogs. Ann Neurol 12: 145–156
28. Hendrickx H, Safar P, Rao GR, Gisvold SE, Miller A (1984a) Asphyxia, cardiac arrest and resuscitation in rats. I. Short-term recovery. Resuscitation 12: 97–116
29. Hendrickx H, Safar P, Rao GR, Gisvold SE, Miller A (1984b) Asphyxia, cardiac arrest and resuscitation in rats. II. Long-term behavioral changes. Resuscitation 12: 117–128
30. Hirsch H, Müller HA (1962) Funktionelle und histologische Veränderungen des Kaninchengehirns nach kompletter Gehirnischämie. Pfluegers Archiv 275: 277–291
31. Hossmann K-A (1988) Resuscitation potentials after prolonged global cerebral ischemia in cats. Crit Care Med 16: 964–971
32. Hossmann K-A, Kleihues P (1973) Reversibility of ischemic brain damage. Arch Neurol 29: 375–384
33. Jastremski MS, Sutton-Tyrell K, Vaagenes P, Abramson NS, Heiselman D, Safar P, Brain Resuscitation Clinical Trial I Study Group (1989) Glucocorticoid treatment does not improve neurological recovery following cardiac arrest. JAMA 262: 3427–3430
34. Kagström E, Smith M-L, Siesjö BK (1983) Local cerebral blood flow in the recovery period following complete cerebral ischemia in the rat. J Cereb Blood Flow Metabol 3: 170–182
35. Kampschulte S, Smith J, Safar P (1969a) Oxygen transport after cardiopulmonary resuscitation. Anaesth Reanimation 39: 95 (in German)
36. Kampschulte S, Morikawa S, Safar P (1969b) Recovery from anoxic encephalopathy following cardiac arrest. Fed Proc 28: 522 (Abstract)
37. Katz L, Vaagenes P, Safar P, Diven W (1989) Brain enzyme changes as markers of brain damage in rat cardiac arrest model. Effects of corticosteroid therapy. Resuscitation 17: 39–53

38. Keough KMW, Davis PJ (1984) Thermal analysis of membranes. In: Kates M, Manson LA (eds) Membrane fluidity. Plenum, New York, pp 55–88
39. Kirimli B, Kampschulte S, Safar P (1968) Cardiac arrest from exsanguination in dogs. Evaluation of resuscitation methods. Acta Anaesthesiol Scand 29 [Suppl]: 183
40. Kochanek PM, Dutka AJ, Hallenbeck JM (1987) Indomethacin, prostacyclin, and heparin improve postischemic cerebral blood flow without affecting early postischemic granulocyte accumulation. Stroke 18: 634–637
41. Kramer RS, Sanders AP, Lesage AM, Woodhall B, Sealy WC (1968) The effect of profound hypothermia on preservation of cerebral ATP content during circulatory arrest. J Thorac Cardiovasc Surg 56: 699–709
42. Krischer JP, Fine EG, Weisfeldt ML, Guerci AD, Nagel E, Chandra N (1989) Comparison of prehospital conventional and simultaneous compression-ventilation cardiopulmonary resuscitation. Crit Care Med 17: 1263–1269
43. Latchaw RE, Johnson DW, Hecht ST, Tarr R, Safar P, Leonov Y, Sterz F (1989) Multifocal cerebral hypoperfusion after cardiac arrest studied with Xenon-CT blood flow analysis. Methodology and results. Proc Amer Soc Neuroradiol. Orlando/Fl, March 1989, p 39
44. Leonov Y, Sterz F, Safar P, Radovsky A, Oku K, Tisherman S, Stezoski SW (1990) Mild cerebral hypothermia during and after cardiac arrest improves neurological outcome in dogs. J Cereb Blood Flow Metabol 10: 57–70
45. Leonov Y, Sterz F, Safar P, Johnson DW, Tisherman Sa, Oku K (1992) Hypertension with hemodilution prevents multifocal cerebral hypoperfusion after cardiac arrest in dogs. Stroke 23: 45–53
46. Levy DE, Caronna JJ, Singer BH, Lapinski RH, Frydman H, Plum F (1985) Predicting outcome from hypoxic-ischemic coma. JAMA 253: 1420–1426
47. Lind B, Snyder J, Safar P (1975) Total brain ischemia in dogs: cerebral physiological and metabolic changes after 15 minutes of circulatory arrest. Resuscitation 4: 97–113
48. Longstreth WT, Invi TS, Cobb LA, Copass MK (1983) Neurologic recovery after out-of-hospital cardiac arrest. Ann Intern Med 98: 588–592
49. Michenfelder JD, Van Dyke RA, Theye RA (1970) The effects of anesthetic agents and techniques on canine cerebral ATP and lactate levels. Anesthesiology 33: 315–321
50. National Academy of Sciences – National Research Council (NAS-NRC) (1966) Cardiopulmonary resuscitation. JAMA 198: 372–379
51. National Academy of Sciences – National Research Council (NAS-NRC) (1974) Standards for cardiopulmonary resuscitation (CPR) and emergency cardiac care (ECC). JAMA 227 [Suppl]: 837–868
52. National Academy of Sciences – National Research Council (NAS-NRC) (1980) Standards and guidelines for cardiopulmonary resuscitation (CPR) and emergency cardiac care (ECC). JAMA 244 (Suppl): 453–508
53. National Academy of Sciences – National Research Council (NAS-NRC) (1986) Standards and guidelines for cardiopulmonary resuscitation (CPR) and emergency cardiac care (ECC). JAMA 255: 2905–2985
54. Negovsky VA, Gurvitch AM, Zolotokrylina ES (1983) Postresuscitation disease. Elsevier, Amsterdam
55. Nemoto EM, Bleyaert AL, Stezoski SW, Moossy J, Rao G, Safar P (1977) Global brain ischemia. A reproducible monkey model. Stroke 8: 558–564
56. Nemoto EM, Frinak S (1978) Rat brain tissue PO_2 after 16 min global ischemia and thiopental therapy. Crit Care Med 6: 113–114
57. Pretto E, Kazziha S, Safar P, Choy W (1985) Enhanced myocardial resuscitability by lidoflazine post-ischemia in isolated perfused rat heart preparation. Anesthesiology 63: A118
58. Reich H, Safar P, Angelos M, Basford R, Ernster L (1988a) Failure of a multifaceted anti-reoxygenation injury (RI) therapy to ameliorate brain damage after ventricular fibrillation (VF) cardiac arrest (CA) of 20 minutes in dogs. Crit Care Med 16: 387 (Abstract)
59. Reich H, Safar P, Angelos M, Leonov Y, Sterz F, Stezoski SW, Alexander H (1988b) Reversibility limit for heart and brain of ventricular fibrillation (VF) cardiac arrest (CA) in dogs. Crit Care Med 16: 390 (Abstract)
60. Rupp SM, Severinghaus JW (1986) Hypothermia. In: Miller RD (ed) Anesthesia, 2nd Ed. Churchill Livingstone, New York, pp 1995–2022
61. Safar P (ed) (1978) Introduction: on the evolution of brain resuscitation. Crit Care Med 6: 199–202
62. Safar P (1980) Amelioration of postischemic brain damage with barbiturates. Stroke 11: 565–568
63. Safar P (1985) Effects of the postresuscitation syndrome on cerebral recovery from cardiac arrest. Crit Care Med 13: 932–935
64. Safar P (1986) Cerebral resuscitation after cardiac arrest. A review. Circulation 74 [Suppl IV]: 138–153
65. Safar P (1988) Resuscitation from clinical death: pathophysiologic limits and therapeutic potentials. Crit Care Med 16: 923–941
66. Safar P, Benson DM, Esposito G, Grenvik A, Sands PA (1974) Emergency and critical care medicine: local implementation of national recommendations. Clin Anesth 10: 65–125
67. Safar P, Stezoski SW, Nemoto EM (1976) Amelioration of brain damage after 12 minutes cardiac arrest in dogs. Arch Neurol 33: 91–95
68. Safar P, Elam J (eds) (1977) Advances in cardiopulmonary resuscitation. Springer, New York Wien
69. Safar P, Gisvold SE, Vaagenes P, Hendrickx H, Bar-Joseph G, Bircher N, Stezoski W, Alexander H (1982) Long-term animal models for the study of global brain ischemia. In: Wauquier A, Borgers M, Amery WK (eds) Protection of tissues against hypoxia. Elsevier, Amsterdam, pp 147–170
70. Safar P, Vaagenes P (1986) Systematic search for brain resuscitation potentials after total circulatory arrest. In: Baethmann A, Go KG, Unterberg A (eds) Mechanisms of secondary brain damage. Plenum, New York, pp 349–375
71. Safar P, Breivik H, Abramson N, Detre K, Brain Resuscitation Clinical Trial (BRCT) I Study Group (1987) Reversibility of clinical death in patients: the myth of the 5 minute limit. Ann Emerg Med 16: 496 (Abstract)
72. Safar P, Bircher NG (1988) Cardiopulmonary cerebral resuscitation. An introduction to resuscitation medicine. World Federation of Societies of Anaesthesiologists, 3rd Ed. Laerdal, Stavanger; Saunders, Philadelphia
73. Safar P (1988) Resuscitation from clinical death: Pathophysiologic limits and therapeutic potentials. Crit Care Med 16: 923–941
74. Safar P, Tisherman S, Latchaw R, Alexander H, Hecht S, Johnson D, Leonov Y, Oku K, Sterz F, Stezoski SW, Tarr R, Wolfson S (1989a) Multifocal cerebral hypoperfusion after cardiac arrest and resuscitation in dogs. Potentials for therapeutic mitigation. Cerebral Resuscitation Research Symposium, Moscow, USSR, March 1989 (Abstract)
75. Safar P, Abramson NS, Angelos M (1990) Emergency cardiopulmonary bypass for resuscitation from prolonged cardiac arrest. Am J Emerg Med 8: 55–67
76. Siesjö BK (1981) Cell damage in the brain: a speculative synthesis. J Cereb Blood Flow Metabol 1: 155–185
77. Siesjö BK (1988) Mechanisms of ischemic brain damage. Crit Care Med 16: 954–963
78. Snyder BD, Tabbaa MA (1988) Assessment and treatment of neurological dysfunction after cardiac arrest. Current concepts of cerebrovascular disease. Stroke 19: 269–273

79. Snyder JV, Nemoto EM, Carroll RG, Safar P (1975) Global ischemia in dogs: intracranial pressure, brain blood flow and metabolism. Stroke 6: 21–27
80. Steen PA, Gisvold SE, Milde JH, Newberg LA, Scheithauer BW, Lanier WL, Michenfelder JD (1985) Nimodipine improves outcome when given after complete cerebral ischemia in primates. Anesthesiology 62: 406–414
81. Sterz F, Leonov Y, Safar P, Johnson D, Oku K, Tisherman SA, Latchaw R, Obrist W, Stezoski SW, Hecht S, Tarr R, Janosky JE (1992) Multifocal cerebral blood flow by Xe-CT and global cerebral metabolism after prolonged cardiac arrest in dogs. Reperfusion with open-chest CPR or cardiopulmonary bypass. Resuscitation 24: 27–47
82. Sterz F, Leonov Y, Safar P, Radovsky A, Stezoski SW, Reich H, Shearman GT, Greber TF (1989) Effect of excitatory amino acid receptor blocker MK-801 on overall, neurologic, and morphologic outcome after prolonged cardiac arrest in dogs. Anesthesiology 7: 907–918
83. Thorn W, Scholl H, Pfleiderer G (1958) Metabolic processes in the brain at normal and reduced temperatures and under anoxic and ischaemic conditions. J Neurochem 2: 150 (in German)
84. Tisherman S, Chabal C, Safar P (1985) Resuscitation of dogs from coldwater submersion using cardiopulmonary bypass. Ann Emerg Med 14: 389–396
85. Vaagenes P, Cantadore R, Safar P, Moossy J, Rao G, Diven W, Alexander H, Stezoski W (1984) Amelioration of brain damage by lidoflazine after prolonged ventricular fibrillation cardiac arrest in dogs. Crit Care Med 12: 846–855

Correspondence and Reprints: P. Safar, International Resuscitation Research Center, University of Pittsburgh, 3434 Fifth Avenue, Pittsburgh, PA, 15260, U.S.A.

Acta Neurochir (1993) [Suppl] 57: 123–129

Functional Consequences of Cerebral Lesions

Taxonomy of Subjective Phenomena: a Neuropsychological Basis of Functional Assessment of Ischemic or Traumatic Brain Lesions

E. Pöppel

Institut für Medizinische Psychologie der Ludwig-Maximilians-Universität, München, Federal Republic of Germany

Abstract

A proper evaluation of functional competence after central lesions has to be based on a classification of functions that one can agree upon. It is a sad fact in neuropsychology that such a classification is not available. An attempt will be made to discuss such a classification (or taxonomy) that might be useful.

The basic idea is that elementary psychological functions are evolutionary products whose availability is dependent on the functional integrity of neuronal modules. Such modules are embedded neuronal mechanisms that are linked to localized structures or distributed neuronal algorithms. Constancy of interindividual loss of psychological functions associated with lesions of modules can be used to define a catalogue of functions. Using this principle one can differentiate four areas of psychological functions that are represented in a modular fashion. These areas are stimulus representations ("perception"), processing of information ("learning and memory"), evaluation of information (for instance by emotions), and finally action or reaction. Functional competence is, however, not only described by the potential availability of elementary psychological functions, but also by formal aspects, i.e. how functions are made available. Such formal aspects refer to activation and in particular to temporal problems of neuronal processing. A particular "time machine" will be discussed which is essential for functional competence.

Central lesions may either effect the *what* of functions or the *how* of functions. A differentiation between these *material* and *formal* aspects of functional competence are essential with respect to recovery or restitution of function.

Keywords: Taxonomy of cerebral functions; neuronal modules und algorithms; temporal neuronal processing.

Introduction

Local injuries of the brain may lead to specific loss of functions, like color vision, movement perception, face recognition, referential memory, emotional disturbance, semantic competence in language, etc. I want to use these observations as building-blocks for a general classification of subjective phenomena. This classification – or taxonomy – will provide a more precise description of neuropsychological and psychopathological phenomena. Only if such a taxonomy is available it is possible to provide a reliable assessment of functions after brain lesions.

The basic hypothesis underlying this taxonomy is that psychological functions are based on neuronal programs that have developed during evolution and that these functions are necessarily dependent on the integrity of neuronal structures or neuronal algorithms. Obviously, I am not the sole proponent of this thesis. At the end of his famous work "The origin of species" (1859), Darwin proposed the following view of subjective phenomena: "In the future I see open fields for far more important research. Psychology will be securely based on the foundation already well laid by Mr. Herbert Spencer, that of the necessary acquirement of each mental power and capacity by gradation. Much light will be thrown on the origin of man and his history." Popper (1982) recently presented a similar idea: "It seems reasonable to assume, in spite of the metaphysical character of the assumption, that the human mind evolves, that it can be regarded as a product of evolution – of an evolution in which the emerging mind plays a very active part."

Mental activity as mentioned here by Popper is a particular feature of the subjective. Pointing out that psychological functions depend on sensory processes does not imply that organisms are passively exposed to their environment. In evolutionary theory the term "Baldwin-Effect" applies, which was characterized by Popper (1982) in the following way: "With the emergence of exploratory behavior, of tentative behavior, and of trial and error behavior, mindlike behavior plays an increasingly active part in evolution. This does not mean that Darwinian selection is transcended, but it means that active Darwinism, the search for a friendly environment, the selection of a habitat by the organism, becomes important."

The thesis expressed here refers to one basic feature of the subjective that has recently been stressed by Searle (1983) – intentionality. Intentionality describes the fact that our mental states are directed at something, or about something or that they refer to something, e.g., "I see something; I believe in something; I expect something; I am afraid of something." However, not all mental states appear to be intentional. Examples for non-intentional subjective states are "I am nervous; I am tired; I am depressed." In this case, the subjective state is not related to a particular object or state, it just describes a general state of being.

After these preliminary remarks on some basic features of mental functions as I see them, I now want to explore the taxonomy of the subjective in some detail. Four classes of elementary psychological functions are discriminated – those of perception, of stimulus processing, of stimulus evaluation, and of response (action and reaction). Let us examine the first, the perceptual domain.

Since more than 100 years it has been known that particular lesions in the brain result in specific functional losses. A patient who has suffered a local injury in the occipital lobe may exhibit a circumscribed blindness in his visual field, e.g., a homonymous hemianopsia (Teuber *et al.*, 1960). Another patient with an injury to a different occipital site may no longer perceive colors (Pöppel *et al.*, 1978). A third patient may no longer be able to recognize faces; he suffers from a so-called prosopagnosia (Meadows 1974). Although extremely rare, such cases are very instructive.

On the basis of numerous studies on perceptual functions, one can argue that functions are locally represented. This observation allows a more general law on elementary psychological functions to be deduced: the fact that psychological functions are lost with inter-individual constancy after circumscribed lesions of the brain provides proof of the existence of these functions, i.e., psychological functions can be defined as those that, in principle, can be lost after circumscribed injuries of the brain.

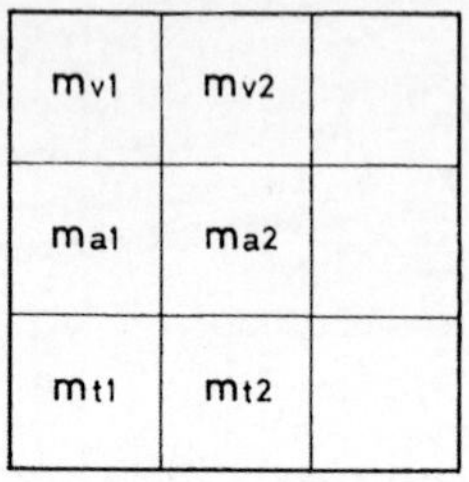

Perceptual Algorithms (A)

Fig. 1. Scheme of the modular representation of the perceptual systems. Each square is meant to indicate a neuronal structure or a neuronal algorithm representing an elementary psychological function

Using a term from technology, we may refer to those structures (or neuronal algorithms) that implement specific functions as modules. Of the different sensory systems in Fig. 1, only three such systems or modalities are shown: m_v refers to the visual system, m_a to the auditory system, and m_t to the tactile system. Each sensory system is characterized by a number of modules; in this example, only three such modules are shown for each modality. The different modules may be viewed as the structural or algorithmic correlates of particular perceptual qualities. Such qualities in the visual domain are, for example, color, movement, or, perhaps, faces. For taste, these qualities are sweet, sour, salty, or bitter. In case a module is lost, this particular quality is no longer available; for instance, if the module m_{v2} is missing the patient might be color-blind.

The next functional domain is that of stimulus processing. Here memory and learning functions are summarized. For this domain it can also be assumed that the integrity of local neuronal structures is essential for the availability of specific psychological functions. A well-known illustration is provided by the case of H. M., a patient who suffered a selective memory loss following bi-lateral surgical ablation of the hippocampus. Since the operation, the patient has suffered from a particular loss of memory, i.e., he is no longer capable of storing information in his referential mnemonic system (Scoville and Milner, 1957). His short-term memory is apparently quite normal, and he does not seem to have any problems remembering events prior to the operation. Thus, those functions are still available that are necessary to use information from his long-term memory. Only one particular aspect of his memory has been lost, the capacity to store new information. In conclusion, the functions of stimulus processing also appear to be represented in a modular way in the brain, as is indicated by numerous neuropsychological observations. Different learning strategies and different aspects of memory are selectively represented, either in a structural or in a functional (algorithmic) module.

Apart from functions of stimulus acquisition and stimulus processing, subjective phenomena are characterized by functions of stimulus evaluation. This functional domain refers to our emotions. Each perceptual contact with the world is instantaneously accompanied by an emotional evaluation. In particular, pain and pleasure are such basic, essential dimensions (Pöppel, 1982, 1985). For this functional region, it also seems true that different evaluative functions are represented in a modular fashion (Fig. 6). The local diencephalic or limbic representation of different emotions has been proven by neuro-ethological and neurological observations (e.g. Ploog, 1980). Lately the integrity of the right hemisphere has been stressed for the availability of negative emotions, like sorrow (Sackheim *et al.*, 1982). A convincing example has been provided by Olds' experiments (1977), which have resulted in a new paradigm for studying learning. By electrical stimulation in the hypothalamus, an area could be localized that apparently leads to a positive evaluation of the present situation. If, in the paradigm of operant conditioning, a rat accidentally pushes a lever leading to an electrical stimulation of the so-called pleasure centre in the hypothalamus, the animal will form an association between lever and positive emotion and will activate the lever continuously. Analogous stimulations of homologous regions of the human brain have led to verbal utterances of pleasure.

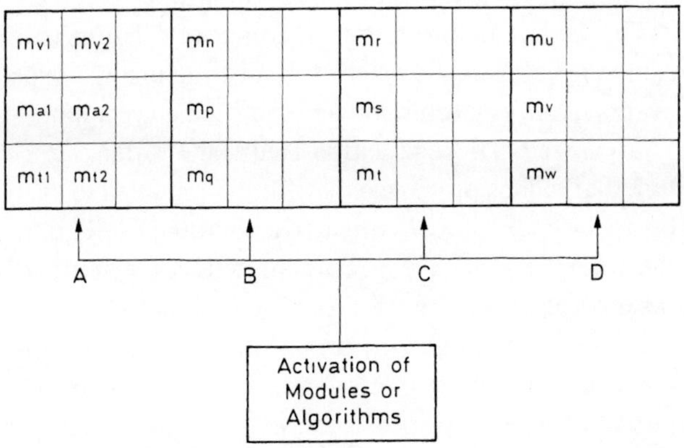

Fig. 2. Scheme indicating activation of the various modular systems: *(A)* Perception modules; *(B)* Processing modules; *(C)* Evaluation modules; *(D)* Response modules

It should be stressed, of course, that different areas are not of equal size. Furthermore, it should not be concluded that every functional region comprises the same number of modules. At present, nothing can be said about the number of such modules, although the number is certainly limited. As the number of modules is not yet known, an open taxonomy with its basic structure is described here. The final functional domain comprises the functions of motor response. The history of the dogma of the localization of functions started with this domain. The French neurologist Broca (1865) was probably the first to stress that spoken language is dependent on the integrity of a circumscribed region in the frontal lobe of the left hemisphere. If this area is destroyed, for instance by stroke or trauma, a particular form of aphasia named after Broca is observed. On the basis of many other observations particularly from work on apraxia it can be concluded that this functional region of motor response is also constructed in a modular fashion.

However, sensory input alone is not sufficient to trigger mental functions. For the brain to function or for the subjective to be available, a certain level of activation is necessary (Fig. 2). A–D refer to the four domains of subjective functions just mentioned. The energetic support of the brain is schematically shown. Only one reservoir supports the different functional regions. However, it is also possible that different regions working in parallel provide the neuronal energy for the different functional regions of the brain. A disturbance in activation can result, in the extreme case, in coma – in less severe cases, in a reduction of vigilance (von Cramon, 1979).

It is now claimed that each intentional act is characterized by the activity of modules in all functional domains. Let me give an example: we may see a beautiful face toward which we orient ourselves. We evaluate this stimulus according to its novelty. We may evaluate this face with respect to its aesthetic value. Without memory, each situation would be completely new to us. This is certainly not the case. Mnemonic information is included in this perceptual act as we compare this face to others, or as we refer this face to our mental image of "faciness". As we react to the stimulus, modules from the functional domains of response are in action. Thus, each act is based on the participation of modules from different functional domains. Furthermore, a certain level of activation of the modules is required.

In the graphic system used here, the mental state of seeing a beautiful face is characterized by the accentuation of certain modules (Fig. 3). The activity of specific modules leads to a new problem of organization of brain states. The hypothesis that different subjective functions are represented in different areas or with different algorithms leads to the question of how the activity of these different regions is temporally coordinated. The problem of such a temporal organization in the brain has been treated by Lashley (1951). The following question arises here: How can a subjective experience in which different modules are active be created by the appropriate local activities ? What is the underlying mechanism providing that evaluation of an event is actually related to the associated perceptual stimulus and not to another one which comes later or earlier?

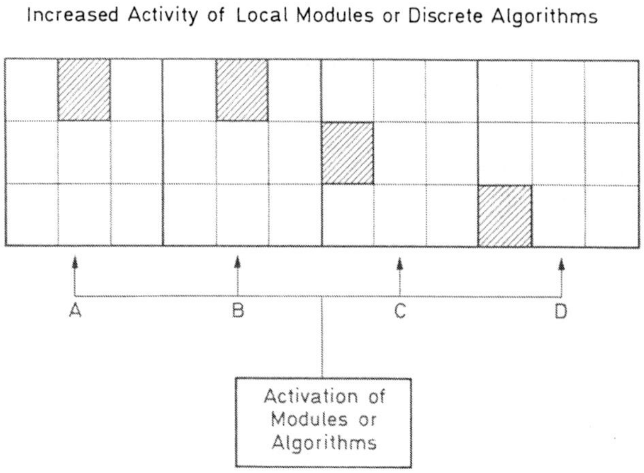

Fig. 3. Scheme indicating a particular mental state in which the higher neuronal activity of specific modules is shown (grey)

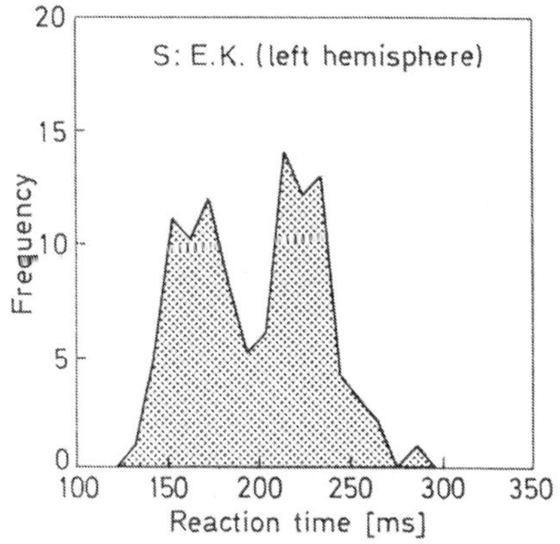

Fig. 4. Histogram of acoustic choice-reaction time of a subject that had to discriminate between optic and acoustic stimuli before the proper response was made. (Only the acoustic response times are shown). Note the bimodal distribution of the response times

As a solution of this intracerebral timing problem that arises because of the modular representation of functions, the following suggestion is made: temporal coordination of psychological functions is provided by neuronal coordination mechanisms, which are expressed as oscillatory processes in neuronal populations. On the basis of such neuronal oscillations, the brain can provide itself with independent temporal states. It can be proposed that, within a period of neuronal oscillation, all physically non-simultaneous events are considered as simultaneous. There are a number of psychophysical observations indicating that, in fact, such high frequency oscillatory processes do exist.

Oscillatory processes can be seen in experiments on choice-reaction time (Pöppel, 1968, 1970, 1978, 1985; Pöppel *et al.*, 1978; Madler and Pöppel, 1987). If a subject or patient has to react to stimuli from the visual or auditory modality in a choice-reaction paradigm, quite often multi-modal histograms of the reactions are observed. In Fig. 4, such a histogram is shown of a subject who participated in a choice-reaction time experiment where visual or auditory stimuli had to be discriminated before the reaction. Here only the reactions to the auditory stimuli are shown. If the stimulus were continuously processed in the brain, one would expect a uni-modal distribution of choice-reaction time. The multi-modality of the histogram, which was observed under stationary conditions, is a clear indication of an oscillatory process which underlies stimulus processing. In a technical sense, one can refer to relaxation oscillations that are stimulus dependent. In Fig. 5, the idea of an oscillatory process is graphically illustrated (Pöppel 1970). Accordingly, a stimulus (S) triggers neuronal oscillation, and the reactions are triggered at particular phases of this oscillation, resulting in multi-modalities of the reaction time.

The author believes that neuronal oscillation of a frequency of 30–40 Hz is the formal structure by which events in the environment are identified and temporally organized. This oscillation is also the formal structure that relates processes in different modules to each other (Pöppel 1985). With such a neuronal oscillation, the brain has a clock (Fig. 6) that provides the synchronization of activity in the different modules. However, it should be stressed that, at present, nothing is known about the structural implementation of this neuronal clock. There are two possible ways to reason about the implementation of a brain clock. One is that such a clock is implemented at one particular locus in the brain (as the reticular formation).

This structure would digest information from all different sensory organs and then secondarily trigger the different thalamo-cortical regions via fast-conducting pathways. The other idea is that each sensory system has its own neuronal oscillation on the basis of a thalamo-cortical feedback. In the latter case an additional synchronization mechanism between different domains would be necessary.

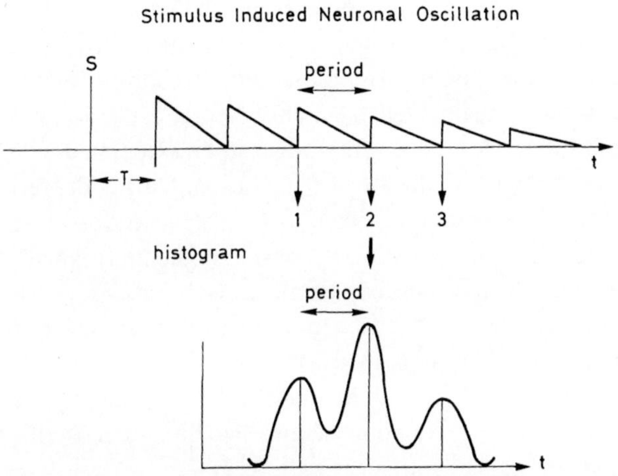

Fig. 5. Model of a neuronal relaxation oscillator explaining multimodal response histograms as in Fig. 4. After a constant transduction time *(T)* a central periodic process is triggered. The period of this central neuronal oscillation is probably close to 30 or 40 ms. Motor responses are only triggered at certain phases of this oscillation. This phase dependency of the response leads to the multimodal distributions in response histograms. The multi-modality therefore is an expression of a temporal information processing that is quantal in nature

In order to complete the taxonomy of the subjective, some observations from chronobiology must be added. It has been shown that mental or physiological functions can considerably vary (Aschoff, 1981) depending on the time of day. In Fig. 7, this fact has been considered, i.e., that activation of the modules underlies a long-term circadian modulation. Different circadian oscillators may be responsible for the circadian modulation of mental and physiological functions. Besides a circadian modulation, infra- or ultradian modulations may also exist, even a circannual modulation of activation is conceivable for the human.

Finally, a further phenomenon of temporal organization must be considered – namely the phenomenon of temporal integration (Pöppel, 1985). It has been demonstrated already by Wundt (1911) that events which follow each other are integrated up to a certain temporal limit into a perceptual unit. This period of integration consists of approximately 3 sec, as indicated by a number of observations (Fig. 8). It has previously been suggested (Pöppel 1985) that a subjective representation of this 3-sec-integration interval could be considered as the subjective present, the "now", or as the content of consciousness at a given moment. This "now", which is conceived here, at first may appear as a mental island within the stream of time. Temporal islands, however, are not typical for the subjective reality of healthy people. The subjective stream of time is characterized by the fact that the islands of "now" are connected with each other. On the basis of the connection of different temporal islands, we obtain the im-

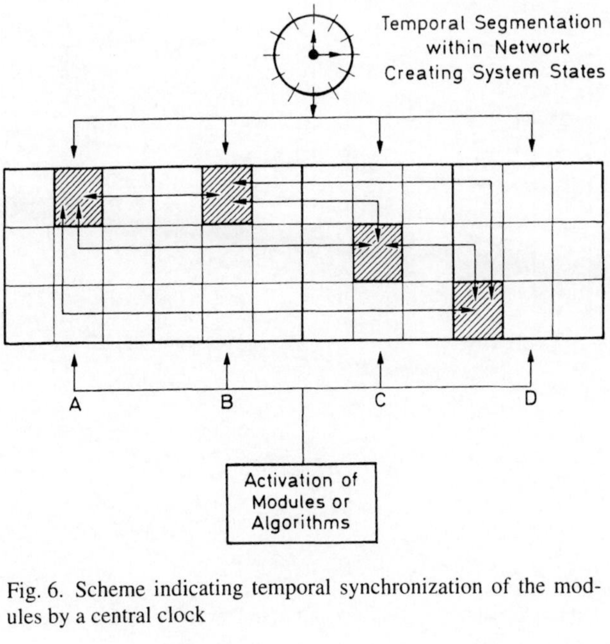

Fig. 6. Scheme indicating temporal synchronization of the modules by a central clock

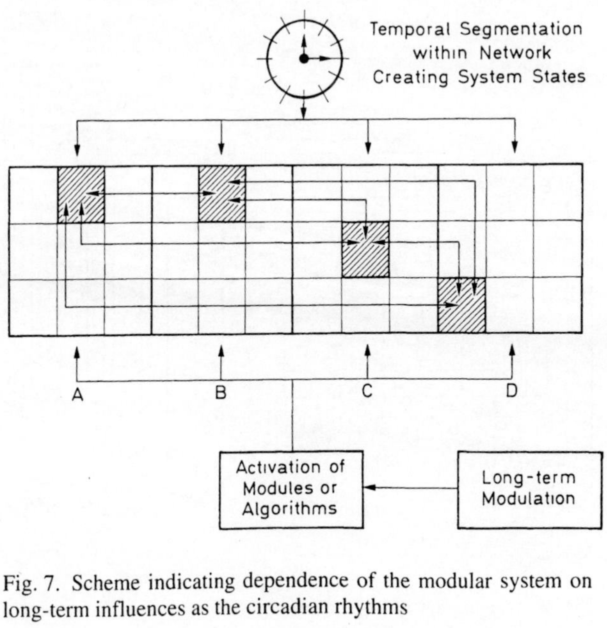

Fig. 7. Scheme indicating dependence of the modular system on long-term influences as the circadian rhythms

pression of flowing time. The connection of different windows of presence is possible on the basis of the content within consciousness. What we have now and what we have next in consciousness is always dependent on what is represented. Each subsequent content of consciousness is determined by the preceding one. On the basis of the semantic connection of contents in consciousness, we do not realize that each act or each separate content of consciousness is limited to only a few seconds. For our subjective reality, it is important what we are aware of and not how. Because of this, we do not realize the temporal structure of consciousness, which is the formal background for the content of consciousness. The temporal structure of consciousness itself is not a content of consciousness.

The taxonomy of subjective phenomena discussed here is biologically oriented. Biological systems can be disturbed. There are four areas within which problems can arise and which can be used to test this taxonomic system. This taxonomy allows the construction of pathological phenomena that can be experimentally tested.

1. Local modules can be lost.
2. Disturbances may be found in the area of activation and its long-term modulations.
3. Disturbances may be found in the temporal short-term organization and synchronization.
4. Problems may arise in the area of temporal integration and the semantic connection of integrated intervals.

The local loss of a module results in a functional loss. Functions that are based on neuronal modules or on discrete neuronal algorithms are then no longer available. Without the integrity of these structures, the psychological repertoire lacks certain essential elements. The modules carry the "whatness", i.e., the content of mental life.

Disturbances or lesions can also be observed in the area of activation. Severe disturbances result in coma or a reduced state of vigilance. Disturbances in the activation system may even result in depression. In a depression, reasoning may be slowed down, the patient is characterized by sleepiness, reduced ability to concentrate, physical lethargy, and general psycho-motor inhibition. All these symptoms can be associated with a disturbance in the area of activation. During depression, typical diurnal changes can also be observed, characterized by severe depression in the morning and an improvement in the afternoon. Some psychiatrists believe that a disturbance in the circadian organization is actually causally related to depression, and certain manic phenomena can be interpreted as an increase in activation due to a poorly controlled circadian oscillator.

Disturbances in the area of temporal organization should theoretically result in completely different pathological phenomena. Certain schizophrenic symptoms could be discussed on the basis of the taxonomy presented here. Is it conceivable that the disturbance of temporal organization may actually lead to schizophrenia? It has been proposed that the 30–40 Hz oscillator

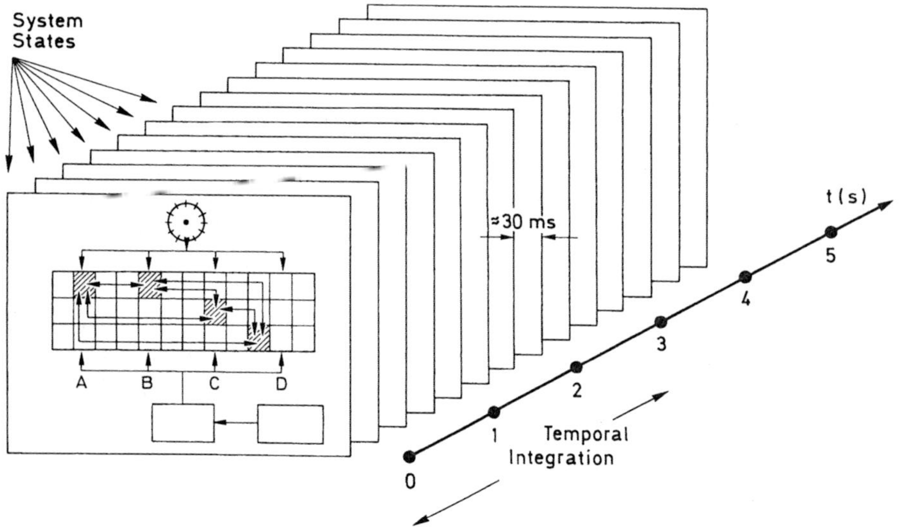

Fig. 8. Scheme indicating the temporal integration of successive mental events (system states) into a unit of not more than a few seconds

is essential for event identification. A disturbance in the temporal connection of different modules could conceivably result in defective event identification so that physically defined stimuli no longer can be given an identity. Personal identity might be endangered if the temporal relation between perception, processing, and evaluation of stimuli is disconnected.

Temporal integration and possible disturbances in this area may lead to different pathological phenomena. This is particularly true for thought disorders. The psychiatrist Bleuler (1969) described such disturbances characteristic of schizophrenic patients in the following way: it appears that the normal connection which is essential to keep thoughts together is no longer available, and that other thoughts suddenly appear. It is suggested that a disturbance of the semantic connection of successive integrative units leads to a discontinuity of mental life, which results in this particular form of schizophrenia.

The taxonomy presented here can only be applied if it can be tested. I believe that such a test is possible through neuropsychological and psychiatric phenomena. The potential that it can be tested is an advantage of this taxonomy. It provides a conceptual basis for the description of the subjective on the basis of evolutionary reasoning. On the other hand, the taxonomy allows an integrated view of neurological and psychiatric phenomena.

References

1. Aschoff J (1981) Biological rhythms. Handbook of of behavioral neurobiology. Plenum, New York
2. Bleuler M (1969) Lehrbuch der Psychiatrie, 11th Ed. Springer, Berlin Heidelberg New York
3. Broca P (1865) Sur le siège de la faculté du langage aniculé. Bull Soc Anthropol 6: 337–393
4. Cramon D von (1979) Quantitative Bestimmung des Verhaltensdefizits bei Störungen des skalaren Bewußtseins. Thieme, Stuttgart
5. Darwin Ch (1963, orig. 1859) Die Entstehung der Arten. Reclam, Stuttgart
6. Lashley K (1951) The problem of serial order in behavior. In: Jeffress LA (ed) Cerebral mechanisms in behavior. Wiley, New York, pp 112–136
7. Madler Ch, Pöppel E (1987) Auditory evoked potentials indicate the loss of neuronal oscillations during general anaesthesia. Naturwissenschaften 74: 42–43
8. Meadows JC (1974) The anatomical basis of prosopagnosia. J Neurol Neurosurg Psychiatry 37: 489–501
9. Olds J (1977) Drives and reinforcements. Raven, New York
10. Ploog D (1980) Emotionen als Produkte des limbischen Systems. Med Psychol 6: 7–19
11. Pöppel E (1968) Oszillatorische Komponenten in Reaktionszeiten. Naturwissenschaften 55: 449–450
12. Pöppel E (1970) Excitability cycles in central intermittency. Psychol Forsch 34: 1–9
13. Pöppel E (1978) Time perception. In: Held R, Leibowitz HW, Teuber H-L (eds) Handbook of sensory physiology, Vol VIII: Perception. Springer, Berlin Heidelberg New York, pp 713–729
14. Pöppel E (1982) Lust und Schmerz. Grundlagen menschlichen Erlebens und Verhaltens. Severin und Siedler, Berlin
15. Pöppel E (1985) Grenzen des Bewußtseins. Zur Wirklichkeit und Welterfahrung. dva, Stuttgart (English edition 1988) Mindswork. Time and conscious experience. Academic Press, Boston)
16. Pöppel E, Brinkmann R, v. Cramon D, Singer W (1978) Association and dissociation of visual functions in a case of bilateral occipital lobe infarction. Archiv für Psychiatrie und Nervenkrankheiten 225: 1–21
17. Popper K (1982) The place of mind in nature. In: Elvee RQ (ed) Mind in nature. Harper and Row, San Francisco, pp 31–59
18. Sackheim HA, Greenberg MS, Weiman AL, Gur RC, Hungerbühler JP, Geschwind N (1982) Hemispheric asymmetry in the expression of positive and negative emotions. Arch Neurol 39: 210–218
19. Scoville WB, Milner B (1957) Loss of recent memory after bilateral hippocampal lesion. J Neurol Neurosurg Psychiatry 20: 11–21
20. Searle JR (1983) Intentionality. An essay in the philosophy of mind. Cambridge University Press, Cambridge
21. Teuber HL, Battersby WS, Bender MB (1960) Visual field defects after penetrating missile wounds of the brain. Harvard University Press, Cambridge
22. Wundt W (1911) Einführung in die Psychologie. Voigtländer, Leipzig

Correspondence and Reprints: E. Pöppel, Institut für Medizinische Psychologie der Ludwig-Maximilians-Universität, Goethestrasse 31, D-W-8000 München 2, Federal Republic of Germany.

Acta Neurochir (1993) [Suppl] 57: 130–135

Blood Flow and Clinical Course in Patients with Ischemic Stroke without Cerebrospecific Therapy

A. Hartmann, C. Dettmers, H. Lagreze, and **Y. Tsuda**

Neurologische Universitätsklinik, Bonn, Federal Republic of Germany

Abstract

To evaluate the course of cerebral tissue perfusion in patients with acute focal cerebral ischemia of the supratentorial compartment regional cerebral blood flow (rCBF) was measured on day 0, 7, 14, 21, and 28 in 132 patients using the 133 Xenon stationary inhalation technique. Ischemic events of the brainstem and hemorrhagic complications were excluded. The clinical status was evaluated using a modified Mathew score.

In 34 patients no hemodilution, anti-edema therapy, or Ca^{++}-antagonists were used but otherwise best medical therapy was applied. These patients represented the so called "natural course" of cerebral ischemia.

In 30/34 patients on day 0 (within 16 hours after onset of symptoms) focal flow abnormalities were found in the involved side. In 9 of these 30 patients and in 1 of the remaining ischemia was observed in the contralateral side. rCBF above normal (relative luxury perfusion) despite pathologic neurologic findings was observed in 8/34 patients on day 3–7. Eight patients presented on day 3–7 with normal flow which later became ischemic again without evidence of another symptomatic episode. Correlation between severity of clinical findings and actual rCBF was low from day 3 to 7 but close on day 0. From day 14–21 hemispheric CBF correlated well with the total neurologic score but focal clinical findings had a lower correlation with focal flow as compared to day 0 and day 28.

Contralateral ischemia was never found after day 14. In 5 other cases with "natural course" described above, a transitory decrease of rCBF below the initial ischemia level was found between day 3 and 14. In all these patients CCT presented with development of severe brain edema requesting antiedema therapy. It was concluded that rCBF measurements reflect focal clinical status in acute cerebral ischemia only at the beginning and at the end of a 4-week period.

Keywords: Cerebral infarction; regional cerebral blood flow; ^{133}Xenon inhalation; clinical correlation.

Introduction

Ischemic stroke resulting in clinical symptoms is due to partial or complete interruption of cerebral blood flow (CBF) in a confined area of the brain. According to the flow threshold theory[1] intensity and duration of neurological symptoms depend on the degree of CBF reduction. After the initial ischemic event both, compensating and aggravating processes evolve, which might affect CBF, cerebral metabolism and, thus, the clinical course. Therapeutic concepts aim at an inhibition of the aggravating factors, at a support of compensating mechanisms, and an improvement of CBF and metabolism. No method of treatment which is routinely used has so far been shown to improve reliably neurological outcome or mortality[2]. Still, the majority of patients suffering from a stroke survive and many return to normal life. This is probably due to the natural course of the stroke.

In order to obtain information on the quality of CBF during the course of an ischemic stroke various methods might be applied. One is the stable Xenon technique, which, using computerized tomography, however, affects cerebral function and, therefore, is of limited value for repeated assessments of CBF[3]. By single photon emission computed tomography (SPECT) tissue perfusion cannot be quantitatively determined, or the methods based on cerebral uptake of IMP or HMPaO do not indicate precisely enough the regional CBF. The 2-dimensional Xenon 133 inhalation technique might be employed for repeated routine measurements of CBF in stroke patients yielding quantitative data on cortical blood flow.

We have made follow-ups of a group of patients with acute ischemic stroke, part of whom (n = 39) has not been subjected to hemodilution or treatment with calcium-antagonists. This appeared to be justified, since no analysis studying these compounds has convincingly proven effectivity of these methods. However, general concepts of treatment, such as stabilization of blood pressure, cardiac function, if necessary reduction of intracranial pressure and normalization of systemic measures, e.g. pulmonary oxygenation were implemented at the best possible level. This group of patients represents therefore cases with a so-called natural course.

The question remains, however, whether measurements of CBF do provide relevant information on the "natural" course. We have attempted to answer this question by repeated measurements of CBF in a group of 27 patients, who entered the hospital on the day of stroke, and who could be followed for a period of at least 4 weeks. Except of the above mentioned measures of general medical care, no hemodilution, rheologically active drugs, or calcium-antagonists were administered.

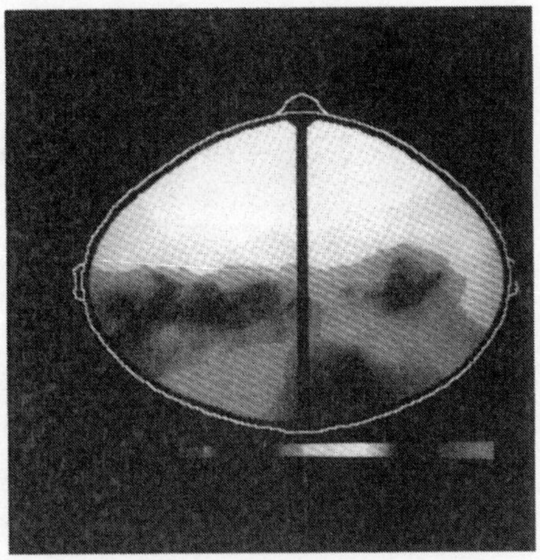

Fig. 1. Regional distribution of CBF in 17 elderly normal individuals without cerebrovascular risk factors, or any other neurological or cardiac disease. The gray tones indicate deviation of CBF in percent from the mean hemispheric CBF, with dark indicating the lowest and bright the highest flow value. A hyperfrontality can be noted also in this group with every individual being older than 50 years

Material and Methods

Patients

From 1979 until 1989 altogether 132 patients with acute ischemic infarction who have entered the neurological service within 24 hours after onset of symptoms have been repeatedly studied in the CBF laboratory. The mean age was 54.1 ± 11.2 years, 83 patients were male. Most of the patients were participating in prospective studies of therapeutical protocols which lasted for several weeks. 39 of those fulfilled the criteria of a "natural" course as outlined above. The mean age of this subgroup of patients was 56.9 ± 9.8 years, 26 were male. The diagnosis was established by angiography, CT, nuclear magnetic resonance, and ultrasound techniques. Extracranial carotid occlusive disease was found in 6, indications of cardiac emboli in 7, while no identifiable cause of stroke in the remaining 12 patients. All but 7 patients underwent measurements of CBF and clinical scoring during a period of at least 4 weeks. In 5 patients violation of the protocol was unavoidable by intermediate treatment of severe brain edema. The remaining 27 patients representing the group with a "natural" course are reported here. No patient was stuporous or comatose, and no patient died during the observation period.

Technique of CBF Measurement

CBF was estimated by NOVO-Cerebrograph 32c with 32 stationary detectors (collimation 40 mm, crystal diameter 3/4 inch), which were positioned in a helmet-like fashion over the skull. Per measurement 25–30 mCi Xenon-133 were inhaled from an 8-ltr-airbag during 60 sec, followed by recording of the clearance curves for 10 min. Endtidal CO_2 (vol%), endtidal Xenon-133, arterial blood pressure (Riva-Rocci) were recorded for correction of recirculation, extracranial contamination, and activity from the nasopharynx. CBF was expressed as initial slope index (ISI, sec^{-1}).

Clinical Scoring

Neurological deficits were scored according to the 100 point-system published by Mathew *et al.*[4] slightly modified by ourselves.

Protocol

CBF measurements and clinical scoring were performed on day 1 (day of onset of symptoms, n = 27), day 3 (n = 27), days 6–8 (n = 27), days 14–16 (n = 24), days 21–23 (n = 23), and days 27–29 (n = 24).

Results

Distribution of CBF in elderly normal individuals is given in Fig. 1. A slight hyperfrontality, albeit less than in normal young volunteers, was constantly noted in all normals without risk factors of cerebrovascular disease or stroke. Test-retest manoeuvers revealing differences of more than 10% between consecutive measurements indicate significant alterations of flow. Interregional CBF differences between hemispheric sides or between ipsilaterally placed detectors did not exceed 15% under physiological conditions.

Two examples of repeated CBF measurements are shown in Fig. 2, indicating a positive correlation between the clinical course and the CBF findings in one case and lack of correlation in the other. Both flow maps demonstrate the relative flow distribution given as ISI seen from the patient's vertex. The right ear is at the right and the nose at the top of the figure. The flow scale with different grey tones is given at the bottom with a bright tone representing the highest and the darkest tone the lowest flow value.

In Fig. 2a blood flow maps of days 1, 7, 14, and 28 of a patient with sudden onset of a right-side homony-

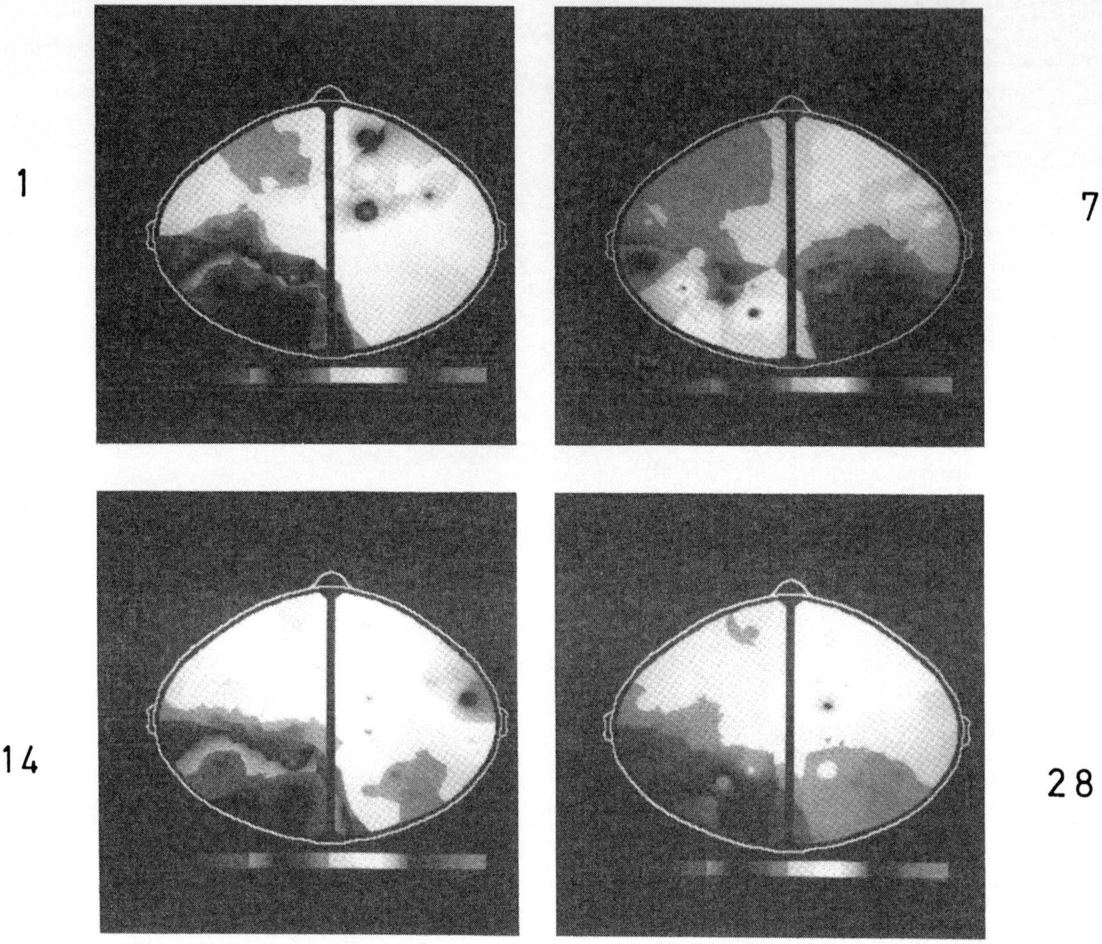

Fig. 2a. CBF mapping in a patient with acute ischemia in the left occipital area. CBF distribution according to the scale at the bottom with dark tone representing low and bright tone representing high flow. The bright tone in the left occipital area on day 7 indicates relative luxury perfusion since clinical symptoms did not change with this transient improvement of CBF

mous hemianopsia. On day 1 the neurological deficit could be associated with severe ischemia in the left occipital lobe. The same CBF pattern was obtained on day 14, while on day 7 blood flow was significantly increased in the left occipital area despite an unchanged visual field defect. Since CBF in this area was higher than in the corresponding contralateral region, the flow finding of the left side may represent luxury perfusion. With the neurological deficit unchanged, CBF on day 28 was almost as low as on days 1 and 14.

Figure 2b demonstrates the flow pattern of a patient with complete left hemiparesis due to middle cerebral artery occlusion. Despite a fast improvement of the neurological deficit during the first week, CBF remained low until day 7. On days 14 and 28 CBF in the right middle cerebral artery territory was slightly lower than in the left side, while the neurological conditions were almost normal. This example indicates that the neurological recovery may be faster than the improvement of CBF. This has already been observed in the case of TIAs, although the CBF course in these cases is running parallel to the neurological symptoms. The fact that in this case (Fig. 2b) in contrast to that as shown in Fig. 2a frontal flow is lower than in the occipital area indicates permanent CBF changes as provoked by general cerebrovascular disease.

The mean CBF course in all 27 patients shown in Fig. 3 is characterized by a slight decrease on day 3 without clinical deterioration, as shown by the mean clinical score. Thereafter CBF improved until day 7, while the clinical conditions remained unchanged. The recovery of the neurological scores started between day 7 and day 14, but an increase of CBF was not noted before day 28. There was a good correlation between the neurological score and mean CBF on days 1 and 28, but not on days 3 and 7 (Fig. 3).

The delayed recovery of neurological function as already observed in other control patients[5] was pre-

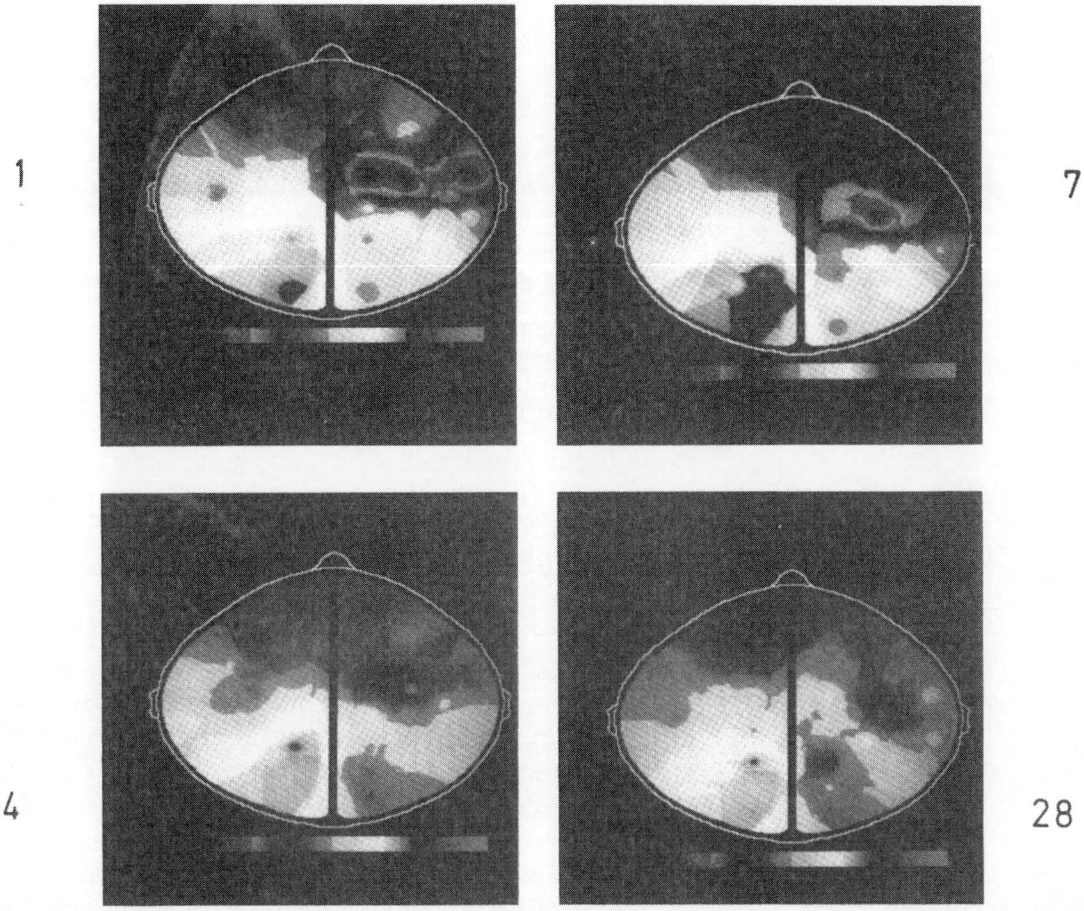

Fig. 2b. Distribution of CBF in a patient with right middle cerebral artery occlusion. CBF recovered slower than clinical symptoms. Low flow is represented by dark and high flow by bright tone (see scale at the bottom)

dominantly due to the improvement of the sub-item "motor power", which started only after the first week. Speech performance according to the Reitan test did not improve in the 4-week observation period, while improvement of cranial nerves started at once but was not complete. However, if one selects those 4 patients who suffered from an isolated motoric dysphasia with some sensory speech involvement without any other neurological deficit, there was an excellent correlation between the focal CBF course of the ischemic area and the recovery of the speech disturbances (Fig. 4). This may indicate that isolated dysphasia as a stroke sign improves more rapidly than dysphasia in patients with additional neurological symptoms probably resulting from a larger area of ischemia.

A particular phenomenon was observed in cases with early reduction of flow in the contralateral side. Principally, a low CBF of the contralateral hemisphere cannot generally be attributed to diaschisis, since a generalized cerebral arteriosclerosis may decrease CBF as well. If, however, the low contralateral CBF improves during the subsequent observation period, diaschisis may be assumed. This occurred in 6 out of 27 patients, in whom CBF of the diseased side increased by about 28% until day 28 and by 18.5% in the contralateral hemisphere obviously with diaschisis (Fig. 5).

There was no correlation with the severity of the decrease of CFB in the stroke-involved side. In the contralateral side CBF reached its final value already on day 7, in the ipsilateral side only on day 21. Focal signs specific of the flow reduction of the contralateral side were not found, however.

Discussion

The study has shown that patients with an acute ischemic insult of the brain may recover to a significant extent without receiving specific therapy. All patients

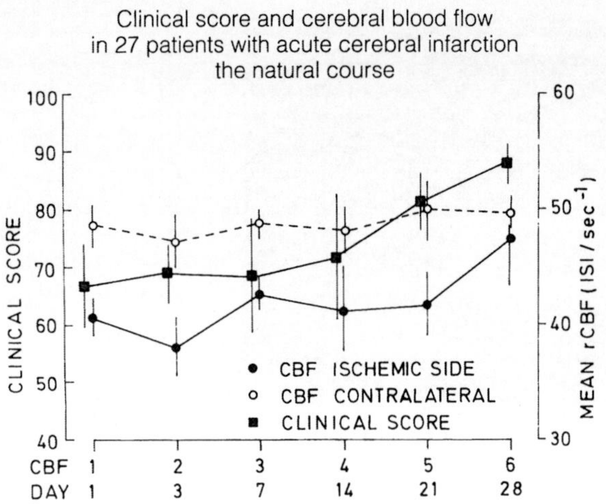

Fig. 3. Mean CBF of the involved hemisphere and clinical score in 27 patients with acute cerebral infarction. Each point represents mean value plus standard deviation. Clinical scale on the left and CBF scale on the right

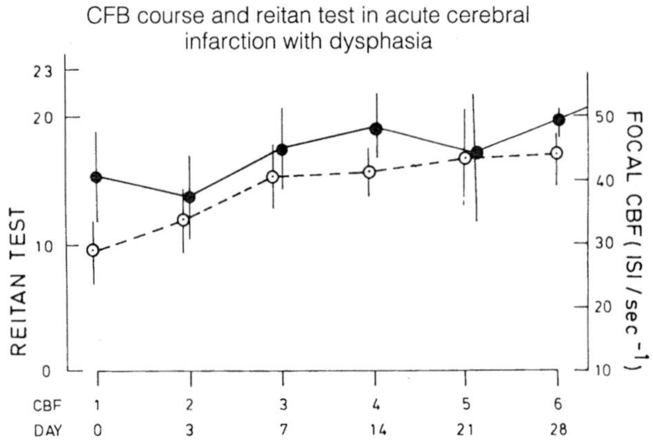

Fig. 4. CBF course and Reitan aphasia test in patients with isolated motor aphasia. Each point represents mean ± standard deviation

were optimally treated by general medical measures except that hemodilution or calcium antagonists were not administered. Despite some beneficial effects of hemodilution in subgroups and some clinical improvement by the calcium antagonist nimodipine[6], the beneficial effect of both therapeutic principles has not been proven yet. Since no other specific therapy of ischemic insults is available – barbiturates and hyperventilation are not applicable for the routine therapy of stroke – it may be justified to administer to such patients only optimal general measures, such as stabilization of blood pressure and of the cardiac conditions, sufficient pulmonary oxygenation, early physical therapy, etc. and antiedematous therapy, if brain edema is present.

In the acute infarct stage CBF correlated well with the neurological symptoms supporting the flow threshold theory. The same correlation was observed after 4 weeks indicating that the blood flow was adjusted to the functional and metabolic needs.

However, there was no close correlation between the hemispheric CBF and neurological score, i.e. the neurological conditions from day 3 to day 21. The slight decrease of mean CBF between the first and third day was not paralleled by a clinical deterioration. Since no clinical evidence of brain edema was present, another CT was not made on day 3. After day 7 the significant clinical improvement in many cases was not reflected by changes of the CBF. Only after day 21 both the CBF and the clinical score were increased. As a cause of this dissociation, first of all the loss of coupling between CBF and metabolism must be considered. The decrease of CBF on day 3 post infarction could be compensated for by an increased O_2- and glucose-extraction providing for an unchanged metabolic rate and, thus, clinical condition. The decrease of CBF on day 3 could have been due to developing brain edema which, however, was not confirmed by clinical signs, by changes of the perfusion pressure in the presence of an impaired autoregulation, or by alterations in the degree and spread of brain tissue lactic acidosis. The recovery of CBF on day 7 may be attributable to various factors, e.g. anastomotic networks, a decrease of brain edema, an increase of lactic acidosis with luxury perfusion, or even recovery of the brain tissue in the penumbra zone.

Another explanation might be that CBF of the peri-infarct area was decreased after the initial event leading to a decrease of the hemispheric CBF altogether without occurrence of new or deterioration of already existing neurological deficits. This assumption is supported by observations that in cases with pure motor aphasia without any other neurological signs the regional CBF of the Broca area paralleled the course of dysphasia as evaluated by the Reitan test (Fig. 4).

An example of a dissociation between CBF or its course and the clinical function is demonstrated by the transient reduction of CBF in the contralateral side opposite to the infarction, i.e. the diaschisis. We did, however, not find any clinical sign resulting from the contralateral flow decrease. The nature of the contralateral flow reduction is not clear. It was present, however, at the day of infarction and recovered faster than the low flow of the involved hemisphere. Brain edema probably may be ruled out as a cause since we have not found any indication in the CT, such as a tissue displacement. Since diaschisis has been observed also in

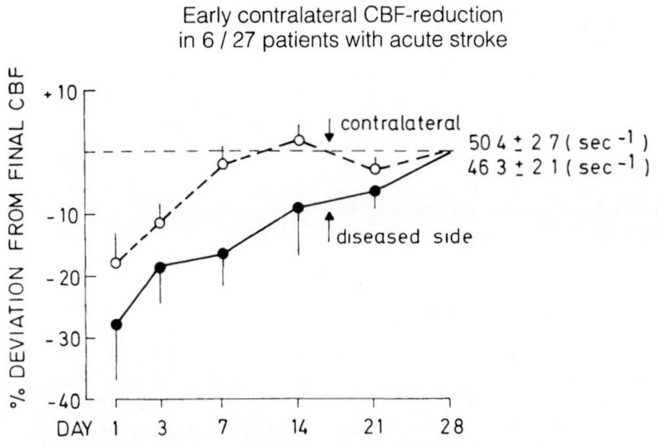

Fig. 5. CBF of the involved (diseased) and contralateral hemisphere in 6 patients with early transient contralateral flow reduction. Each point represents mean value ± standard deviation. CBF is given as percent deviation from the final CBF value which is indicated

cases without cardiac emboli, its development appears to be independent also of embolization.

As shown, CBF was correlated with the clinical deficit at the onset of stroke and 4 weeks later, but not in between. Therefore, the question arises whether measurements of CBF provide prognostic information as to the clinical recovery at an early stage. None of the patients in this series who fully recovered had an initial flow below 40 (sec^{-1}). All cases (n = 5) with an initial CBF below 36 (sec^{-1}) recovered to the maximum score of 90 out of 100. This indicates that initial CBF measurements might indeed provide some clues as to what extent the patient might recover, a conclusion which is drawn by others as well[7].

References

1. Astrup J, Siesjö BK, Symon L (1981) Thresholds in cerebral ischemia – the ischemic penumbra. Stroke 12: 723–725
2. Chopra JS *et al* (eds) (1990) Progress in cerebrovascular disease, Vol 3. Amsterdam, Elsevier
3. Hartmann A, Wassmann H, Czernicki Z, Dettmers C, Schumacher HW, Tsuda Y (1987) Effect of stable Xenon in room air on regional cerebral blood flow and EEG in normal baboons. Stroke 18: 643–648
4. Mathew T, Meyer JS, Rivera VM, Charney JZ, Hartmann A (1972) Double-blind evaluation of glycerol therapy in acute cerebral infarction. Lancet 1327–1329
5. Hsu CY, Norris JW, Hogan EL, Bladin P, Dinsdale HB, Yatsu FM, Earnest MP, Scheinberg P, Caplan LR, Karp HR, Swanson PD, Feldman RG, Cohen MM, Mayman CI, Cobert B, Savitsky JP (1988) Pentoxifylline in acute nonhemorrhagic stroke – a randomized placebo-controlled double blind trial. Stroke 19: 716–722
6. Gelmers HJ, Gorter K, De Weerdt CJ, Wiezer HJ (1988) A controlled trial of nimodipine in acute ischemic stroke. N Engl J Med 318: 203–207
7. Heiss WD, Prosenz P, Herles HJ (1973) Effekt von Hämodilution und Dehydratation auf die regionale Gehirndurchblutung. Nervenarzt 44: 166–169

Correspondence and Reprints: A. Hartmann, Neurologische Universitätsklinik, Sigmund-Freud-Strasse 25, D-W-5300 Bonn 1, Federal Republic of Germany.

Acta Neurochir (1993) [Suppl] 57: 137–140

Emergency Care and Treatment in Acute Cerebral Insults

Assessment of Emergency Care in Trauma Patients

B. Bouillon, M. Schweins, A. Lechleuthner, M. Vorweg, and **H. Troidl**

II. Department of Surgery, Köln-Merheim, Federal Republic of Germany

Abstract

There are many reasons for evaluation of an emergency care system, such as expenses (1.035 Bio. DM in 1985) and quality control.

From January 1, 1987 to December 31, 1987 information on all patients seen by an emergency physician in the field have been recorded prospectively in a standard form by the Cologne emergency medical services. Cologne has 1,000,000 inhabitants and covers an area of 405 km^2.

The patients' status, diagnosis and therapeutic interventions were recorded. Trauma patients were further assessed as to time of accident, cause of accident, and trauma score. All trauma patients with a trauma score < 16 were followed up to their discharge from the hospital.

In 1987, 2,073 trauma patients were treated. Overall mortality at the time of discharge was 9.2%. This result alone, however, is not sufficient for assessment of the trauma system. It is important to provide better information on the patient. The trauma evaluation score already used in the US became also a valid instrument in West-Germany. It shows a high correlation between survival and the patients' physiological status in the field. Standard curves could be established for comparing individual or regional trauma systems.

Keywords: Head injury; emergency care system; trauma evaluation score; quality assessment.

Why should one be interested in assessing emergency care in trauma patients? Henrik Blum[15] once said that "any program worth trying is worth evaluating; and if it's not worth evaluating it's not worth trying". He is probably right when we think of evaluating a trauma system in general but additionally there are a lot of controversies in emergency care that wait for answers such as scoop and run versus stabilization or the discussion whether emergency physicians rather than paramedics should go out into the field. Other questions are the effects of steroids or the benefits of early ventilatory support in brain trauma.

Increasing costs made people ask for the benefit of prehospital emergency care as insurances payed a total of DM 1,035,000,000 in 1985 for these services in West-Germany[2].

Trauma is one of the most important diseases of modern time[12]. In 1965 the American College of Surgeons published an article "Accidental death and disability, the neglected disease"[7]. We feel that in Germany even today the importance of trauma is not clear to everybody. Trauma in 1986 was the leading cause of death up to the age of 45 (Fig. 1)[5]. Calculating the working years of life lost in 1985 in Germany, trauma takes the lead over cancer and cardiovascular disease. This phenomenon can be explained by the age distribution of trauma patients as mostly younger people are affected.

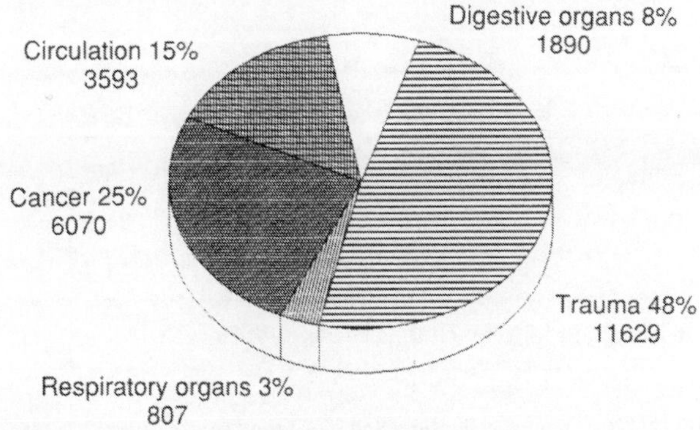

Fig. 1. Causes of death up to the age of 45 in 1986 in West-Germany according to the Bundesamt für Statistik, Wiesbaden

In the literature we find mortality rates from 5%–80% for multiple injured patients. The problem we figure is the problem of definition of a multiple injured patient[4,12,14,16]. The most widely used definition is the one given by Tscherne that multiple injured patients have at least two body regions affected and one injury or the combination of injuries is considered life-threatening[13]. Discussing mortality rates, this definition is not sufficient for comparability of a patient population. Susan Baker described this dilemma very well in 1983, when she said "if you have never felt the need for any type of severity scoring system, then you probably never had to explain how it is that the survival rate of 85% in your trauma center is actually better than the survial rate of 97% in some other hospital were the patients are much less seriously injured"[11].

She and other authors claim that scoring systems are necessary for the purpose of classification of trauma patients and as a common language for comparing study results[6,8,11].

One of the most widely used scoring systems is the Glasgow Coma Scale. It was introduced by G. Teasdale in 1974 and uses eye opening, verbal and motor response for patients classification[9]. The problem using the Glasgow Coma Scale alone for the classification of trauma patients is that the score changes over time. If you only get the score information of e.g. 7 you do not know, whether you talk of a patient having a concussion and waking up, or of a patient with a subdural hematoma who is getting worse. The second problem is that the Glasgow Coma Scale does not reflect impairment of consciousness through haemorrhagic shock, such as in a patient with a rupture of the spleen who has no brain trauma at all.

	GCS	TS	ISS	TRISS
Sensitivity	68	85	81	84
Specificity	62	60	40	69
Accuracy	64	72	66	80

Fig. 2. Sensitivity, specificity and accuracy in percent of the 4 scoring systems Glasgow Coma Scale *(GCS)*, Trauma Score *(TS)*, Injury Severity Score *(ISS)* and *TRISS* in 262 multiple injured patients treated in the Cologne-Merheim hospital from 1984–1987

	Age	TS	ISS
Total	35.3	11.1	36.5
Survivors	32.6	12.5	31.6
Deaths	40.3	7.7	44.4
Predicted Mortality Rate (TRISS) = 35%			

Fig. 3. The mean age, Trauma Score *(TS)* and Injury Severity Score *(ISS)* as well as the predicted mortality rate of 262 multiple injured patients treated in the Cologne-Merheim hospital from 1984–1987

The Trauma Score introduced by Champion in 1981 is a physiologic score using the Glasgow Coma Scale and physiologic parameters of respiratory and circulatory function[6]. It has been shown that the score correlates very well with the probability of survival of the patient and is widely used in the US for classification of trauma patients. The score reflects problems of multiple injured patients, such as mentiond above but has still a problem of change over time. A patient with a rupture of the liver might have a high Trauma Score when seen by a rapid emergency care system while having a low score if there is a relevant delay. A rapid system could even artificially worsen its results by picking up patients who would have been declared dead upon arrival in a slow system.

The Injury Severity Score (ISS) published in 1974 by S. Baker scores six body regions according to the anatomic severity of the injury[1]. It then uses the sum of squares of the three most heavily injured body regions which results in a score of 0–75. In several studies it could be shown that the Injury Severity Score has a high correlation with mortality, the length of stay in hospital, the need for respiratory support and morbidity. It has been further shown that age is one of the most important prognostic factors altering survival.

Combining the anatomic severity of injury described by the Injury Severity Score the physiologic status of the patient at the first point of time when seen by an emergency physician described by the Trauma Score and the most important prognostic factor age using a logistic regression model results in an even better descriptive Score. This score was first introduced by Champion 1981 and is called the TRISS[6]. In a retrospective analysis of 262 multiple injured patients treated in our institution from 1984–1987 we calculated the different scores and analysed their ability in predicting the outcome. The Glasgow Coma Scale showed a sensitivity of 68% meaning correct prediction of death and a specificity of 62% describing the ability of correct detecting those who will survive resulting in an accuracy rate of 64%. The Trauma Scores shows a much better sensitivity of 85% and a comparable specificity. The Injury Severity Score also shows a high

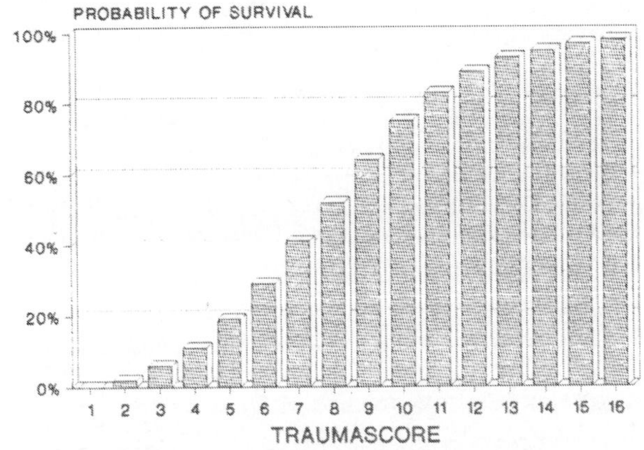

Fig. 4. Correlation of the probability of survival and the Trauma Score in 648 patients in Cologne in 1987. The prehospital treatment was delivered by the Cologne emergency medical service and continued in 26 hospitals

sensitivity of 81% but a much lower specificity of 40%. If we now combine Trauma Score and Injury Severity Score considering age we can calculate the TRISS Score. In our population it shows the highest accuracy with 80% with a sensitivity of 84% and a specificity of 69% (Fig. 2).

If we look at our over all mortality rate of 30% we can define the population we were looking at with a mean age of 35.3 years, a mean Trauma Score of 11.1 and a mean Injury Severity Score of 36.4 (Fig. 3). We can further calculate a predicted mortality rate with the TRISS methodology of 35%. Using this instruments we can much better interpret our own results and compare it with the expected mortality rate.

Obtaining those results in the retrospective analysis we wanted to perform a prospective validation of those scoring systems. The first study we launched was the evaluation of the Trauma Score. Before starting the study we had to view another problem. The American literature used a Trauma Score calculated upon arrival of the patient in the emergency room. This procedure is impossible in a setting where emergency care physicians use aggressive therapeutic interventions in multiple injured patients. In 1987 45% of 310 multiple injured patients seen by the emergency service had i.v. narcotics or analgetics and 43% were intubated in the field. These interventions made a scoring in the emergency room impossible as respiratory rate and effort as well as the level of consciousness could not be assessed any more. Therefore we had to use the score prior to these aggressive interventions in the field. We therefore introduced the score in our emergency physician chart which is completed by the emergency physician before handing over the patient to his colleagues in the hospital. We first did a 4-months pilot study from 1st September 1986 to 31st December 1986 to make sure that emergency physicians get familiar with this new scoring system.

The emergency service area we evaluated was the Cologne rescue area, a mostly urban setting of 405 km^2 and 966.307 inhabitants. There are four emergency cars and one helicopter staffed with physicians. When an emergency call reaches the dispatch center it decides which team is to be sent out into the field according to an algorithm. If a life-threatening emergency is suspected an emergency physician by car or helicopter is sent out. Within 10 minutes 80% of the emergencies are seen by an emergency physician in the field. This figure shows that the system acts fast which is also partly explained by the urban characteristic of this area. In 1987 a total of 11 168 emergencies were taken care of by emergency physicians. More than 50% were medical emergencies (5853), 20% surgical. There were 2136 trauma related emergencies in the Cologne area in 1987. The causes of accident were motor vehicle in more than 50% (1067). The age distribution of the trauma patients shows a peak at the age of 25. Looking at the diagnoses in trauma patients made by the emergency physicians in the field we see that 1764 patients (82.6%) of the trauma patients had minor laceration and 372 (17.4%) were considered as multiple injured. 97 patients were diagnosed as death upon arrival and 62 patients suffered from penetrating trauma.

The distribution of the Trauma Score reveals a distribution similar to the one published in other studies where 85% had a Trauma Score of higher than 14, and 15% a Trauma Score of 14 or lower. All patients with a Trauma Score from 1 to 15 and a randomised sample of 160 patients of a Trauma Score of 16 were followed up in 27 hospitals for survival. The follow up rate was 94%. Only in 38 patients no further information could be gathered. Calculating the survival rate for each Trauma Score shows that similar to the original publication by Champion we can find a strong correlation between the score and survival (Fig. 4). This confirms that the Trauma Score is also valid in Germany. In this study we could show that the Trauma Score has a strong correlation with the survival when assessed in the field. With this validation study we completed the first step for further quality control.

An important point to mention is that as yet the outcome mostly looked at in assessing trauma patients is mortality. In follow up studies we performed on trauma patients treated in our institution we could find that another very important endpoint in those patients who survive is the quality of life. 70% of the trauma patients admitted in our institution in 1987 judged their quality of life after their injury as average or bad[3,10].

In conclusion we state that scoring systems are usefull for classification of trauma patients. Standard scores should be used for comparison of trauma systems. The Trauma Score, Injury Severity Score and the TRISS-Score have been proven effective in trauma patients. On a regular time base national and international standards should be assessed and published. The quality of individual or regional trauma systems could then be evaluated.

References

1. Baker ST, O'Neill B, Hadeon W, Long WB (1974) The injury severity score: a method for describing patients with multiple injuries and evaluating emergency care. J Trauma 14: 187
2. Bouillon B, Gail HE, Huber J, Klingshirn H, Korbmann H, Kühner R, Raftopoulo A (August 1987) Ansatzpunkte für Forschungsarbeiten zum Rettungswesen. Projektgruppenbericht der Bundesanstalt fur Straßenwesen, Bereich Unfallforschung
3. Bouillon B, Hirschel V, Imig R, Tiling T, Troidl H (1989) Lebensqualität – Kriterium in der Behandlungsstrategie Schwerstverletzter. Langenbecks Arch Chir [Suppl] II: 117–122
4. Boyd DR (1973) A symposium on the Illinois trauma program: a systems approach to the care of critically injured. J Trauma 13: 275
5. Bundesamt fur Statistik, Wiesbaden, 1986
6. Champion HR, Sacco WJ, Carnazzo AJ, Copes W, Fouty WJ (1981) Trauma Score. Crit Care Med 9: 672–676
7. National Academy of Science/National Research Council (NAS/NRC) (1966) Accidental death and disability – the neglected disease of modern society. Government Printing Office, Washington, DC
8. Seefelder C, Matzek N, Rossi R (1988) Polytrauma-Bewertungsskalen. Notfallmedizin 14: 227– 236, 317–329
9. Teasdale G, Jennett B (1974) Assessment of coma and impaired consciousness. Lancet II: 81–84
10. Troidl H, Kusche J, Vestweber K-H, Eypasch E, Koeppen L, Bouillon B (1987) Quality of life: an important endpoint both in surgical practice and research. J Chron Dis 40: 523–528
11. Trunkey DD, Siegel J, Baker SP, Gennarelli TA (1983) Panel: Current status of trauma severity indices. J Trauma 23: 185–201
12. Trunkey DD (1983) Trauma. Sci Am 249: 28
13. Tscherne H, Regel G, Sturm JA, Friedl HP (1987) Schweregrad und Prioritäten bei Mehrfachverletzungen. Chirurg 58: 631–640
14. Van Wagoner FH (1961) Died in hospital: a three-year study of deaths following trauma. J Trauma 1: 184
15. Waller JA (1974) The accident, the ugly duckling, and the three preventions: a fable for mature health officers (Editorial). Am J Pub Health 64: 301, 409
16. West JG, Trunkey DD, Lim RC (1979) Systems of trauma care: a study of two counties. Arch Surg 114: 455–460

Correspondence and Reprints: B. Bouillon, II. Department of Surgery, Cologne-Merheim, Ostmerheimerstrasse 200, D-W-5000 Köln 91, Federal Republic of Germany.

Acta Neurochir (1993) [Suppl] 57: 141–144

Current Level of Prehospital Care in Severe Head Injury – Potential for Improvement

P. Sefrin

Institute of Anaesthesiology, University of Würzburg, Würzburg, Federal Republic of Germany

Abstract

The fact that 50–60% of cases with severe head injury result from traffic accidents underlines the great significance of emergency care and of its organization. Many patients with severe head injury are threatened from vital complications diagnosed with delay, or not at all, which plays a major role not only for survival but also for the quality of recovery and regaining of employment capabilities. Thus, the necessity of qualified and trained physicians with experience in emergency care is obvious. Emergency care can be divided into an early resuscitation phase of securing or reestablishment of general vital functions, and a following stabilisation phase with administration of measures directed towards the specific conditions underlying trauma.

1. Prevention and treatment of respiratory complications. In addition to classical emergency care measures, endotracheal suction might be employed. The most effective method for clearance of airways and, thus, securing of the cerebral oxygenation is endotracheal intubation. Early intubation provides also for control of the intracranial pressure by hyperventilation and administration of O_2. Recently assistant ventilation is available as compared to the past when only controlled ventilation was possible.

2. Circulatory support. A major requirement for a sufficient cerebral perfusion is an adequate cerebral perfusion pressure making necessary early fluid substitution. In case the patient is in circulatory shock, shock-specific treatment may compete with adequate positioning of the patient.

3. Pharmacological treatment in the prehospital phase. Although dexamethasone has been reported to directly influence brain edema, its benefits in head injury are not clear. Currently conducted clinical studies using markedly higher doses may provide so far missing information. Benzodiazepines might be given for sedation, while administration of barbiturates for cerebral protection has not been found effective under these circumstances, although barbiturates should be administered in seizures. Patients presenting with psychomotoric unrest or a low blood pressure may be subjected to a combination of analgesia and sedation.

Taken together, early implementation of competent prehospital emergency treatment may not only prevent acute life threatening complications on the scene, but also provide the basis for successful administration of definite conservative or surgical care after admission to the hospital.

Keywords: Traffic accidents; head injury; prehospital emergency care.

Introduction

In 60–70% of all patients with multiple trauma, the clinical prognosis and neurological outcome is determined by the degree of severity of the craniocerebral trauma (CCT) involved[4]. In view that 50–60% of the cases of craniocerebral trauma result from road accidents, the role of the primary care given in the context of accident rescue measures is of particular importance. In recent years the political institutions and cost-carriers have come to realize that investments in a viable rescue system in the long term lead to cost savings in the health system. Corresponding studies in North Bavaria have shown that such a system indeed pays off, with a mathematic benefit of DM 5.30 resulting from each DM invested in the rescue service[2].

The success of such medical measures in the treatment of CCT depends to an overriding extent on the expertise of the emergency doctor and also on the equipment that is available as well as on the time that elapses until primary care can be administered. Appropriate organizational improvements and restructuring means that today life-saving services can arrive at the site of the accident within about 8 minutes of being alerted. The expansion of emergency medical services – both ground-based and also aerial – has been intensified, with the number of facilities and thus also of the density of emergency doctors having increased. In view that special emergency interventions are necessary to prevent acute danger for the life of accident victims, the presence of a doctor is an essential precondition for the success of such primary care. However, of the total number of accidents in which the intervention of emergency services were required, so far an emergency doctor was present in only 34%[9].

A great number of CCT victims are endangered by vital complications that are recognized or treated either not at all or only at a late stage. This means that not only the chances for survival but also the long-term prognosis for the restitution of the victim's capacity to work are affected, as any immediately occurring complication may result in subsequent additional sequelae. There is in the meantime no doubt as to the benefit of high-quality emergency care of accident victims in the sense of primary intensive therapy. A prospective study in 147 neurotraumatized accident victims proved the necessity and efficacy of appropriate preclinical care[8]. It was shown that the survival rate of victims who receive primary medical care at the site of the accident is substantially greater than that of patients delivered to hospital without having received medical first aid and without medical escort. Patients with qualitatively satisfactory preclinical care have a significantly higher chance of surviving their injuries. Whereas only 9.3% of the intubated patients with CCT developed aspiration, the proportion of nonintubated patients who went on to develop aspiration was 46.3%. The consequence was that 35.5% of the intubated patients subsequently suffered from respiratory insufficiency, whereas as many as 72.2% of the nonintubated patients developed this clinical condition.

The typical and specific symptoms form the basis for the essential basic interventions and subsequent intensive-care measures, the aim of which is to prevent secondary brain damage from various pathophysiological approaches. The primary care of emergency patients can generally be divided into an initial phase for the securement/regaining of the vital functions (reanimation phase) on the one hand and the subsequent stabilization phase with trauma-specific measures on the other.

Prevention and Removal of Respiratory Obstructions

The measures primarily suited to remove/prevent an acute danger as presented by obstructions of the respiratory tract are the so-called "primitive" measures (Esmarch maneuver, digital cleaning, stable lateral position). These measures are taught to laypersons in the context of first-aid courses; however, their application in emergency situations is wanting. A previous analysis showed that 38% of all positionings were clearly incorrect, most of these being the complete lack of or extremely incorrect lateral positioning. A similar tendency can be observed among rescue service personnel: 24% of all positionings that were carried out displayed grave mistakes. Most of these were unconscious accident victims who were transported in the dorsal position[9]. A more recent questionnaire distributed among emergency doctors revealed that these consider the first-aid measures carried out in unconscious patients to be suboptimal in 80% of all cases (42% insufficient, 38% barely satisfactory). Great difficulties are perceived in the recognition of respiratory obstructions and in the implementation of mouth-to-mouth resuscitation: 62% of all emergency doctors and 66% of emergency medical personnel consider these aspects to be insatisfactory[10]. The safest way to not only keep the respiratory tract free but also to ensure sufficient oxygenation of the brain is intubation. Its use in CCT should thus be relatively wide-spread: despite the clear indication for its use, 80% of a group of emergency patients that subsequently died had not been intubated prior to the rescue transport leaving the scene of the accident. A procedure currently under discussion – that of mask respiration – can be said to be merely the insufficient attempt of the nonexpert. Although recent studies made in the USA recommend mask respiration with continual drainage during transport as a sufficient prophylactic measure[6], the authors are nevertheless unable to provide unequivocal evidence that this is indeed a better prophylactic measure for the transport of the victim than intubation given at an early stage.

An additional advantage of early intubation is the possibility of O_2-enriched hyperventilating respiration, which leads to a reduction of the intracranial pressure (ICP). A precondition for the implementation of artificial respiration is that the doctor giving first aid has routine experience in the use of ventilators and resuscitators. He must be aware that – beside the positive effects of PEEP respiration – such a method also involves negative effects on the victim's hemodynamics. When ventilating a spontaneously breathing victim with a bellows resuscitator, attention must be paid that the assisted respiration is synchronous with the victim's own breathing rhythm. Asynchronic respiration – like the head-down dorsal position – can lead to considerable and above all recurrent increases in ICP. For this reason, the use of automatic respirators in the past has been problematic in cases in which it was not possible or advisable to paralyse the patient for respiration. An automatic respirator with which it is possible to select assisted respiration with either 50 or 100% O_2 is available now. This is reason enough for us to maintain our restrictive attitude toward relaxation in emergency rescue and completely dismiss long-term relaxation.

Stabilization of the Circulatory System

The precondition for sufficient cerebral perfusion is a sufficiently high mean arterial pressure. The therapy of shock, which must still be considered as one of the severest complications that can arise in the accident victim, consists of the initiation of volume substitution. After all, 54% of the patients in our collection of accident victims with manifest symptoms of shock died in the further course of treatment. This finding is confirmed by American analyses that detected a mortality rate of 73% in neurotraumatized accident victims who displayed manifest shock at the site of the accident as compared with a rate of 41% in victims without shock[8]. This reinforces the importance of early administration of adequate volume substitution measures. Even the attempts of primary care helpers to staunch bleeding as a cause of shock can be considered as failures: more than one third of all measures taken by laypersons to staunch bleeding were either insufficient or else incorrect (36%), and the share of incorrect measures taken by rescue personnel was still 10%. The mistakes made most frequently by emergency doctors were in shock therapy: no volume substitution for transportation at all was made in 18% of the cases, and insufficient in a further 6%[9].

Medical Preclinical Therapy

The most essential medical measure in the preclinical area is the prophylaxis or reduction of an intracranial pressure increase due to cerebral oedema, especially in cases in which primary traumatic damage can no longer be influenced. A capability to directly influence cerebral oedema has long been ascribed to dexamethasone. Initial positive therapeutical results are now subject to controversial discussion regarding the efficacy of dexamethasone in CCT. Recent findings, however, indicate that favourable effects of relatively high doses of initially 500–1000 mg are to be expected. A prospective multicenter randomized double-blind study, in which we and our patients in the emergency rescue services are also involved, is currently underway. In cases in which the emergency doctor opts for steroid therapy, this should be administered as early on as possible, i.e. at the accident site and in a sufficiently high dose.

Due to the wide therapeutical range and the rapid onset of action, benzodiazepines in particular offer themselves to the doctor for sedation. The benefit of barbiturates, previously recommended for their "brain protection" effect, has not been confirmed in preclinical practice. However, barbiturates may be used as a sedative and above all – if necessary – as a potent narcotic. Nevertheless, aspects that must be considered when administering this class of drugs include not only their blood-pressure-decreasing effect, but also the fact that their hypnotizing effect gives rise to a masking of the actual neurological symptoms. These side-effects make it a precondition that the user has sufficient experience in the application of these drugs and in the neurological evaluation.

Convulsions occur relatively early on and require immediate countermeasures. Accident victims diplaying restlessness, increased muscle tone and hyperventilation in the sense of a "hypertonic syndrome"[5] should be sedated with diazepam or midazolam. Under the precondition that respiration is sufficient, the use of barbiturates in manifest convulsions can be left at the discretion of the emergency doctor. In the event of exogenously caused psychomotor restlessness and hypertonic blood pressure reactions, e.g. due to additional peripheral injuries, then – in addition to sedatives – analgetics must also be administered, with preference being given to those with only slight effects on the circulation, such as piritramide or fentanyl. Despite this clear indication, the emergency service of the City of Munich has used this therapy concept in only 18% of the accidents in which craniocerebral trauma was involved[3].

The early administration and implementation of the therapeutical measures described here may not only afford avoidance of an acute threat to the life of an accident victim, but it also provides the basis for the initiation of conservative and operative therapy concepts in the clinic. Unfortunately, this cannot be perceived as being the case in general, as analyses of various emergency services show that therapeutical measures are adequate in only 28–32% of all cases[8]. In a study on the situation in Munich, Dorsch found that one half of the patients with more severe craniocerebral trauma must be considered as having been inadequately treated[3]. On the basis of our own findings[9], we think that the medical measures taken can be described as "sub-optimal" in about one half of the cases. At least 10% of the accident victims that die could be saved, if the opportunities provided by the preclinical first aid system were exploited to the full by the appropriate emergency-medical qualification of the doctors. The preconditions regarding the presence of rescue services have been fulfilled almost entirely. Singbartl has calculated that a statistically higher number of survivors receive adequate preclinical medical aid at the site of the accident than those who die[8].

Improvements in the primary care of accident victims can be achieved by both training in first-aid courses, starting at school and subsequently intensified by special training courses for doctors in rescue services.

References

1. Bettag W (1970) Erstversorgung und Transport Schädel-Hirn-Verletzter. Zschr Allg Med 46: 863
2. Deutscher Verkehrssicherheitsrat Bayer. Staatsministerium des Inneren (1985) Notfallrettung Unterfranken, Dokumentation Band II. Zur Wirksamkeit und Wirtschaft des Rettungsdienstes 1985
3. Dorsch A (1987) Abhängigkeit der Qualität notärztlicher Erstversorgung von der Fachzugehörigkeit des Notarztes. Inaugural-Dissertation, Technische Universität München
4. Gordon E (1979) Non-operative treatment of acute head-injuries: the Karolinska experience. Int Anaesth Clin 17: 181
5. Karimi-Nejad A (1988) Traumatische Schädel-Hirn- und Rükkenmarksschädigungen. In: Lehmann H-U (ed) 8. Tagung der Sektion "Rettungswesen" der Deutschen Interdisziplinären Vereinigung für Intensivmedizin (DIVI). Demeter, Gräfelfing
6. Rhee KJ, O'Malley RJ, Turner JE, Ward RE (1988) Field airway management of the trauma patient: the efficacy of bag mask ventilation. Am J Emerg Med 6: 33
7. Richard KE (1984) Die präklinische Versorgung von Verletzten mit schwerem Schädel-Hirn-Trauma. Intensivmed 21: 46–50
8. Singbartl G (1985) Die Bedeutung der präklinischen Notfallversorgung für die Prognose von Patienten mit schwerem Schädel-Hirn-Trauma. Anaesth Intensivther Notfallmed 20: 251–260
9. Sefrin P, Albert M, Schulz E (1980) Konsequenzen für die Primärversorgung von Notfallpatienten aus einer prospektiven Studie an 106 tödlichen Verläufen. Anaesth 29: 667–672
10. Sefrin P (1988) Schwerpunkte der Ersten-Hilfe-Ausbildung aus notfallmedizinischer Sicht. FP 8528/2 der Bundesanstalt für Straßenwesen, 1988 (not published)
11. Unfallverhütungsbericht (1988) Straßenverkehr 1987 – Übersicht: Rettungswesen. Drucksache 11/2364 vom 25. 5. 1988

Correspondence and Reprints: P. Sefrin, Institute of Anaesthesiology, University of Würzburg, Josef-Schneider-Strasse 2, D-W-8700 Würzburg, Federal Republic of Germany.

Acta Neurochir (1993) [Suppl] 57: 145–151

Prehospital Management of Head Injuries: International Perspectives

J. H. Garcia

Department of Pathology, Division of Neuropathology, Henry Ford Hospital, Detroit, Mi., U.S.A.

Abstract

The extent of disability and, therefore, the management of head injured patients before their arrival to the hospital should be influenced by a number of factors including:

- the age of the victim (children with comparable severity of head injury have less disability and lower mortality than persons over the age of 40)
- the type of injury (motor-vehicle accidents are the most frequent cause of head trauma in the U.S. and Canada)
- the decision to transfer the patient to either the nearest hospital or to a designated neurotrauma center (at several communities in the U.S. and Canada, recent analysis has demonstrated that the extent of disability and the mortality among head injured victims can be significantly decreased by their admission to specialized trauma units)
- elapsed time as another important factor (at least in one type of traumatic injury – acute subdural hemorrhage – it has been shown than an interval of less than 2 hours between the time of injury and the time of the craniotomy can significantly decrease both the mortality and the extent of neurological impairment among the survivors).

A number of acute injury effects on endothelial, neuronal or glial cells could potentially be influenced by compounds that may be administered before the victim arrives to the hospital. Evidence based on experimental and human observations will be reviewed concerning the potential ability to intervene in one or more of these areas: generation of free radicals, effect of opioid peptides, release of platelet-derived growth factors and their effects on astrocytic proliferation, alterations in thyroid hormones, hypoalbuminemia as a reflection of the effect of cytokines, and changes in vascular permeability and reactivity.

The application of methods with a statistically proven value to prognosticate the outcome of a head injury, such as developed by Choi *et al.* (J Neurosurg, 1988), may allow evaluation of the effects that free radical scavengers, as an example, may have on the patient's recovery. More standardized and controlled animal experiments are needed before any of these drugs can be used in clinical trials.

Keywords: Head injury; prognostic factors; hospital admission policy; delay of treatment.

Introduction

In 1966 the U.S. National Research Council issued a white paper labeling trauma, "the neglected disease of modern society", (NCTCS). Since that time several attempts have been made to implement new methods or to improve management methods that may decrease the number of deaths and increase the functional recovery of survivors. The management of head injured patients before they are admitted to the hospital, should take into consideration:

1) the prevalent causes of traumatic injury at the specific community or locality,

2) the age of the patient,

3) the time elapsed between accident occurrence and beginning of surgical treatment,

4) the type of facility where medical treatment is to be provided, and

5) the need (or the lack thereof) to start any form of medical therapy before the patient's arrival to the hospital.

Causes of Traumatic Injury

According to Siegel (1987) reliable demographic data on traumatic injury in the United States as well as in other Western countries are difficult to obtain; the task is next to impossible to complete in less developed countries. The number of fatal traumatic injuries in the United States (for the period 1980–1984) averaged almost 150,000/year, or over 2.5 times the number of casualties the U.S. armed forces experienced in the 10-year period of the Vietnam war (Siegel, 1987). Of particular concern is the fact that in the United States almost 33,000 deaths per year are caused by firearms; only 5% of these were unintentional, 50% were classified as the result of suicide and 43% were entered as homicides (Siegel, 1987). In 1979 the mortality rate for all accident victims in the United States was reported at

48/100,000 with a mortality of 23/100,000 for motor vehicle victims. In the age group of 1–34 years, traumatic injury kills more Americans than all other diseases combined (Hawkins, 1987). Annual mortality rates in England and Wales have been reported at 12/100,000 and in Australia at 27/100,000 (Hardman, 1985).

While motor vehicle accidents are the prevalent cause of traumatic head injury in the United States, and Canada (Siegel, 1987; Burns, 1985), bicycle accidents were the main cause of brain injury in a well-defined population of the People's Republic of China, where the incidence rate of head trauma was 56/100,000/year (Wang, 1986). Annual mortality from motor vehicle accidents in the United States declined from 51,091 in 1980 to 42,584 in 1983, a fact that has been partly linked to changes put into effect in the speed-limit laws (Siegel, 1987).

The Age of the Victims

Among 8,814 patients with head injuries, who were treated at 41 hospitals in three separate metropolitan areas, prospective analyses have shown that the mortality and the degree of neurologic disability are markedly lower in the pediatric age group, namely those under the age of 14 years, except when acute subdural hematoma or severe hypotension were the main diagnoses (Luerssen, 1988). A similar type of finding was reported by Alberico and associates (1987) who conducted a prospective analysis of 330 head-injured patients (100 were under the age of 19) treated at the same hospital. The analysis showed not only a higher percentage of favorable outcome for the under-19-year-old group (43 vs. 28%; $p < 0.01$) but they also found a 24% mortality among head injured patients under the age of 19, compared to 45% ($p < 0.001$) in the older group of patients

Time Elapsed Between Injury and Beginning of Treatment

Possibly because of the savings in time, the mortality among brain-injured patients transported by an aeromedical service was lower than that recorded for patients transported to the hospital by land; in a study that analyzed the result of treatment given to 232 patients with comparable types of brain injury (Glasgow Coma Scale ≤ 8), even though all patients received the same type of treatment at the same facility, over the same period of time, mortality was 9% lower among the group of patients transported by air; also the degree of disability was less severe among those transported by air compared with the group transported by the land advanced system (Baxt, 1987b).

The influence of time as a factor in altering mortality was forcefully shown in a restrospective study of 82 patients with severe head injury who were admitted to the same hospital and who underwent craniotomy for treatment of acute subdural hematoma. Patients who had the blood clot removed within the first 4 hours of the injury had a mortality of 30% compared to 90% among those operated after 4 h ($p < 0.001$). This observation is especially meaningful since it has been determined that about 25% of all head trauma victims, who are admitted to the hospital in coma, have acute subdural hematomas (Seelig, 1981).

Among 42 trauma victims with severe head trauma there were 25 fatally injured patients; in this group dynamic CT scan of the head showed (2 h from the time of impact) a remarkable increase in brain bulk which was mainly the reflection of brain edema accompanied in a significant percent of patients by diffuse cerebral ischemia. In cases with fatal outcomes the development of brain edema occurred very early and was massive (Yoshino, 1985).

Trauma Center: Regionalization

Beginning in the early 1970s several United States and Canadian studies began showing that areas endowed with regional trauma centers had experienced a dramatic reduction in the number of preventable deaths (West, 1988).

The impact that medical care at a trauma system may have on the survival of patients with a Trauma Score (TS) of 8 or less was examined by making comparisons between the *observed* rate of survival and the *predicted* one using a method that calculates the probability of survival based on age, physiologic score, and anatomic sites of the injury. Two hundred and eighty-three patients (8.3%) out of 3,394 treated at trauma centers in a 12 month period had a TS of 8 or less. The calculated probability of survival from these patients was 18% whereas the observed survival was 29%. Sixty patients with penetrating trauma had a probability of survival of 8% compared to the actual survival of 20%. The improved survival was attributed to the integration of prehospital and hospital care (Shackford, 1987).

A multidisciplinary audit conducted by an independent panel demonstrated that preventable deaths in non-trauma hospitals constituted 7.6% of all trauma-related deaths, whereas in trauma centers the figure was only 2.0%. At the former site most errors were made in diagnosis, whereas at the trauma centers preventable deaths were mostly the result of errors in technique (Shackford, 1987). The mortality from motor vehicle accidents was evaluated in two California counties; in one all victims were taken to a designated trauma center, whereas in the other patients were transported to the nearest hospital. About two-thirds of the non-CNS related deaths and one third of the CNS related deaths in the non-trauma centers were judged preventable. Only one-fourth were so judged among those admitted to the trauma center (West, 1979).

Regional studies that analyzed mortality/morbidity figures at the same community have shown in two Florida counties (Kreis, 1986) and a third one in California (Shackford, 1986) the significant decrease in the percent of preventable deaths secondary to trauma, as a reflection of establishing specialized and coordinated regional trauma centers. In a California county, preventable deaths occurred at a rate of 13.6% before regionalization compared to 2.7% after implementing the regional system. Moreover, the care of motor-vehicle victims was considered suboptimal in 32% of patients before regionalization, compared to 2.7% after its implementation ($p < 0.01$). This study was based on an independent audit of 591 trauma-related admissions (before regionalization) that was compared to the care given to 1,366 motor vehicle victims admitted to the same group of hospitals after the system was implemented (Shackford, 1986).

The American College of Surgeons issued, in 1986, a set of criteria defining trauma centers with various levels of capability to handle traumatically injured patients (West, 1988) and endorsements of these criteria as well as the concept of establishing regional neurotrauma centers, have been issued by the main Neurosurgical Societies in the United States (Pitts, 1987).

A national survey of the regional trauma systems in the U.S. showed that in 1988 regrettably only 2 states were in full compliance with the 8 essential components based on criteria set by the American College of Surgeons. In that year as many as 29 states had yet to initiate the process of trauma center designation (West, 1988). Sanctioned trauma center designation began in the U.S. in the late 1970s in attempts to improve hospital capabilities to care for injured patients. Guided by criteria issued by the American College of Surgeons Committee on Trauma and led by surgeons dedicated to improving trauma care, many states focused early on the quality of care issued only to struggle later with the political consequences that followed the designation process (Maull, 1986).

The American Association of Neurosurgeons and the College of Neurological Surgeons support the concept of organizing neurosurgical trauma units that include the appropriate combination of prepared communities and institutions staffed by adequate numbers of neurosurgeons. The AANS and CNS also support the guidelines of the American College of Surgeons regarding designation of institutions qualified to receive trauma patients and also support the concept of pre-hospital triage of trauma victims based on judgements made by Emergency Medical Service personnel guided by criteria selected by the local neurosurgical community (Pitts, 1987).

Reduction in trauma morbidity and mortality depends on early identification of severely injured patients, proper initial stabilization and safe transfer to the hospital. The Trauma Score and the American College of Surgeons Anatomical Injury Categorization (AIC) have been judged valuable triage tools (Moylan, 1985).

The California Association of Neurological Surgeons also supports the concept of neurological trauma care consisting of an appropriate amalgamation of prepared communities, institutions and adequate numbers of neurosurgeons. The CANS further supports the guidelines of the American College of Surgeons regarding institutions designated to receive trauma cases (Smith, 1986). The logic for seeking to develop specialized neurotrauma centers stems partly from demonstrated increased morbidity and mortality that head trauma adds to the outcome of traumatic injury, and the specialized nature required in the management of these patients; in a study of 441 non-CNS and 104 CNS injured trauma patients, who were all treated at the same medical center over a period of 12 consecutive months, mortality was 0.9% for the former group and 30.8% for those with head injury ($p < 0.001$) (Baxt, 1987a).

Guidelines for the initial management of head injury victims including indications for hospital admission were formulated by neurosurgeons in the United Kingdom and a multidisciplinary group of clinicians and administrators attending a seminar convened by the U.K. Department of Health and Social Security analyzed appropriate data to show that admitting practices varied among centers and were very different from those recommended in the guidelines. It was concluded that implementation of the guidelines could have a substantial effect on clinical practice and could reduce hospital admissions by as much as 46% (Fowkes, 1986).

Mechanism of Cell Injury in Head Trauma

In recent years the attention of researchers working in this field has centered on attempts to modify possible effects of acute cell injury by administering medical treatment immediately after the traumatic event; application of this approach is something that in humans would require therapeutic intervention either at the site of the accident or en route to the hospital. A number of these potential therapies have been tried in standardized models of head injury; most studies have been conducted on the fluid percussion injury to the brain in the cat or the rat. A survey of some of the more recent reports based on these experiments is included below.

Menkin introduced in 1940 the concept that *endogenous* mediators, released from areas of tissue necrosis or inflammation, trigger generalized defensive body responses. We now know that the release of these mediators is functionally independent of mechanisms controlled by the CNS, the endocrine system or the immune system. These mediators are responsible for triggering fever, initiating leukocytosis, accelerating metabolic activity and oxygen consumption as well as activating lymphocytes (Beisel, 1987). The observations that these mediators activate and modulate both T- and B-lymphocytes led to the selection of the name of Interleukin-1 (IL-1) mediators. A major breakthrough emerged when T-cell growth factor was isolated and characterized; this lymphokine is called Interleukin-2 (IL-2) (Beisel, 1987). Known previously as endogenous pyrogen, LAF, leukocytic endogenous mediator, or leukocytosis inducing factor, these mediators constitute a unique mechanism whose characteristics include: activation of IL-1 release in proportion to the magnitude of the stimulus, stimulation of fever apparently by initiating PGE synthesis within hypothalamic neurons, stimulation of production and release of neutrophils by the bone marrow, activating lymph cells and causing T-lymphocytes to produce IL-2, assisting thymocytes and T-lymphocytes in proliferative responses and stimulating production of immunoglobulins by B-lymphocytes and plasma cells (Beisel, 1987). Severely head injured patients are hypermetabolic/hypercatabolic and exhibit many aspects of the *post-injury acute-phase* response. Increased interleukin-1 (IL-1) in the ventricular fluid may be responsible for these metabolic abnormalities. Serum albumin levels were evaluated throughout an 18 day period in 62 head injured patients receiving aggressive nutritional support. Hypoalbuminemia, documented upon hospital admission, persisted until 2 weeks post injury despite aggressive nutritional support. In endothelial-cell cultures, albumin transfer across monolayers increased by 459% under the influence of IL-1 and the increases were both dose and time dependent. Such findings suggest that hypoalbuminemia after severe head trauma may be explained by the enhanced endothelial permeability induced by IL-1 (McClain, 1988). Delayed injury after CNS trauma may be caused by release of *endogenous factors*. Endogenous opioid peptides have been proposed as one such class of injurious factors, based on pharmacological studies demonstrating a therapeutic effect of naloxone and other opiate receptor antagonists, following CNS trauma. Regional alterations in opiate receptors have been observed in several brain areas following fluid percussion brain injury in cats. Significant decrease in endorphin (ir-End) was observed in the hypothalamus at 2 h following high but not low level injury; similar results were observed at other sites. In the anterior pituitary a significant increase in ir-End and significant decrease in dynorphin (ir-Dyn) was observed at 2 h following both types of injury. Damage to brain tissues was most pronounced in regions showing significant increases in ir-Dyn but not in other opioids. In the medulla, the increase in ir-Dyn correlated significantly with a fall in systemic mean arterial pressure (McIntosh, Head, 1987).

Regional changes in brain opioid immunoreactivity and CBF were studied after fluid percussion brain injury in the cat. The effect of an opiate antagonist was compared with that of its dextroisomer, which is inactive at opiate receptors. Administration of win-(–) but not of win-(C) or saline at 15 min. after injury significantly improved mean arterial pressure, EEG amplitude, regional CBF, and reduced the severity and incidence of intracranial bleedings.

Win-(–) also improved survival following experimental brain injury. These findings suggest that dynorphin, through its actions on opiate receptors, may contribute to the mechanisms of secondary injury following head trauma and suggest that selected opiate-receptor antagonists may be useful in the treatment of traumatic brain injury (McIntosh, Hayes, 1987).

Acute illness affects thyroid function; few studies have correlated the severity of the underlying medical problem with indexes of thyroid function. Traumatically brain injured patients were chosen for this study as a relatively homogeneous, previously healthy group whose illness was readily quantifiable by the Glasgow Coma Scale (GCS). Triiodothyronine *(tri)* and thyroxine *(thyr)* levels fell significantly within 24 hours of the head injury. Patients with the greatest neurologic dysfunction had the lowest *tri* and *thyr* level on day 4; significant correlations existed between the GCS on day 4 and the concommitant *thy* and *tri* levels. Patients who died or remained vegetative had *thy* and *tri* levels 30% to 50% lower than those who had a good recovery ($p < .015$). Highly significant correlations existed also on day 4 between thy and tri levels and *norepi* (norepinephrine) and *epi* (epinephrine) concentrations. Thus, head trauma induces a gradient of thyroid dysfunction that occurs promptly, is proportional to the degree of neurologic impairment and also reflects the ultimate outcome. The significant association with catecholamine levels suggests a role for the sympathetic nervous system in its causation (Woolf, 1988).

McIntosh *et al.* (1988) have shown significant decrease in mortality, lessened neuronal deficit and restoration of brain phosphate metabolism in rats subjected to brain percussion injury and given an analog of the thyrotropic releasing hormone.

Recent experiments utilizing NMR spectroscopy have demonstrated that reperfusion of ischemic myocardium is associated with a burst of free radical production. The evidence supporting the free radical hypothesis is suggestive but not conclusive. Definitive demonstration of the role of oxygen radicals will require studies measuring the production of these species in conscious models (Bolli, 1988).

In the model of fluid percussion injury to the rat or cat brain the following abnormalities have been documented: transient decrease in PCr and increase in Pi; these and other effects were considerably accentuated by concomitant hypotension (DeWitt, 1986).

After experimental brain trauma or acute hypertension the brain produces superoxide anion radicals, and brain arterioles display endothelial lesions, dilatation of the lumen and loss of reactivity to CO_2. These abnormalities are prevented by pretreatment with free radicals (FR) scavengers or inhibitors of the cyclooxygenase component of prostaglandin H (PGH) synthase; thus, arachidonic acid metabolism by PGH synthase with concomitant formation of tissue injureing oxygen radicals may be the cause of the vascular damage. Experiments were conducted to determine whether kinins, which are known to stimulate arachidonate metabolism and induce brain arteriolar dilation, may be involved in initiating cerebrovascular abnormalities produced by brain trauma in cats. Before fluid percussion to the brain was induced, arterioles on both brain hemispheres constricted normally in response to CO_2, whereas after the percussion injury arterioles pretreated with the kinin antagonist dilated less and displayed normal reactivity to CO_2. These results suggest that a specific kinin receptor stimulates PGH synthase-dependent factor that is responsible for free radical-mediated cerebrovascular injury. Given the ubiquitous distribution of this system, kinins may be an important mediator of vascular injury (Ellis, 1988).

Sequential changes in brain lactate level and brain pH were studied *in vivo* by NMR spectroscopy over a period of 8 h following fluid percussion brain injury. A transient fall in intracellular pH occurred in animals subjected to moderate and high but not low level injury; intracellular pH returned to baseline by 90 mins. Transient rises in brain lactate level were temporally parallel and correlated significantly with the changes in pH.

Post injury alterations in pH and lactate levels were identical in magnitude in animals subjected to either moderate or high level injury. However, animals subjected to moderate injury experienced moderate chronic neurologic deficit that disappeared after 4 weeks, whereas animals subjected to high level injury showed greater histopathologic damage and more severe chronic neurologic deficit. These data suggest that the extent of post traumatic intracellular cerebral acidosis in this model of head injury is not directly related to the severity of functional neurologic deficit (McIntosh, Faden, 1987). In an earlier study, the level of lactate in the cerebrospinal fluid, obtained from the ventricles, was found to be a reliable indicator of improvement or deterioration among patients with severe head injury (DeSalles, 1986).

The recovery of certain neurons, whose axons are physically interrupted, is said to be delayed (or inhibited) by the impairment of the retrograde axonal flow that results in a lack of nerve growth factor in the perikaryon. Continuous intraventricular injection of nerve growth factor in rats whose cholinergic axons had been surgically severed, resulted in a remarkably improved preservation of those neuronal perikarya whose axons had been interrupted (Kromer, 1987).

Takamiya (1988) applied immunocytochemical methods to study the chronology and type of astroglial reactions that follow injury to the brain in young rats. Two days after the injury, GFAP-positive cells had increased in number and then spread to the entire ipsilateral cortex by 3 days after the injury. Astrocytes in layers II–VI of the cortex became "reactive" without mitosis. The association between platelet-derived growth factor (PDGF) and the appearance of reactive astrocytes following brain injury was investigated using a specific antagonist of PDGF, Trapidil, a coronary vasodilator that was injected both locally and intraperitoneally. Following injury to the cerebral cortex in 4-week old rats, Trapidil dramatically suppressed the appearance of reactive astrocytes in areas distant from the wound (Takamiya, 1986). Nieto-Sampedro (1988) has shown that a soluble astrocyte mitogen inhibitor, which is normally present in rat brain, is destroyed by brain injury at the same time that receptors for epidermal growth factor (EGF), a peptide that stimulates astrocytic division, appear on the surface of the astrocytes. Intracerebral injection of an antagonist capable of binding the inhibitor resulted in the appearance of numerous reactive astrocytes. Thus, modulation of astroglial responses may become possible as part of therapeutic trials.

Specificity of Prognosis

Sixty percent of deaths among 159 severely head injured patients occurred by day three and 12% of the patients remained in coma for more than two weeks. Two years after the injury, 51% of the patients were dead, 7% were severely disabled or vegetative, and 42% had a good to moderate recovery. Of the patients in prolonged coma only one-third made a good or moderate recovery. The high proportion of poor outcomes among patients in prolonged coma suggests that this group should be specifically targeted for research. This type of research may require an index of the severity of injury with better precision than the Glasgow Coma Scale (Lyle, 1986).

Increased specificity in prognosis was the objective of analyzing data from 523 patients admitted to the same hospital with severe head injury and with known 6-month outcomes. A combination of the patient's age, the best motor response and the pupillary response was the most accurate prognostic indicator. The model correctly predicted outcome in 78% of the cases. When predictions into an outcome category adjacent to the actual outcome were accepted, the model was accurate in 90% of the cases (Choi, 1988).

References

1. Alberico AM, Ward JD, Choi SC, Marmarou A, Young HF (1987) Outcome after severe head injury. Relationship to mass lesions, diffuse injury, and ICP course in pediatric and adult patients. J Neurosurg 67: 648–656
2. Baxt WG, Moody P (1987a) The differential survival of trauma patients. J Trauma 27: 602–606
3. Baxt WG, Moody P (1987b) The impact of advanced prehospital emergency care on the mortality of severely brain-injured patients. J Trauma 27: 365–369
4. Beisel WR (1987) Humoral mediators of cellular response and altered metabolism. In: Siegel JH (ed) Trauma emergency surgery and critical care. Churchill Livingstone, New York, pp 57–78
5. Bolli R (1988) Oxygen-derived free radicals and postischemic myocardial dysfunction ("stunned myocardium"). J Am Coll Cardiol 12: 239–249
6. Burns CM (1985) Accident-injury organization: Canadian overview. Can J Surg 28: 482–486
7. Choi SC, Narayan RK, Anderson RL, Ward JD (1988) Enhanced specificity of prognosis in severe head injury. J Neurosurg 69: 381–385
8. DeSalles AA, Kontos HA, Becker DP, Young MS, Ward JD, Moulton R, Gruemer HD, Lutz HL, Maset AL, Jenkins L, Marmarou A, Muizelaar (1986) Prognostic significance of ventricular CSF lactic acidosis in severe head injury. J Neurosurg 65: 615–624
9. DeWitt DS, Jenkins LW, Wei EP, Lutz H, Becker DP, Kontos HA (1986) Effects of fluid-percussion brain injury on regional cerebral blood flow and periarteriolar diameter. J Neurosurg 64: 787–794
10. Ellis EF, Holt SA, Wei EP, Kontos HA (1988) Kinins induce abnormal vascular reactivity. Am J Physiol 255: H397–400
11. Fowkes FG, Ennis WP, Evans RC, Roberts CJ, Williams LA (1986) Admission guidelines for head injuries: variance with clinical practice in accident and emergency units in the UK. Br J Surg 73: 891–893
12. Hardman JM (1985) Cerebrospinal trauma. In: Davis RL, Robertson DM (eds) Textbook of neuropathology. Williams and Wilkins, Baltimore, pp 842– 882
13. Hawkins ML, Treat RC, Mansberger Jr AR (1987) Trauma victims: field triage guidelines. South Med J 80: 562–565
14. Kreis Jr DJ, Plasencia G, Augenstein D, Davis JH, Echenique M, Vopal J, Byers P, Gomez G (1986) Preventable trauma deaths: dade county, Florida. J Trauma 26: 649–654
15. Kromer LF (1987) Nerve growth factor treatment after brain injury prevents neuronal death. Science 235: 214–216
16. Luerssen TG, Klauber MR, Marshall LF (1988) Outcome from head injury related to patient's age. A longitudinal study of adult and pediatric head injury. J Neurosurg 68: 409–416
17. Lyle DM, Pierce JP, Freeman EA, Bartrop R, Dorsch NW, Fearnside MR, Rushworth RG, Grant JM (1986) Clinical course and outcome of severe head injury in Australia. J Neurosurg 65: 15–18
18. Maull KI, Schwab CW, McHenry SD, Leavy P, Carl L, Woo P, Overholt S, Sinclair T, Aprahamian C (1986) Trauma center verification. J Trauma 26: 521–524
19- McClain CJ, Henning B, Ott LG, Goldblum S, Young AB (1988) Mechanisms and implications of hypoalbuminemia in head-injured patients. J Neurosurg 69: 386–392
20. McIntosh TK, Vink R, Faden AI (1988) An analogue of thyrotropinreleasing hormone improves outcome after brain injury ^{31}P-NMR studies. Am J Physiol 254: R785–792
21. McIntosh TK, Faden AI, Bendall MR, Vink R (1987) Traumatic brain injury in the rat: alterations in brain lactate and pH as characterized by ^{1}H and ^{31}P nuclear magnetic resonance. J Neurochem 49: 1530–1540
22. McIntosh TK, Hayes RL, DeWitt DS, Agura V, Faden AI (1987) Endogenous opioids may mediate secondary damage after experimental brain injury. Am J Physiol 253: E565–574
23. McIntosh TK, Head VA, Faden AI (1987) Alterations in regional concentrations of endogenous opioids following traumatic brain injury in the cat. Brain Res 425: 225–233
24. Moylan JA (1985) Trauma injuries. Triage and stabilization for safe transfer. Postgrad Med 78: 166–171

25. NCTCS (1966) National Committee of Trauma and Committee of Shock: accidental death and disability. The neglected disease of modern society. Washington DC, National Academy of Science, National Research Council
26. Nieto-Sampedro M (1988) Astrocyte mitogen inhibitor related to epidermal growth factor. Science 240: 1784–1786
27. Pitts LH, Ojemann RG, Quest DO (1987) Neurotrauma care and the neurosurgeon: a statement from the Joint Section of Trauma of the AANS and CNS. J Neurosurg 67: 783–785
28. Roy PD (1987) The value of trauma centres: a methodologic review. Can J Surg 30: 17–22
29. Seelig JM, Becker DP, Miller JD, Greenberg RP, Ward JD, Choi SC (1981) Traumatic acute subdural hematoma. Major mortality reduction in comatose patients treated within four hours. N Engl J Med 304: 1511–1518
30. Shackford SR, Hollingworth-Fridlund P, Cooper GF, Eastman AB (1986) The effect of regionalization upon the quality of trauma care as assessed by concurrent audit before and after institution of a trauma system: a preliminary report. J Trauma 26: 812–820
31. Shackford SR, Mackersie RC, Hoyt DB, Baxt WG, Eastman AB, Hammill FN, Knotts FB, Virgilio RW (1987) Impact of a trauma system on outcome of severely injured patients. Arch Surg 122: 523–527
32. Siegel JH, Dunham CM (1987) Trauma, the disease of the 20th century. In: Siegel JH (ed) Trauma emergency surgery critical care. Churchill Livingstone, New York, pp 1–32
33. Smith RW (1986) California Association of Neurological Surgeons' Emergency Services Committee report: guidelines for establishment of trauma centers. J Neurosurg 65: 569–571
34. Takamiya Y, Kohsaka S, Toya S, Otani M, Mikoshiba K, Tsukada Y (1986) Possible association of a platelet-derived growth factor (PDGF) with the appearance of reactive astrocytes following brain injury *in situ*. Brain Res 383: 305–309
35. Takamiya Y, Kohsaka S, Toya S, Otani M, Tsukada Y (1988) Immunohisto-chemical studies on the proliferation of reactive astrocytes and the expression of cytoskeletal proteins following brain injury in rats. Brain Res 466: 201–210
36. Wang CC, Schoenberg BS, Li SC, Yang YC, Cheng XM, Bolis CL (1986) Brain injury due to head trauma. Epidemiology in urban areas of the People's Republic of China. Arch Neurol 43: 570–572
37. West JG, Williams MJ, Trunkey DD, Wolferth Jr CC (1988) Trauma systems. Current status -future challenges. JAMA 259: 3597–3600
38. West JG, Trunkey DD, Lim RC (1979) Systems of trauma care: a study of two counties. Arch Surg 114: 455–460
39. Woolf PD, Lee LA, Hamill RW, McDonald JV (1988) Thyroid test abnormalities in traumatic brain injury correlation with neurologic impairment and sympathetic nervous system activation. Am J Med 84: 201–208
40. Woolf PD, Hamill RW, Lee LA, Cox C, McDonald JV (1987) The predictive value of catecholamines in assessing outcome in traumatic brain injury. J Neurosurg 66: 875–882.
41. Yoshino E, Yamaki T, Higuchi T, Horikawa Y, Hirakawa K (1985) Acute brain edema in fatal head injury: analysis by dynamic CT scanning. J Neurosurg 63: 830–839

Correspondence and Reprints: Julio H. Garcia, Department of Pathology, Division of Neuropathology, Henry Ford Hospital, 2799 West Grand Boulevard, Detroit, Mi., 48202, U.S.A.

Acta Neurochir (1993) [Suppl] 57: 152–159

Management of Intracranial Hypertension in Head Injury: Matching Treatment with Cause

J. D. Miller, I. R. Piper, and **N. M. Dearden**

Department of Clinical Neurosciences, University of Edinburgh, Scotland, U.K.

Abstract

Raised intracranial pressure (ICP) is common after head injury and strongly associated with mortality and morbidity. Empirical and prophylactic therapy with steroids and barbiturates has proved unsuccessful. Ideally, therapy should be targeted at the predominant cause of the increase in ICP. In head injury these may be

(1) an increase in cerebral blood volume best treated by hyperventilation and hypnotic drugs,
(2) an increase in brain water content best treated by osmotherapy and
(3) increased CSF outflow resistance best treated by CSF drainage. This last cause seldom predominates in head injury.

To determine whether it is possible to identify the best therapy in individual head injured patients, we are comparing osmotherapy (mannitol) and hypnotic drugs (thiopentone and gamma-hydroxybutyrate) in selected patients with severe head injury where it is possible to maintain standard conditions of ventilation and stable blood pressure and to measure ICP, CPP, brain electrical activity, PR ratio of the ICP wave form and cerebral $AvDO_2$ before and during each of the two forms of therapy. Effective therapy means that ICP has been reduced to 20 mm Hg with preservation or improvement in CPP.

17 patients have been studied so far and 4 groups identified. Osmotherapy was superior to hypnotic in 5 cases, hypnotic superior to mannitol in 3 cases, both were effective in 5 cases and neither effective in 4 cases. Patients in whom hypnotics were superior tended to be younger, with diffuse rather than focal brain injury, had the highest levels of brain electrical activity prior to treatment and a higher PR ratio. During successful therapy $AvDO_2$ decreased and patients made a good recovery suggesting that in the patients who responded best to hypnotic drugs the primary brain damage was not severe.

Keywords: Head injury; intracranial hypertension; cerebral monitoring; osmotherapy and hypnotics.

Introduction

Raised intracranial pressure (ICP) defined as a pressure in excess of 20 mmHg for 5 minutes or more, is common following severe head injury, even after intracranial hematomas have been evacuated and patients are artificially ventilated to moderate hypocapnia with full paralysis and sedation. In a consecutive series of 215 patients who were comatose after head injury, raised ICP occurred in 53% of cases (Miller *et al.*, 1981).

Not only is raised ICP common, it is associated with a poor outcome from injury. The higher the level of ICP the greater is the mortality and morbidity (Miller *et al.*, 1977, 1981). In its most extreme form, ICP can rise to the level of the arterial pressure and the cerebral circulation ceases, resulting in brain death. This is the mode of death in 50% of fatal head injuries (Becker *et al.*, 1977).

When ICP rises, it may be in the form of a steady increase in baseline ICP over a period of hours or minutes, attributable to cerebrovascular engorgement, brain edema or development of a space-occupying lesion, or as a series of graduated pressure waves in which cerebrovascular engorgement causes ICP to rise to 30, 40 or 50 mmHg in a matter of seconds (Lundberg, 1960). Such a pressure wave can trigger a vasopressor response, resulting in a sharp increase in arterial pressure and a further rise in ICP. Such major increases in ICP soon act as a factor limiting cerebral blood flow, promoting transtentorial herniation and culminating in cerebral infarction and brain death. Thus while raised ICP can be caused by traumatic brain damage and subsequent swelling, if left unabated it can also produce ischemic brain damage as high tissue pressure and vascular distortion compromise regional cerebral blood flow. This is one of the most important secondary insults to the already injured brain.

The rapidity with which intracranial hypertension develops has important implications for the choice of therapy for raised ICP. There may be insufficient time for a trial and error approach to the choice of therapy. In this short review we state the case for a more selective approach to therapy for raised ICP following head injury, based upon efforts to define the predominant cause for the increase in ICP operating in a particular patient at the time that treatment is needed.

Another reason for favouring a fresh approach to the management of intracranial hypertension after head injury is that current forms of therapy for raised ICP are neither sufficiently nor consistently successful. Possible reasons for this are considered next.

Limitations of Current Therapy for Raised ICP After Head Injury

Hyperventilation causes cerebral vasoconstriction, produces a fall in cerebral blood volume, and hence in ICP. To be successful, cerebrovascular reactivity to CO_2 must be retained, and this is not always the case following severe head injury. Furthermore, the effect of hypocapnia on ICP does not last. Within 24 hours, constricted cerebral vessels have returned to their normocapnic caliber, and the effect on ICP will be lost (Muizelaar *et al.*, 1988). During hyperventilation, there is a progressive increase in the lactate content of CSF. This reverses the potential increase in pH in CSF that would result from the respiratory alkalosis. This rise in CSF lactate is abolished if hyperventilation is conducted during hyperbaric oxygenation, and is therefore probably caused by a degree of cerebral ischemia, induced by the hypocapnic cerebral vasoconstriction (Miller and Ledingham, 1971).

Steroid therapy was introduced to clinical neurosurgery more than 20 years ago and soon resulted in major improvements in the management of patients with brain tumors and other focal brain lesions. Administration of dexamethasone, betamethasone or methylprednisolone to suitable patients produces clinical improvement in 6 to 12 hours, increases brain compliance with loss of waves of increased ICP in 24 to 36 hours, then gradually reduces baseline ICP levels over 48 to 72 hours (Brock *et al.*, 1976; Miller and Leech, 1975; Miller *et al.*, 1977). In acute head injury, the results are quite different. Despite early encouraging reports from Gobiet *et al.* (1976) and Faupel *et al.* (1976), later reports all seem to indicate that steroids neither reduce the mortality nor the incidence or severity of raised ICP in patients with acute severe head injury (Gudeman *et al.*, 1979; Cooper *et al.*, 1979; Pitts and Kaktis, 1980; Saul *et al.*, 1981; Braakman *et al.*, 1983). These trials seemed to show, at least, that steroid therapy was not harmful. However, in the randomised double blind trial of high dose steroid therapy reported by Dearden and his associates (1986), steroid treated patients who had intracranial hypertension fared significantly worse than similar placebo-treated patients. It is of interest that steroid treated patients tend to be hyperglycemic, and hyperglycemia increases the brain damage caused by incomplete cerebral ischemia, the situation that applies in the head injured patient with raised ICP (DeSalles *et al.,* 1987).

One reason for the apparent failure of steroid therapy in acute head injury is that the mechanism responsible for increased ICP may be quite different from that which applies in peritumoral edema. Even when edema is present in head injury, it may be of a type that is different from the chronic vasogenic edema that surrounds a brain tumor.

Mannitol has been employed in the reduction of ICP for nearly 30 years. Surprisingly, disagreement remains concerning the means by which this osmotic agent reduces ICP. It is suggested that withdrawal of water from the brain across the osmotic gradient may be greatest in normal brain, where the blood-brain barrier is intact (Pappius and Dayes, 1965). Recent studies of the effect of mannitol on brain water content *in vivo* using magnetic resonance imaging show, however, that the water withdrawal is maximal in areas of blood-brain barrier abnormality (Bell *et al.*, 1987). Head injured patients have not yet been studied in this way.

Neither of these mechanisms may account for rapid reduction in ICP. Mannitol reduces blood viscosity and cerebral vasoconstriction follows as cerebral vessels reduce in caliber to maintain constant cerebral blood flow, but only if autoregulation is preserved. Early reduction in ICP is produced by a fall in cerebral blood volume according to this hypothesis, advanced by Muizelaar and Kontos and their colleagues (1983, 1984; Rosner and Coley, 1987). In addition to this mechanism, Takagi and his co-workers (1983) have proposed that mannitol can also induce a loss of fluid from the lateral ventricles. It is likely that more than one of these mechanisms operates at any time.

When repeated doses of mannitol are required, it is important to ensure that baseline serum osmolality is not allowed to rise above 320 mOsm/L. If further mannitol is given at this point, not only is ICP unlikely to fall but renal failure may follow (Becker and Vries, 1972).

Hypnotic therapy takes the form of short-acting anesthetic agents such as thiopentone or gammahydroxybutyrate (gammaOH). These drugs depress cerebral metabolism, cerebral blood flow and volume are reduced and ICP falls. Unfortunately these agents also possess cardiodepressant properties and tend to reduce arterial pressure so that there may be little improvement in cerebral perfusion pressure even when ICP is substantially reduced. Systemic hypotension is somewhat unpredictable but is more likely in the presence of hypovolemia.

After initial enthusiastic reports of the efficacy of barbiturate therapy in reducing ICP and improving outcome in head injured patients by Shapiro, Marshall and their associates (Rockoff *et al.*, 1979), three prospective randomised trials of barbiturate therapy in head injury have been reported. Schwartz and colleagues (1984) found that mannitol was superior to barbiturate in controlling raised ICP after head injury and there was no difference in out-come between treatment groups. Patients treated with barbiturates had lower levels of cerebral perfusion pressure. Ward *et al.* (1985) found no difference between barbiturate-treated patients and those on a standard regimen that included mannitol with respect to overall mortality, number of patients dying with severe intracranial hypertension, and prevalence of raised ICP. Episodes of arterial hypotension were significantly more frequent in the barbiturate treated patients. The most recent report by Eisenberg *et al.* (1988) shows no benefit for barbiturates in terms of mortality of severe head injury, but confirms valuable reduction of ICP in selected cases. While all hypnotic drugs may cause arterial hypotension, especially if given in the presence of hypovolemia, gammaOH appears to cause this problem less than thiopentone (Leggate *et al.*, 1986).

Current Position of Therapy for Raised ICP After Head Injury

No single therapy seems to be universally applicable, and treatments may have complications. Nevertheless, there seems to be evidence of benefit for barbiturates or mannitol in individual cases. Other approaches to therapy, such as calcium channel blockade or antagonism of NMDA receptors are in early stages of evaluation and development. For the time being, at least, we have to define more precisely the indications for the agents we have currently available in the light of the knowledge we now have about their possible mechanisms of action, and after trying to determine in individual patients the causal mechanism of the rise in ICP.

Before resorting to any form of therapy it is essential to identify and correct any of the simple factors that can produce an increase in ICP. The patient should be at 10 to 15 degrees of head-up tilt, the position of the head should be checked and excessive flexion or rotation corrected, the airway should be secured and sufficient sedative and/or relaxant medication provided to ensure that the patient does not make respiratory efforts against the ventilator. Inspection of ventilator inspiratory pressure can indicate this, as well as increases in intrathoracic pressure due to bronchospasm or increase in central venous pressure related to pneumothorax of lung collapse. Any elevation of body temperature should be corrected, blood gases should be checked to exclude hypoxemia or hypercapnia. The possibility of epileptic activity should be considered and treated if necessary, serum sodium should be checked to exclude hyponatremia (Na < 120 mEq/L). Finally, if ICP remains high despite these measures, repeat CT scan should be obtained to exclude late or recurrent intracranial hemorrhage as the explanation for the rise in ICP.

Causes of Raised ICP Following Head Injury

Once the above factors have been excluded or corrected, there are three groups of causes for post-traumatic intracranial hypertension. The first is congestive brain swelling due to a rapid increase in cerebral blood volume. This may be the result of arterial dilatation, that may be active or passive, or venous obstruction, such as raised intrathoracic pressure or, rarely, sagittal sinus occlusion. The second group is the brain edemas. Any or all of the recognised forms of edema can occur after head injury (Miller, 1979). Perifocal, vasogenic edema forms around brain contusions; hydrostatic edema occurs when previously compressed brain is perfused at high pressure after removal of an intracranial hematoma. Cytotoxic edema occurs in areas of ischemic brain, and osmotic edema occurs following hyponatremia. The third main cause of raised ICP is increased CSF outflow resistance, but this rarely occurs in the early post-traumatic phase.

Identifying the Cause of Raised ICP

It is important to identify the nature of the brain injury, to determine whether it is more likely to be focal or diffuse. The CT scan is a major help in this determination. We believe that congestive brain swelling is more likely to complicate diffuse brain injury, and edema is more likely after focal injury. The most direct way to measure cerebral blood volume is by PET or SPECT, but these are not bedside techniques and are not yet widely available. Estimates of blood flow velocity in the major cerebral arteries can be obtained at the bedside by Transcranial Doppler but information about vascular caliber is lacking and statements about blood volume must be inferential. The same applies to inferences drawn from measurements of the arterial-jugular bulb oxygen content difference, where it must be recognised that these reflect global changes in CBF and/or metabolism and may fail to identify regional differences.

Evidence is slowly accumulating that analysis of the waveform of the ICP recording and comparison with the arterial wave may yield data that permit inferences about whether the ICP is increased predominantly because of either vascular or non-vascular, compliance-related, factors. It appears that the if the ICP wave is resolved into a number of harmonic waves, the amplitude of the slower harmonics reflects the vascular contribution, while the higher frequency harmonics reflect factors related to compliance, including brain edema (Bray *et al.*, 1986; Piper *et al.*, 1989).

It is always valuable, in patients who are inaccessible to neurological evaluation because of relaxant drugs, to obtain a measure of brain electrical activity. In relation to therapy for raised ICP, it appears that if brain electrical activity has already disappeared, benefit is never obtained from therapy with hypnotic drugs.

Paired Comparison of Osmotic and Hypnotic Therapy

Since it may be possible to identify whether raised ICP in individual patients is predominantly due to vascular factors or to brain edema, it is important to test whether it is true that hypnotic drugs are in fact more effective in the former case, and mannitol in the latter. We are therefore conducting a comparison study in severely head injured patients with raised ICP in which each patient receives both mannitol and one or two hypnotic drugs (thiopentone and/or gammaOH) to compare their efficacy in reducing ICP below 20 mmHg while preserving or increasing cerebral perfusion pressure.

Sixteen patients have been studied, one twice. All patients were managed in a standardised way, using artificial ventilation. To be eligible for this study constant values for inspiration, pause, expiration time, respiratory rate and inflation pressure had to apply. In all patients continuous monitoring of ICP, SAP, CPP, CVP, body temperature (core and peripheral) and brain electrical activity (Cerebral Function Monitor) was obtained. Intermittent samples of arterial and jugular bulb blood were drawn for measurement of blood

Table 1. *Characteristics of Patients in Four Treatment Groups*

Treatment group		Patient	Age	Brain injury	GCS on admission	Therapy	Outcome
A	Mannitol superior to hypnotic	GM	24	F	4	M, T	MD
		JM	14	F	4	M, G	D
		KL	21	F	9	M, T	MD
		JB	31	F	5	M, T	SD
		GM	44	F	11	M, T	D
B	Hypnotic superior to mannitol	DG	10	D	5	M, G	MD
		GG	3	D	7	M, T	MD
		WG	25	D	6	M, T	GR
C	Both effective	JM	14	F	4	M, T	D
		CS	8	F	7	M, T	GR
		HW	22	F	7	M, T	GR
		HM	54	F	7	M, T	D
		DM	17	F	4	M, T	D
D	Neither effective	JR	41	F	6	M, T	SD
		CG	62	F	5	M, T	D
		FS	43	D	3	M, T	D
		LP	24	F	4	M, T	D

Brain injury classified as focal *(F)* or diffuse *(D)*; Glasgow Coma Score *(GCS)*; Therapy – mannitol *(M)* thiopentone *(T)* or gammahydroxybutyrate *(G)*. Outcome classified as good recovery *(GR)*, moderate disability *(MD)*, severe disability *(SD)*, or dead *(D)*.

gases and pH and calculation of the cerebral arterio-venous oxygen content difference ($AvDO_2$). The ICP waveform was examined under these constant conditions of ventilation and the ratio of the pulse pressure to the respiratory wave amplitude was calculated (P:R ratio).

Increases in ICP above 25 mmHg in the first 24 hours and 30 mmHg thereafter that had no correctable simple cause were treated in randomised order by mannitol (0.5 g/kg body weight) then by hypnotic drugs. These consisted of thiopentone (5 mg/kg) in 6 cases, thiopentone followed on a subsequent occasion by gammaOH in 9 cases, or gammaOH alone (60 mg/kg) in 2 cases. Based upon the capacity of each agent to reduce ICP while maintaining CPP above 60 mmHg in adults or 50 mmHg in children under 10 years, we identified four response groups. These are mannitol superior to hypnotics (n = 5), hypnotics superior to mannitol (n = 3), both effective (n = 5), and neither effective (n = 4).

Data from the study are shown in Tables 1, 2 and 3, and can be summarised as follows. Patients in whom hypnotic drugs were most effective had diffuse brain injuries and tended to make a better recovery. Mannitol

Table 2. *Changes in Intracranial Pressure (ICP) and Cerebral Perfusion Pressure (CPP) Following Mannitol or Hypnotic Drugs in Four Treatment Groups (Mean ± SEM)*

Treatment group		Treatment	ICP pre	ICP post	CPP pre	CPP post
A	Mannitol superior to hypnotic (n = 5)	mannitol	37.8 ± 5.1	20.4 ± 2.2	61.2 ± 4.6	77.6 ± 6.1
		hypnotic	39.8 ± 6.5	26.4 ± 4.2	54.6 ± 4.8	51.2 ± 3.5
B	Hypnotic superior to mannitol (n = 3)	mannitol	44.7 ± 2.3	33.7 ± 4.6	50.7 ± 6.0	62.3 ± 8.0
		hypnotic	42.7 ± 5.0	14.3 ± 3.8	46.7 ± 1.4	68.3 ± 2.4
C	Both effective (n = 5)	mannitol	36.8 ± 2.7	23.6 ± 1.4	52.8 ± 4.7	71.4 ± 5.4
		hypnotic	33.2 ± 4.3	20.2 ± 5.1	57.8 ± 5.6	67.4 ± 7.8
D	Neither effective (n = 4)	mannitol	58.7 ± 10.6	44.2 ± 10.2	50.0 ± 3.5	55.0 ± 2.4
		hypnotic	56.2 ± 10.7	45.8 ± 13.8	53.7 ± 5.1	52.3 ± 4.6

Table 3. *Values for Brain Electrical Activity – Cerebral Function Monitor (CFM) Lower Border Voltage, Pulse:Respiratory Ratio (P:R) of the Intracranial Pressure Waveform, and the Cerebral Arteriovenous Oxygen Content Difference Before and During Therapy*

Treatment group		Treatment	CFM uV	P:R ratio	AvO_2 Difference pre	(ml O_2/dl) post
A	Mannitol superior to hypnotic (n = 5)	mannitol	3.8 ± 0.5	0.98 ± 0.23	6.95 ± 1.66	4.81 ± 1.63
		hypnotic	4.4 ± 0.4	0.99 ± 0.34	6.66 ± 1.47	8.63 ± 2.43
B	Hypnotic superior to mannitol (n = 3)	mannitol	8.0 ± 0.6	2.9 ± 1.15	6.95 ± 3.72	6.44 ± 3.61
		hypnotic	8.7 ± 0.3	3.4 ± 0.92	7.11 ± 3.55	5.36 ± 2.48
C	Both effective (n = 5)	mannitol	5.7 ± 0.5	0.9 ± 0.12	5.87 ± 0.29	4.58 ± 0.35
		hypnotic	5.6 ± 0.5	1.1 ± 0.55	6.53 ± 0.50	6.12 ± 0.20
D	Neither effective (n = 4)	mannitol	2.6 ± 0.5	2.9 ± 1.40	7.14 ± 2.05	6.96 ± 2.32
		hypnotic	3.6 ± 0.6	2.6 ± 1.18	8.02 ± 3.11	8.34 ± 3.06

was more effective in patients with focal brain injury (contusion or after hematoma evacuation). Hypnotics were more effective when the P:R ratio was higher and when brain electrical activity was well preserved. Indeed the extent of the ICP reduction afforded by hypnotics is a linear function of the CFM lower border voltage (Dearden and Miller, 1989).

While the pre-treatment value of $AvDO_2$ was not of value in discriminating between preferred therapies, successful treatment of raised ICP with good preservation of CPP was accompanied by a reduction in $AvDO_2$. In a small number of cases, unsuccessful treatment of raised ICP was accompanied by an increase in $AvDO_2$ into the ischemic range of values (> 9 ml/dl).

This is a continuing study. Because of the stringent conditions needed for valid comparison between two sequential treatments only a small proportion of our severely head injured patients can be included in the study, but already some tentative conclusions can be drawn. A selective approach to therapy for raised ICP seems to be feasible and appropriate. Analysis of ICP waveform and brain electrical activity are helpful. Measurement of cerebral $AvDO_2$ can indicate when therapy is not only unsuccessful but is proving harmful to the patient. While the numbers of observations are still limited, it is beginning to appear that the patients who respond best to hypnotic drugs are those who have sustained less severe degrees of primary traumatic brain injury.

Overall Experience of Therapy with Mannitol and Hypnotic Drugs

In the period 1986/1987 we have managed 208 patients with severe head injury (GCS 8 or less with no eye opening after resuscitation) in our Intensive Care Unit (Table 4). Artificial ventilation and ICP monitoring were not employed in 85 patients (41%). This was either because the injury was not considered sufficiently severe to warrant such intensive measures (CT scan normal), or because a decision not to treat had been made on grounds of patient age (over 70 years) and/or poor neurological status (probable brain death). In the remaining 123 patients who were ventilated and had ICP monitoring, we used mannitol and/or hypnotic drugs in 85 cases (69%) and needed no therapy for raised ICP in 38 patients, although in 9 of these cases a single dose of mannitol had been given prior to arrival in the ICU. Mortality in patients who required therapy for intracranial hypertension (44%) was significantly higher than mortality in patients who did not require treatment for raised ICP (8%), a confirmation of the adverse significance of intracranial hypertension after head injury (Miller *et al.*, 1977, 1981).

In the 85 patients who required treatment for raised ICP in the ICU, mannitol alone was regarded as adequate therapy in 52 cases, while barbiturates were used in 33 patients. However in 31 of these mannitol was also given at some time, although 16 of these patients were in the comparison study. Thus in our everyday practice of management of severe head injury, barbiturate therapy alone was regarded as sufficient therapy for raised ICP in 2 plus 3 study patients of 208 severely head injured patients seen during a two-year period. From this data it can be concluded that if a blind choice of therapy for raised ICP has to be made urgently in a severely head injured patient, then mannitol would be our chosen agent. However, our ongoing paired comparison study does already show that hypnotic drugs will prove to be a superior therapy to mannitol in a small number of carefully selected cases. Correct indi-

Table 4. *Results of Management of Severely Head Injured Patients – Edinburgh 1986/87*

Patient group	Outcome number	GR/MD	SD/veg.	dead
All patients	208	104 (50%)	25 (12%)	79 (38%)
ICP not monitored	85	42 (49%)	4 (5%)	39 (46%)
ICP monitored	123	62 (50%)	21 (17%)	40 (33%)
no therapy used	29	23 (79%)	4 (14%)	2 (7%)
early mannitol only	9	5 (56%)	3 (33%)	1 (11%)
mannitol in ITU	52	22 (42%)	6 (12%)	24 (46%)
hypnotics ± mannitol	33	12 (36%)	8 (24%)	13 (39%)

cations for hypnotic therapy are, diffuse head injury, P:R ratio greater than 1.5, preserved brain electrical activity (Bingham *et al.*, 1985) and preserved cerebrovascular reactivity to carbon dioxide (Nordstrom *et al.*, 1988).

Conclusions

Raised intracranial pressure, brain swelling and edema continue to present a major challenge in the management of the patient with severe brain injury, despite prompt detection and evacuation of intracranial hematomas and artificial ventilation. If raised ICP persists after all easily remediable causes have been dealt with, the most effective approach to therapy seems to be to identify in the individual patient whether the predominant cause of the intracranial hypertension is congestive brain swelling or brain edema (an increase in brain water content). Imaging of the brain by CT, analysis of the ICP waveform, monitoring of brain electrical activity and measurement of cerebral arteriovenous oxygen content difference all provide information of value in the selection and evaluation of therapy for raised ICP in the head injured patient.

References

1. Becker DP, Vries JK (1972) The alleviation of increased intracranial pressure by the chronic administration of osmotic agents. In: Brock M, Dietz H (eds) Intracranial pressure. Springer, Berlin Heidelberg New York, pp 309–315
2. Becker DP, Miller JD, Ward JD, Greenberg RP, Young HF, Sakalas R (1977) The outcome from severe head injury with early diagnosis and intensive management. J Neurosurg 47: 491–502
3. Bell BA, Smith MA, Kean DM, MacDonald HL, Barnett GH, Douglas RHB, Smith MA, McGhee CNJ, Miller JD, Tocker JL, Best JJK (1987) Brain water measured by magnetic resonance imaging: correlation with direct estimation and change following mannitol and dexamethasone. Lancet 1: 66–69
4. Bingham RM, Procaccio F, Prior PF, Hinds CJ (1985) Cerebral electrical activity influences the effects of etomidate on cerebral perfusion pressure in traumatic coma. Br J Anaesth 57: 843–848
5. Braakman R, Schouten HJA, Blauw-van Dishoeck M, Minderhoud JM (1983) Megadose steroids in severe head injury: results of a prospective double-blind clinical trial. J Neurosurg 58: 326–330
6. Bray RS, Sherwood AM, Halter JA, Robertson C, Grossman RG (1986) Development of a clinical monitoring system by means of ICP waveform analysis. In: Miller JD, Teasdale GM, Rowan JO, *et al* (eds) Intracranial pressure VI. Springer, Berlin Heidelberg New York, pp 260–264
7. Brock M, Wiegand H, Zillig C, Zywietz C, Mock P, Dietz H (1976) The effect of dexamethasone on intracranial pressure in patients with supratentorial tumors. In: Pappius HM, Feindel W (eds) Dynamics of brain edema. Springer, Berlin Heidelberg New York, pp 330–336
8. Cooper PR, Moody S, Clark WK, Kirkpatrick J, Maravilla K, Gould AL, Drane W (1979) Dexamethasone and severe head injury: a prospective double-blind study. J Neurosurg 51: 307–316
9. Dearden NM, Gibson JS, McDowall DG, Gibson RM, Cameron MM (1986) Effect of high dose dexamethasone on outcome from severe head injury. J Neurosurg 64: 81–88
10. Dearden NM, Miller JD (1989) Paired comparison of hypnotic and osmotic therapy in the reduction of intracranial hypertension after head injury. In: Hoff JT, Betz AL (eds) Intracranial pressure VII. Springer, Berlin Heidelberg New York, pp 471–481
11. DeSalles AAF, Muizelaar JP, Young HF (1987) Hyperglycemia, cerebrospinal fluid lactate and cerebral blood flow in severely head injured patients. Neurosurgery 21: 45–50
12. Eisenberg HM, Frankowski RF, Contant CF, Marshall LM, Walker MD and the Comprehensive Central Nervous System Trauma Centers (1988) High dose barbiturate control of elevated intracranial pressure in patients with severe head injury. J Neurosurg 69: 15–23
13. Faupel G, Reulen HJ, Müller D, Schürmann K (1976) Double blind study on the effects of steroids on severe closed head injury. In: Pappius HM, Feindel W (eds) Dynamics of brain edema. Springer, Berlin Heidelberg New York, pp 337–343
14. Gobiet W, Bock WJ, Liesegang J, Grote W (1976) Treatment of acute cerebral edema with high dose of dexamethasone. In: Beks JWF, Bosch DA, Brock M (eds) Intracranial pressure III, Springer, Berlin Heidelberg New York, pp 231–235
15. Gudeman SK, Miller JD, Becker DP (1979) Failure of high dose steroid therapy to influence intracranial pressure in patients with severe head injury. J Neurosurg 51: 301–306
16. Leggate JRS, Dearden NM, Miller JD (1986) The effects of gammahydroxybutyrate and sodium thiopentone on intracranial pressure in severe head injury. In: Miller JD, Teasdale GM, Rowan JO, *et al* (eds) Intracranial pressure VI. Springer, Berlin Heidelberg New York, pp 754–757
17. Lundberg N (1960) Continuous recording and control of ventricular fluid pressure in neurosurgical practice. Acta Phychiatr Neurol Scand 36 [Suppl 149]: 1–193
18. Miller JD (1979) Clinical management of cerebral oedema. Br J Hosp Med 20: 152–166
19. Miller JD, Ledingham IM (1971) Reduction of increased intracranial pressure: comparison between hyperbaric oxygen and hyperventilation. Arch Neurol 24: 210–216
20. Miller JD, Leech PJ (1975) Assessing the effects of mannitol and steroid therapy on intracranial volume/pressure relationships. J Neurosurg 42: 274–281
21. Miller JD, Becker DP, Ward JD, Sullivan HG, Adams WE, Rosner MJ (1977) Significance of intracranial hypertension in severe head injury. J Neurosurg 47: 503–516
22. Miller JD, Sakalas R, Ward JD (1977) Methylprednisolone treatment in patients with brain tumors. Neurosurgery 1: 114–117
23. Miller JD, Butterworth JF, Gudeman SK, Faulkner JE, Choi SC, Selhorst JB, Harbison JW, Lutz HA, Young HF, Becker DP (1981) Further experience in the management of severe head injury. J Neurosurg 54: 289–299
24. Muizelaar JP, Lutz HA, Becker DP (1984) Effect of mannitol on ICP and CBF and correlation with pressure autoregulation in severely head injured patients. J Neurosurg 61: 700–706
25. Muizelaar JP, van der Poel HG, Li Z, Kontos HA, Levasseur JE (1988) Pial arteriolar diameter and CO_2 reactivity during prolonged hyperventilation in the rabbit. J Neurosurg 69: 923–927
26. Muizelaar JP, Wei EP, Kontos HA, Becker DP (1983) Mannitol causes compensatory cerebral vasoconstriction and vasodilatation to blood viscosity changes. J Neurosurg 59: 822–828
27. Nordström CH, Messeter R, Sundbarg G, Schalin W, Werner M, Ryding E (1988) Cerebral blood flow, vasoreactivity and

oxygen consumption during barbiturate therapy in severe traumatic brain lesions. J Neurosurg 68: 424–431
28. Pappius HM, Dayes LA (1965) Hypertonic urea: its effect on the distribution of water and electrolytes in normal and edematous brain tissues. Arch Neurol 13: 395–402
29. Piper IR, Dearden NM, Miller JD (1989) Can waveform analysis of ICP separate vascular from non-vascular causes of intracranial hypertension? In: Hoff JT, Betz AL (eds) Intracranial pressure VII. Springer, Berlin Heidelberg New York, pp 157–163
30. Pitts LH, Kaktis JV (1980) Effect of megadose steroids on ICP in traumatic coma. In: Shulman K, Marmarou A, Miller JD, *et al* (eds) Intracranial pressure IV. Springer, Berlin Heidelberg New York, pp 638–642
31. Rockoff MA, Marshall LF, Shapiro HM (1979) High-dose barbiturate therapy in man: a clinical review of sixty patients. Ann Neurol 6: 194–199
32. Rosner MJ, Coley I (1987) Cerebral perfusion pressure: a hemodynamic mechanism of mannitol and the pre-mannitol hemogram. Neurosurgery 21: 147– 156
33. Saul TG, Ducker TB, Salcman M, Carro E (1981) Steroids in severe head injury: a prospective randomised clinical trial. J Neurosurg 54: 596–600
34. Schwartz ML, Tator CH, Rowed DW (1984) The University of Toronto head injury treatment study: a prospective randomised comparison of pentobarbital and mannitol. Canad J Neurol Sci 11: 434–440
35. Takagi H, Saito T, Kitahara T, Morii S, Ohwada T, Yada K (1983) The mechanism of the ICP-reducing effect of mannitol. In: Ishii S, Nagai H, Brock M (eds) Intracranial pressure V. Springer, Berlin Heidelberg New York, pp 729–733
36. Ward JD, Becker DP, Miller JD, Choi SC, Marmarou A, Wood C, Newlon PG, Keenan R (1985) Failure of prophylactic barbiturate coma in the treatment of severe head injury. J Neurosurg 62: 383–388

Correspondence and Reprints: Douglas Miller, Department of Clinical Neuroscience, Western General Hospital, Edinburgh EH4 2XU, U.K.

Acta Neurochir (1993) [Suppl] 57: 160–164

Traumatic Brain Tissue Acidosis: Experimental and Clinical Studies

A. Marmarou, R. Holdaway, J. D. Ward, K. Yoshida, S. C. Choi, J. P. Muizelaar, and **H. F. Young**

Division of Neurosurgery, Medical College of Virginia, Richmond, Virginia, U.S.A.

Abstract

We have been focusing on potential metabolic derangement associated with severe head injury and a clinical trial directed toward treating brain tissue acidosis is currently underway. More specifically, we based this study on the hypothesis that following brain trauma brain tissue acidosis develops which may contribute to the prolongation of coma and neurologic deficit. Tromethamine (THAM), a safe and low toxicity agent which buffers in major part by causing a hypocapnic alkalosis, was selected for trial. Patients admitted with GCS < 8 were randomized into one of three arms: control; THAM plus hyperventilation; hyperventilation alone. Each regimen was maintained for 5 days post injury. Our analysis of 3 and 6 months Glasgow outcome score showed that prophylactic hyperventilation retards recovery, and the use of THAM overcomes the apparent deleterious effects of hyperventilation. One explanation is that the reduced ICP instability observed in THAM treated patients may account for this improvement. Is THAM effective in buffering traumatized brain tissue? What factors account for improvement in ICP stability?

We addressed these questions in experimental studies utilizing MR spectroscopy to measure brain lactate production and tissue pH in fluid percussed anaesthetized cats. The protocol was designed to match our clinical trial, and brain injured animals were randomized into control, THAM, and hyperventilated groups. We observed that brain lactate production increased with trauma and remained above control at 8 hrs post injury. Lactate production in THAM treated animals was not elevated. Highest lactate production was associated with injured animals treated with sustained hyperventilation. We also observed that the PCr/P_i ratio in trauma alone was reduced to 78% of control at 1 hr and to 65% in the hyperventilated group. The reduction of PCr/P_i of the THAM treated group was similar to the reduction seen in trauma alone and equaled 77.8%. Brain tissue pH of THAM treated animals was slightly alkalotic compared to trauma controls.

These data suggest that THAM treatment moderates the pH disturbances following trauma and the slightly alkalotic milieu may improve ICP stability in patients by helping to maintain vasoconstriction.

Keywords: Head injury; brain tissue acidosis; lactate production; THAM buffering.

Introduction

The deleterious effects of secondary insults superimposed upon the initial structural damage following traumatic head injury have clearly been shown to be correlated with outcome (Miller and Becker, 1982). Many of these secondary insults have been linked to metabolic derangements associated with the injury, and the presence of lactate in the cerebrospinal fluid has been considered a marker of this metabolic damage. Under normal conditions, the brain produces lactate as a by-product of energy metabolism, formed by the cytosolic reduction of pyruvate by lactate dehydrogenase (LDH). With aerobic conditions, lactate concentrations in CSF are low, however, with traumatic injury, CSF lactate is increased. The underlying mechanism for increased lactate production associated with mechanical injury is unclear. With complete ischemia, oxidative phosphorylation ceases immediately and stores of glucose and glycogen are utilized to supply anaerobic glycolysis resulting in lactate accumulation. Thus, investigators have theorized that an ischemic insult occurring soon after injury may be responsible for the elevated CSF lactate observed in head injured patients (Graham *et al.*, 1978). Others have discounted this theory based upon measures of CBF within 12 hours of admission which were not significantly depressed (DeSalles *et al.*, 1987).

This report briefly summarizes the theoretical and experimental studies conducted at our institution to shed further light upon the significance of lactic acidosis in head injury. One objective was to study the time course of lactate clearance from CSF obtained from head injured patients and using a mathematical model, predict the time relative to moment of injury when lactate seepage into CSF occurred. This would help clarify if the lactate production was shortlived or if the metabolic perturbation was sustained.

Predictive Value of CSF Lactate

Clinical studies have shown that the level and rate of clearance of CSF lactate is correlated with outcome. (Crockard and Tailor 1972; DeSalles *et al.*, 1986; King *et al.*, 1974; Seitz and Ocke, 1977). In the first 24 hours after head injury, CSF lactate levels are invariably high. A decrease to the normal range within 48 hours of injury has been shown to be consistent with good outcome, while sustained or increasingly high levels of CSF lactate generally correlate with poor outcome. Observations of CSF lactate represent the net accumulation of lactate production less clearance. A general equation can be derived describing the amount of lactate present in CSF as a function of time for an arbitrary rate of entry hereafter termed the "seepage function".

Briefly, the time rate of change in lactate is given by the difference in mass flow rate of lactate entry less mass flow rate of lactate exit. In terms of differentials,

$$dQ / dt = K(t) - r / V Q(t), \qquad (1)$$

where $K(t)$ equals lactate seepage rate (mMol/min), $Q(t)$ the amount of lactate in CSF at time t (mMol), V the CSF volume (ml) and r, the rate of influx/efflux of formed CSF (ml/min).

This equation was solved to yield the general solution for an arbitrary seepage function $K(t)$. In this report, three seepage functions were studied: an impulse function; an exponentially decaying function of variable decay rate and initial value, and a combined pulse ramp function of variable magnitude and descending slope. These mathematical functions simulated different processes which could conceivably account for the observed lactate concentration in head injured patients. In the case of the impulse function, it simulated a "bolus" input of lactate suggesting a brief period of lactate production which might occur soon after injury followed by a rapid return to normal. The exponential function of variable decay rate and initial value simulated a rapid lactate production followed by a more gradual reduction toward normal level. Finally, the combined pulse ramp simulated an initial rapid lactate production combined with a continuous rate of lactate entry into CSF. The seepage functions were selected to test whether or not the presumed metabolic derangement was brief with rapid recovery (impulse) or sustained during the first 72 hours post injury (pulse ramp).

The results of this analysis indicated that a lactate "generator" simulated by a rapid pulse followed by a gradual exponential decay invariably led to an increasing slope of CSF lactate concentration peaking between 6 and 12 hours followed by a gradual return toward normal. The fact that our patient lactate values were always descending with time indicated that this type of seepage function was not possible.

The Optimal Seepage Function

The seepage function which best fits the clinical data was described by a rectangular seepage pulse. The exact features, however, could not be determined for the following reason. Our earliest measure of CSF lactate from head injured patients was approximately 8 hours. Thus, it was not possible to determine the maximum lactate concentration. However, from this theoretical analysis it was possible to determine the range of values of seepage function which matched the temporal lactate course. The minimum duration of lactate production that could account for our clinical lactate profile was a rectangular pulse of 2.52 hours duration and 1.0 mMol/hr in magnitude. The maximum lactate production that could fit the clinical profile was a rectangular pulse of 5.45 hours duration and magnitude 0.5 mMol/hr.

Thus, from this analysis, we theorized that whatever the mechanism accounting for lactate production, it occurred soon after injury and could be sustained for up to 5.45 hours.

Thereafter, lactate seepage into CSF ceased thereby discounting the notion of a sustained lactate production.

Comparison of Blood and CSF Lactate Profiles

It is interesting to compare the temporal course of blood and CSF lactate in the severely head injured patient. Lactate levels were measured in 114 severely head injured patients (GCS 8 or less) admitted to the neuroscience intensive care unit. The temporal profile of lactate values were analyzed using a regression model of the form

$$[\log(\text{lactate}) = a + b\,(\text{time})]$$

Table 1. *Lactate Levels in CSF and Blood*

	CSF	Arterial
a	1.63 ± 0.063	1.48 ± 0.88
b	−0.015 ± 0.002	−.026 ±.003

The temporal course of lactate was analysed according to the equation log (lactate) = a + b (time).

The values of a + b for blood are indicated above.

where time was truncated at 48 hours. The values of *a* and *b* for CSF and arterial lactate levels are given in Table 1.

Each of the two slope values are significant at $p < .001$ indicating that the lactate level of an average patient decreases over the 48 hour period. The differences in slopes between CSF are significant at $p < .01$ suggesting that blood lactate levels decrease faster than CSF lactate. When these data were analyzed with respect to outcome (GOS), we found a weak correlation between slope and GOS. However, the degree of correlation at 6 months was significant at 0.05 indicating that good outcome was characterized by a relatively rapid reduction in lactate level in 48 hours. Of interest is the observation that the same information could be derived from the rate of fall in arterial lactate.

Experimental Studies of CSF and Arterial Lactate

It was clear from the clinical measurements that both CSF and arterial lactate levels were strongly related. The observation of elevated CSF lactate was consistent with earlier studies suggesting a metabolic derangement associated with brain injury (DeSalles *et al.*, 1986). However, the evidence of brain tissue acidosis by CSF analysis is indirect and the severity of brain tissue acidosis following traumatic injury and the relationship of injury time relative to appearance of lactate in CSF and blood were unclear. We studied lactate dynamics in brain, CSF and serum following fluid percussion trauma in the anaesthetized adult cat (Inao *et al.*, 1988). In these studies, brain tissue lactate and pH were measured using magnetic resonance techniques. In the mild trauma group, brain and CSF lactate values were elevated moderately although brain pH and serum lactate remained at control values. With severe trauma, brain lactate index (ratio of actual to control) increased by 82 % and pH reduced by .05 units indicating brain tissue acidosis. In the brain injured cat, brain lactate reached a peak at 1.5 hours after trauma and steadily decreased to normal levels by 8 hours post trauma. In contrast, maximal CSF and arterial lactate were observed 15 minutes after trauma and decreased during the next two hours.

From these experimental studies, we concluded the following. First, the proportion of lactate in brain tissue, CSF and blood increased in proportion to the severity of injury. Secondly, the observation that lactate levels in blood and CSF are maximal immediately post injury while brain tissue lactate peaks at 1.5 hours suggests that brain tissue production could not account for the rapid appearance of lactate in CSF and blood.

This led to the speculation that the initial elevation of CSF lactate reflected a systemic response to trauma while the secondary rise of CSF lactate was due to slow seepage of lactate produced by the brain tissue. What factors account for lactate elevation in serum? Others have shown that within seconds of fluid percussion trauma, circulating catecholamines increase dramatically and epinephrine rises 500 fold with increases in norepinephrine of nearly 100 fold (Rosner *et al.*, 1984). Studies of surgical trauma and shock have shown that epinephrine and norepinephrine concentrations increase serum glucose and lactate up to 4 fold by stimulating glycogenolysis in liver and muscle. (Håkanson *et al.*, 1984; Liddell *et al.*, 1979).

Elevated Blood Lactate in Non CNS Injured Patients

Considering the possibility that circulating levels of arterial lactate observed in head injured patients may not entirely be of CNS origin, we studied patients admitted to our general surgical intensive care unit for injuries other than CNS. Serial blood samples were obtained and analyzed for lactate level. The temporal blood lactate course of a patient with injuries confined to the neck and abdomen are shown in Fig. 1. The initial sample level measured 4.4 mM/l and gradually approached normal levels within 48 hours. This blood lactate course is very close to that observed in severe head injury.

Effect of THAM on CSF and Blood Lactate in Head Injured Patients

These studies described above were in technical support of an ongoing clinical trial which treated severely head injured patients with tromethamine (THAM), a systemically applied buffer for treatment of brain tissue acidosis (Ward *et al.*, 1989). In this

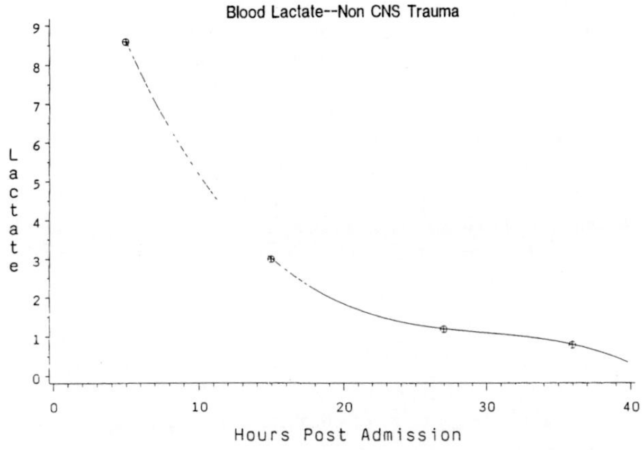

Fig. 1. The temporal course of arterial lactate (mMol/l) of a patient with neck and abdominal injuries and no CNS involvement. The temporal course is very similar to that observed in the severely head injured patient

study, severely head injured patients were randomized into either control, hyperventilation ($PaCO_2$ 24 ± 2 mmHg) or THAM plus hyperventilation adjusted to achieve an equal reduction in arterial $PaCO_2$ as the hyperventilation group.

The differences in lactate profile of arterial, venous and CSF were not significantly different among the three treatment groups indicating that THAM or the combination of THAM and hyperventilation, although raising the arterial blood to slightly alkalotic levels, did not alter the production or clearance rate of lactate. It was observed that the ICP profile of the THAM treated patient was less variable and more easily managed, particularly in the period beyond 48 hours post injury.

Effect of THAM on Brain Tissue Lactate

Although experimental studies have shown that use of THAM buffer for treatment of brain tissue acidosis improved mortality in brain injured cats (Rosner and Becker, 1984), the direct influence of THAM upon brain lactate and pH has not been explored. Currently, we are proceeding with an experimental study utilizing the fluid percussion model of brain injury in which we have simulated the clinical trial in the laboratory. Animals were divided into three groups and treated with sustained hyperventilation, THAM, and THAM plus hyperventilation and followed for 8 hours post injury.

We observed that brain tissue lactate in hyperventilated animals were similar to lactate profiles measured in untreated controls. However, the brain lactate rise seen normally and peaking at 1.5 hours was completely ameliorated in THAM treated animals. Moreover, the PCr/P_i ratio which remained suppressed at 8 hours in all groups returned to control levels in THAM treated animals. These studies provide evidence of a direct effect of THAM upon lactate production in brain injured animals and offer a possible explanation for the reduced swelling and less variable ICP course in head injured patients treated with THAM.

Summary and Conclusions

In ischemic injury, with cessation of phosphorylation and depletion of aerobic energy stores, the production of lactate is more clearly understood than the metabolic perturbations associated with traumatic brain injury. In the clinical setting, the CBF in the greater majority of head injured patients is not at ischemic levels. Investigators have argued that the brain may be relatively ischemic based upon traumatically induced increased energy demand (Andersen *et al.*, 1988; Unterberg *et al.*, 1988). Another possible explanation is that early ischemia may be undetected by CBF measures in the clinical setting which are usually conducted several hours after injury. Still another explanation is that the lactate seen in CSF is a normal response to trauma as a result of compartmentalization of energy where glycolysis is triggered preferentially to restore ionic homeostasis (Andersen and Marmarou, 1989). Finally, the systemic response to trauma must be considered as serum levels of lactate are increased and with structural damage and hemorrhage, CSF samples may reflect serum contents.

Thus, although these studies have provided additional information to better understand the conditions under which brain tissue acidosis develops, the underlying mechanism for lactate production in the head injured patient is not completely resolved and further studies in the clinical and laboratory setting are necessary.

References

1. Andersen BJ, Unterberg AW, Clarke GD, Marmarou A (1988) Effect of posttraumatic hyperventilation on cerebral energy metabolism. J Neurosurg 68: 601–607
2. Andersen BJ, Marmarou A (1989) Energy compartmentalization in neural tissue. J Cereb Blood Flow Metabol 9 [Suppl 1]: 386
3. Crockard HA, Taylor AR (1972) Serial CSF lactate/pyruvate values as a guide to prognosis in head injury coma. Eur Neurol 8: 151–157
4. DeSalles AF, Kontos HA, Becker DP, Yang MS, Ward JD, Moulton R, Gruemer HD, Lutz HL, Maset AL, Jenkius L, Marmarou A, Muizelaar JP (1986) Prognostic significance of ventricular CSF lactic acidosis in severe head injury. J Neurosurg 65: 615–624

5. DeSalles AAF, Muizelaar JP, Young HF (1987) Hyperglycemia, CSF lactic acidosis and CBF in severely head-injured patients. Neurosurgery 21: 45–50
6. Graham DI, Adams JH, Doyle D (1978) Ischaemic brain damage in fatal non-missile head-injuries. J Neurol Sci 39: 213–234
7. Håkanson E, Rutberg H, Jorfeldt L (1984) Endocrine and metabolic responses after standardized moderate surgical trauma: influence of age and sex. Clin Physiol 4: 461–473
8. Inao S, Marmarou A, Clarke GD, Andersen BJ, Fatouros PP, Young HF (1988) Production and clearance of lactate from brain tissue, cerebrospinal fluid, and serum following experimental brain injury. J Neurosurg 69: 736–744
9. King LR, McLaurin RL, Knowles Jr C (1974) Acid-base balance and arterial and CSF lactate levels following human head injury. J Neurosurg 40: 617–625
10. Liddell MJ, MacLean LD, Shizgal HM (1979) The role of stress hormones in the catabolic metabolism of shock. Surg Gynecol Obstet 149: 822–830
11. Miller JD, Becker DP (1982) Secondary insults to the injured brain. J R Coll Surg Edinb 27: 292–298
12. Rosner MJ, Newsome HH, Becker DP (1984) Mechanical brain injury: the sympathoadrenal response. J Neurosurg 61: 76–86
13. Rosner MJ, Becker DP (1984) Experimental brain injury: successful therapy with a weak base tromethamine with an overview of CNS acidosis. J Neurosurg 60: 961–971
14. Seitz HD, Ocker K (1977) The prognostic and therapeutic importance of changes in the CSF during the acute phase of head injury. Acta Neurochir (Wien) 38: 211–231
15. Unterberg AW, Andersen BJ, Clarke GD, Marmarou A (1988) Cerebral energy metabolism following fluid-percussion brain injury in cats. J Neurosurg 68: 594–600
16. Ward JD, Choi S, Marmarou A, Moulton R, Muizelaar JP, DeSalles AAF, Becker DP, Kontos HA, Young HF (1989) Effect of prophylactic hyperventilation on outcome in patients with severe head injury. In: Hoff JT, Betz AL (eds) Intracranial pressure VII. Springer, Berlin Heidelberg New York Tokyo, pp 630–633

Correspondence and Reprints: A. Marmarou, Division of Neurosurgery, Medical College of Virginia, MCV Station Box 508, Richmond, Virginia 23298, U.S.A.

Subject Index

Adenosine 80
Antiaggregatory therapy 53
Axonal damage 49
Axonal injury 41

Blood-brain barrier 30, 64
Brain tissue acidosis 160

Capillary perfusion 35
Cardiac arrest 110
Cerebral acid-base regulation 57
Cerebral infarction 53, 130
Cerebral ischemia 21
Cerebral metabolic rates 35
Cerebral monitoring 152
Cerebral resuscitation 110
Cerebral water content 89
Clinical correlation 130
Contrast enhancement 30
Contusion index 41

Delay of treatment 145
Delayed neuronal death 80
Denervation 94
Dextran magnetite particles 30

Electron microscopy 49
Emergency care system 137
Energy metabolism 21
Excitotoxicity 94

Focal cerebral ischemia 73, 102
Forebrain ischemia 80

Glial neuronal pH 57

Head injury 41, 49, 137, 141, 145, 152, 160
Hippocampus 80
^{1}H-observed ^{13}C-decoupled spectroscopy 9
Hospital admission policy 145
^{1}H, ^{31}P-spectroscopy 21
^{1}H-spectroscopy 1
Hyperammonemia 9

Infarct formation 73
In vivo metabolism 1
Intracellular pH 9, 21
Intracranial hypertension 41, 89, 152
Ischaemia 1, 21, 35, 53, 57, 73, 94
Ischemic flow thresholds 102
Ischemic neuronal damage 94
Isotonic saline infusion 89

Label flow 9
Lactacidosis 57
Lactate production 160

Magnetic resonance imaging 30
Mediator compounds 64
Microthrombosis 53
MK-801 73
Multimodality treatment after cardiac arrest 110

Neuronal modules und algorithms 123
NMDA-receptor 73, 94
Nodal blebs 49

Osmotherapy and hypnotics 152

PET scanning 35
Prehospital emergency care 141
Prognostic factors 145

Quality assessment 137

Recovery from ischemia 102
Regional cerebral blood flow 130

Secondary ischemia 41
Spectral editing 1
Stroke 35

Taxonomy of cerebral functions 123
Temporal neuronal processing 123
THAM buffering 160
Traffic accidents 141
Trauma evaluation score 137
Triflusal 53

Vasogenic brain edema 64, 89
Vasomotor response 64

^{133}Xenon inhalation 130